THERMAL CONDUCTIVITY 15

A Continuation Order Plan is available for this series. A continuation order will bring delivery of each new volume immediately upon publication. Volumes are billed only upon actual shipment. For further information please contact the publisher.

THERMAL CONDUCTIVITY 15

Edited by

Vladimir V. Mirkovich

Canada Centre for Mineral and Energy Technology
Ottawa, Ontario, Canada

SPRINGER SCIENCE+BUSINESS MEDIA, LLC

Library of Congress Cataloging in Publication Data

International Conference on Thermal Conductivity, 15th, Ottawa, 1977.
  Thermal conductivity 15.

  "Proceedings of the Fifteenth International Conference on Thermal Conductivity
held in Ottawa, Ontario, Canada, August 24–26, 1977."
  Includes index.
  1. Heat – Conduction – Congresses. I. Mirkovich, Vladimir V. II. Title.
QC320.8.I57   1977   536'.2012   78-12943
  ISBN 978-1-4615-9085-9   ISBN 978-1-4615-9083-5 (eBook)
  DOI 10.1007/978-1-4615-9083-5

Proceedings of the Fifteenth International Conference on Thermal
Conductivity held in Ottawa, Ontario, Canada, August 24–26, 1977

© 1978 Springer Science+Business Media New York
Originally published by Plenum US 1978
Softcover reprint of the hardcover 1st edition 1978

Plenum Press, New York is a division of Plenum Publishing Corporation
227 West 17th Street, New York, N.Y. 10011

FOREWORD

Once again, it gives me a great pleasure to pen the Foreword
to the Proceedings of the 15th International Conference on Thermal
Conductivity. As in the past, these now biannual conferences provide a broadly based forum for those researchers actively working
on this important property of matter to convene on a regular basis
to exchange their experiences and report their findings. As it is
apparent from the Table of Contents, the 15th Conference represents
perhaps the broadest coverage of subject areas to date. This is
indicative of the times as the boundaries between disciplines become increasingly diffused. I am sure the time has come when Conference Chairmen in coming years will be soliciting contributions
not only in the physical sciences and engineering, but will actively
seek contributions from the earth sciences and life sciences as
well. Indeed, the thermal conductivity and related properties of
geological and biological materials are becoming of increasing importance to our way of life.

As it can be seen from the summary table, unfortunately,
proceedings have been published only for six of the fifteen conferences. It is hoped that hereafter this Series will become
increasingly well known and be recognized as a major vehicle for
the reporting of research on thermal conductivity.

Conference and Year	Title of Volume	Publisher and Year
7th (1967)	THERMAL CONDUCTIVITY Proceedings of the Seventh Conference	Superintendent of Documents/GPO (1968)
8th (1968)	THERMAL CONDUCTIVITY Proceedings of the Eighth Conference	Plenum Press (1969)
9th (1969)	NINTH CONFERENCE ON THERMAL CONDUCTIVITY	USAEC (1970)
13th (1973)	ADVANCES IN THERMAL CONDUCTIVITY Papers Presented at XIII International Conference on Thermal Conductivity	University of Missouri, Rolla (1974)
14th (1975)	THERMAL CONDUCTIVITY 14	Plenum Press (1976)
15th (1977)	THERMAL CONDUCTIVITY 15	Plenum Press (1978)

I wish to take this opportunity to congratulate Dr. Vladimir Mirkovich, General Chairman of the Conference and Editor of these Proceedings, for his untiring efforts in the planning and conducting of a conference under difficult circumstances. His full cooperation with CINDAS (Center for Information and Numerical Data Analysis and Synthesis) as the Permanent Sponsor of the International Thermal Conductivity Conferences is also acknowledged.

                                    Y. S. Touloukian
                                    Director, CINDAS
                                    and
June 1978   Distinguished Atkins Professor
West Lafayette, Indiana   Purdue University

PREFACE

The 15th International Thermal Conductivity Conference (ITCC) was held in Ottawa, Ontario, Canada, August 24-26, 1977, hosted by the Canada Centre for Mineral and Energy Technology (CANMET) and by the Department of Energy, Mines and Resources (EMR). This, the 15th of a now well-established series of conferences, set a precedent by being held in Ottawa for the second time. The conditions, however, are quite different from those of 1962. Then, petroleum was abundant and energy shortages, or shortages of any kind of material for that matter, were of concern to no one. At that time, a forum was needed for researchers in the field of heat transport to discuss experimental techniques, to establish reference materials and to enhance the theory of heat transport. Despite all the advances to date this forum is still required, as many problems still remain to be solved and new ones are being added. As most forms of energy are derived from heat and all types of energy eventually convert into heat, it should not be surprising that in the present period of energy shortages, interest in thermal transport properties should intensify. In fact, some 25 per cent of the papers presented at the 15th ITCC dealt with generation, conversion or conservation of energy, undoubtedly reflecting an increased research activity in the energy field. The number of papers on technical applications appears to be on the increase and among them a novel topic was introduced to ITCC: problems of heat transport in materials associated with storage and disposal of radioactive nuclear waste.

The Thermal Conductivity Conferences were started in 1961 on the initiative of Mr. C.F. Lucks, who was at that time with the Columbus Laboratories of the Battelle Memorial Institute. They were held annually until 1973 when a biennial schedule was adopted, in part to coordinate work and to cooperate with our European counterpart, the European Conferences on Thermophysical Properties. The Thermal Conductivity Conferences have been self-perpetuating without the support of a professional society or some governmental agency, - a clear indication of their value to science and technology as well as of the dedication of the past chairmen. The list of hosts to date are as follows:

Conference and Year	Host Organization and Site	Chairman
1st 1961	Battelle Memorial Institute Columbus, OH	C.F. Lucks
2nd 1962	National Research Council (Canada) Ottawa, Canada	M.J. Laubitz
3rd 1963	Oak Ridge National Laboratory Gatlinburg, TN	D.L. McElroy
4th 1964	U.S. Naval Radiological Defense Lab. San Francisco, CA	R.L. Rudkin
5th 1965	University of Denver Denver, CO	J.D. Plunkett
6th 1966	Air Force Materials Laboratory Dayton, OH	M.L. Minges G.L. Denman
7th 1967	National Bureau of Standards Gaithersburg, MD	D.R. Flynn B.A. Peavy
8th 1968	Thermophysical Properties Research Center/Purdue University W. Lafayette, IN	D.Y. Ho R.E. Taylor
9th 1969	Ames Laboratory & Office of Naval Research Ames, IA	H.R. Shanks
10th 1970	Arthur D. Little, Inc. & Dynatech R/D Co. Boston, MA	A.E. Wechsler R.P. Tye
11th 1971	Sandia Laboratories, Los Alamos Scientific Laboratories and University of New Mexico Albuquerque, NM	R.U. Acton P. Wagner A.V. Houghton, III
12th 1972	Southern Research Institute & University of Alabama Birmington, AL	W.T. Engelke S.G. Bapat M. Crawford
13th 1973	University of Missouri - Rolla Lake of the Ozarks, MO	R.L. Reisbig H.J. Sauer, Jr.
14th 1975	University of Connecticut Storrs, CT	P.G. Klemens
15th 1977	Dept. of Energy, Mines & Resources Ottawa, Canada	V.V. Mirkovich

As in previous conferences, a feature of the 15th ITCC was the presentation of the Thermal Conductivity Award. This time the Conference was faced with a somewhat unexpected, but nevertheless not unpleasant, dilemma of having to select among no less than three equally meritorious and distinguished candidates. The board of governors solved the problem by agreeing unanimously that in this exceptional case each of the three candidates should receive the award. For their outstanding contributions to the field of thermal conductivity, the Thermal Conductivity Award was conferred on (alphabetically): Dr. R.E. Taylor of CINDAS and University of Purdue; Dr. Y.S. Touloukian, also of CINDAS; and Mr. R.P. Tye of Dynatech R/D Co.

The 15th ITCC is in debt to the Session Chairmen for conducting the sessions and giving their valuable time in reviewing the manuscripts. The support and encouragement received from the EMR/CANMET staff through all stages of this conference is greatly appreciated. The arrangements made for publication of these Proceedings as well as the help and guidance received from the continuing co-sponsor, CINDAS/Purdue University, is gratefully acknowledged. Invaluable advice was obtained from all members of the ITCC Governing Board, particularly from Mr. R.P. Tye and from the members of the local Scientific Committee, Drs. M.J. Laubitz and J.G. Cook. Thanks are due to Mr. R.M. Buchanan of the CANMET staff for his efficient handling of the finances, and, while it is impossible to mention individually all of those who have so much contributed toward smoothing the operations of the Conference, Mr. A. George McDonald, also of the CANMET staff, must be singled out for his exceptional efforts in helping with every aspect of the 15th ITCC.

For a conference such as ours to receive external financial support is certainly pleasant, but more significantly, such support is also a recognition of the importance of the conference by the leaders of the scientific and technological society. It therefore gives the chairman of the 15th ITCC a great deal of pleasure to acknowledge the generous support given by the following Canadian Companies: Aluminium Company of Canada, Ltd.; Fiberglas Canada Limited; Gulf Oil Canada Limited; Canadian Refractories, Ltd.; Domtar Construction Materials, Ltd.; and particularly by Imperial Oil Limited (ESSO).

The 16th International Thermal Conductivity Conference will be held in the latter part of 1979. It will be chaired by Mr. D.C. Larsen under the sponsorship of the IIT Research Institute, Chicago, Ill.

Ottawa, Ontario, Canada   V.V. Mirkovich
May, 1978   Chairman, 15th ITCC

CONTENTS

SESSION B

SOLIDS AT HIGH TEMPERATURE

SESSION C

REACTOR MATERIALS

SESSION D

TECHNIQUES, DATA ANALYSIS

SESSION E

THEORY, SOLIDS AT LOW TEMPERATURE

SESSION F

GASES AND LIQUIDS

SESSION G

ROCKS AND SOILS

SESSION H

INSULATION

PLENARY SESSION II

SESSION I

NUCLEAR WASTE DISPOSAL, COAL

SESSION J

DIFFUSIVITY, CONTACT CONDUCTANCE

PLENARY SESSION I

Keynote address by: P.G. Klemens
                       Dept. of Physics and
                       Institute of Materials
                       Science
                       Univ. of Connecticut
                       Storrs, Connecticut

KEYNOTE ADDRESS: 15TH INTERNATIONAL THERMAL CONDUCTIVITY CONFERENCE

P. G. Klemens

Dept. of Physics and Institute of Materials Science

University of Connecticut, Storrs, Connecticut 06268

As we come together for the 15th Thermal Conductivity Conference to continue what has now become an established tradition, it is not inappropriate to remind ourselves of what brings us together and what we hope to gain from these conferences.

Obviously the common interest is Thermal Conductivity - measurements, results of measurements, interpretation and comparison, applications both to technology and to other sciences, and especially the continuing endeavor to extend the measurement techniques and our understanding to materials and to conditions which were previously inaccessible.

The motivation for our work falls broadly into two groups:-

(1) To study the mechanism of thermal conduction within the framework of the atomic theory of matter, and thus to relate and reconcile the observed conductivities to the existing microscopic models and to other material properties.

(2) To measure or estimate by other means the thermal conductivity of specific materials under specified conditions, because this information is needed either for various technological purposes or in some other science.

We talk to each other because we believe that these activities can be mutually supportive. Basic theory is needed to give some assurance that the results of a particular measurement are not unreasonable, and to help estimate thermal conductivity values when no measurements have been made; on the other hand the measurement

3

techniques and the data generated for applied purposes have also relevance to basic studies. Finally we have a common interest to see our subject area flourish, **and in order to bring this about** we are eager to hear of new areas of application.

In spite of this common interest there has been an unfortunate tendency for the two groups of workers to go their separate way. The theorist publishes his theory in a way which may appeal to other theorists, who after all are his peer-group to referee his papers and his funding proposals, but the results of his theory are not always put into a usable form. The basically oriented experimentalist performs measurements to test or illustrate a particular principle, and when he publishes the results, he often concentrates on this test, and reports some derived quantity like a Lorenz number, a lattice component or an ideal thermal resistivity, rather than thermal conductivity data. Admittedly one can reconstitute the data from information given in the paper, but few potential users will do so, and the people at CINDAS will tend to disregard such data. On the other hand, there exists much data which was acquired for applied purposes but which contains information of basic interest. Often this data is not utilized because it is published where basic scientists do not look, and also because of insufficient material characterization or lack of auxiliary data. One result is the uneven distribution of measurements amongst materials. I need not elaborate this point: Dr. Touloukian has vividly illustrated the difficulties encountered by the user of data;[1]   the basic scientist experiences corresponding difficulties. At this point one should mention that the work of CINDAS has led to a considerable improvement of the situation in recent years.

Let me return to the basic motivation for thermal conductivity studies. We note with satisfaction that these studies have been significant on several occasions in the development of physical concepts. Let me give some examples from the area with which I am familiar - solid state physics:-

The ratio of thermal to electrical conductivities or Lorenz number has provided a test of the electron theory of metals, culminating in Sommerfeld's theory of the degenerate electron gas.

The need to understand the thermal conductivity of insulators led Debye to his model which describes the vibration of solids in terms of elastic waves.

The thermal conductivity of superconductors does not parallel the electrical conductivity but is often less in the superconducting state.[2] This formed a powerful argument for the two-fluid concept and its derivative models.

The energy gap in superconductors was first seen through the exponential temperature dependence of the thermal conductivity.[3] Evidence from specific heat and from infrared absorption came later.

The topology of the Fermi surface of simple metals, in contrast to original expectations, was seen to be more complex from the low temperature Lorenz number. The thermal conductivity of insulators, particularly its reduction by simple point defects, had a powerful influence on the development of lattice dynamics.

We can summarize these developments as follows:- Thermal conductivity is sufficiently different in nature from the other physical properties, so that it provides important clues concerning the nature of matter and about microscopic interaction processes.

Note that many of the developments which I have cited, and all the recent ones, referred to thermal conductivity at low temperatures. This leads us to consider another trend in the separation between basic and applied programs: they have tended to emphasize different temperature regimes. In basic research the emphasis has been on low temperatures. After all, on a logarithmic temperature scale there is more range below room temperature than between room temperature and the melting point; also the interesting quantum effects occur at low temperatures, and the transport properties become more sensitive to defects the lower the temperature. In contrast, the technological needs are mainly at room temperatures and above; it is only recently, and on a modest scale, that liquified gas technology and progress in high field superconductors has led to a technical need for thermal conductivities at low temperatures.

Furthermore, the high temperature regime poses particular challenges in the problems of measurement. With a few notable exceptions the basic scientists had abandoned this region to their applications-oriented colleagues.

Of course, the isolation between basic and applied research was never an absolute one, and in any case was certainly not characteristic of our field alone. Basic science is not being supported for its own sake, but because it underlies the technical developments which are needed to solve urgent problems. Forty years ago, when science was smaller, the basic scientists were well aware of this fact of life, and took an active interest in applications. When the scale of scientific and technical effort was expanded, this need for interaction was downgraded or overlooked. Now that the expansion has stopped, we need to get the most out of the available resources, and this can only be done by bringing basic and applied science together. This is so in all fields, and certainly in our own.

The format of the Thermal Conductivity Conferences has therefore been very much in tune with the needs of the time and this may
explain why an informal group, acting outside the scientific and
professional societies, could keep the conferences going during a
period which has certainly not been an easy one for the individual
worker, but has been characterized by frequent project changes and
dislocations.

Currently, the dominant challenge for technology arises from
the energy crisis. The resulting technical efforts will have a
need for thermal conductivity information. This need will be found
in every major project, in connection with better heat exchangers,
with nuclear reactor materials, with insulators for liquified gas
storage and other energy storage schemes, with building materials,
with rock properties for geothermal energy, and so on. There is,
of course, basically nothing new here, since nearly every technical project requires thermal conductivity information, sometimes
in a traditional manner, often as an unexpected facet of a wider
problem.

In this universality lies both our strength and our weakness.
Our strength, because thermal conductivity knowledge is so widely
needed. Our weakness, because this need is often only one in a
chain of technical problems that form the project. By the time
this need is recognized and the work initiated, the entire project
may be in trouble, and the work proceeds under a cloud of uncertainty. The individual worker and his group suffer from frequent
changes in direction, even though the actual scientific problem
has a habit of appearing under different guises in different projects. This is a problem in management, but it affects all of us
in a very real way.

If thermal conductivity research is to make an optimum contribution to the technological needs of the future, this management problem must be solved, especially as it is the applications-
oriented research groups which suffer the most from frequent
changes in objectives. However, stability and reasonable continuity of support would be good for all of us. There are many general
and basic problems, with a direct bearing on applied work, which
would benefit from a continuity of effort. Let me mention a few,
drawn from my own area of solid state research.

(a) We still lack the ability to separate the total thermal con-
     ductivity reliably into separate carrier components: elec-
     tronic and lattice, or lattice and radiative. This is needed
     not only for basic studies, but without this knowledge it is
     not possible to interpolate or extrapolate existing data.

(b) Internal radiation as a conduction mechanism is still poorly understood; this lack is particularly acute in geophysical studies and thermal and thermoplastic models of the earth, but affects our understanding of nearly all dielectric solids at high temperatures.

(c) The pressure dependence of the thermal conductivity and the related problem of the effect of thermal expansion on the temperature dependence of the thermal conductivity is waiting for experimental results.

(d) Theories of electronic and lattice thermal conductivities have been developed in the limits of low and high temperatures, but are particularly crude at intermediate temperatures – yet it is precisely this range of ordinary temperatures where we have the biggest demand for data and for reliable interpolation schemes.

(e) In lattice thermal conductivities the role of low-frequency longitudinal phonons, four-phonon processes and the dependence of conductivity on crystal size are still not well understood.

(f) We understand the ballistic flow of heat carriers, i.e. a direct rectilinear transmission from body to body, and the diffusive flow, where the mean free path is short. We are ill at ease in the intermediate situation, where the mean free path is comparable to external dimensions. This problem occurs at very low temperatures in phonon flow, but also at high temperatures in radiative heat transfer.

(g) After a great deal of effort, conduction through inhomogeneous media, very important in practice, still presents special difficulties.

Of course, this is a very personal list, and I am sure that many of you could add to this list of problems. And this, after all, is one of the reasons why we are here at this Conference.

REFERENCES

(1) Y. S. Touloukian "The Impact of Physical Properties Research on Technological Advancement" in Seventh Symposium on Thermophysical Properties, Gaithersburg, Maryland, May 1977.
(2) H. Bremmer "Onderzoeking over de Warmtegeleiding bij lage Temperaturen" (Leiden Dissertation), De Voorpost, Rotterdam, 1934.
(3) B. B. Goodman, "The Thermal Conductivity of Superconducting Tin below 1°K" Proc. Phys. Soc. (London) A66, 217 (1953).

SESSION A   GENERAL SUBJECTS

Session Chairmen: Y.S. Touloukian
                         Purdue University
                         Lafayette, Indiana

                         I.D. Peggs
                         Atomic Energy of Canada
                         Pinawa, Manitoba

THE FEASIBILITY OF USING INFRARED SCANNING TO TEST FLAWS IN CERAMIC MATERIALS*

C. K. Hsieh, Mechanical Engineering Department,

University of Florida, Gainesville, Florida 32611

W. A. Ellingson, Materials Science Division, Argonne

National Laboratory, Argonne, Illinois 60439

ABSTRACT

The feasibility of using infrared scanning to test the integrity of refractory materials was established through the use of a 60% aluminum oxide refractory. Heat was injected from the back side of the sample and scanning with the infrared system was performed on the front surface. To minimize the lateral heat dissipation from the sample, a nichrome heating wire was embedded in a narrow groove cut around the view field; the scanning surface was painted black to minimize the background-reflection effects. Prior to tests the ground level surface energy was scanned in a two-dimensional array. Follow-up scans were made after various defects were simulated on the back side of the sample. By means of an image-processing technique, the ground-level surface energy and noise effects were removed from the scanned data. This also permitted the processed data to reveal the true effects due to material defects. In the experiment, the scanned thermal patterns were analyzed to obtain six features for comparison. The study has shown that, based on an analysis of these features, a quantification of material defects can be made. The study also reveals the limitations of certain extracted features to distinguish material flaws.

---

*Work supported by the U.S. Energy Research and Development Administration.

INTRODUCTION

One of the energy-related research activities supported by the U.S. Energy Research and Development Administration and carried out at Argonne National Laboratory deals with the development of nondestructive test methods for application to coal-conversion systems. Part of this project addresses the need to detect the erosive wear of refractory linings in coal-gasifier vessels and associated ash and char transfer lines. Because of the continuous exposure of the refractory to fluidized beds of highly abrasive coal particles at elevated pressures and temperatures, the erosion of the material is excessive and often a rapid process. Infrared scanning is one of the methods being developed at Argonne to detect and quantify the erosion.

The use of infrared scanning to test material integrity has been applied to bonds [1,2] and near subsurface defects.[3-7] More recently, a pattern-recognition technique has also been applied to the analysis of infrared scanning data and has been used to test the integrity of metal and refractory-metal laminates.[8-9] With the exception of these latter studies, nearly all of the existing literature on infrared scanning deals with qualitative studies. The change of color or the shade of gray in the infrared picture has been used as an indication of temperature nonuniformity and as clues to the presence of material flaws. This nonquantitative approach is inadequate to fill the present need.

The purpose of the present study is to establish the feasibility of using infrared scanning to test flaws in ceramic materials. A ceramic tile was used for study, and the tests were carried out under steady-state conditions. The selection of a plane slab simplified the studies to the extent that the observed infrared data were free from surface curvature effects. The study is intended as an exploratory effort to identify the value of the method and to provide inputs for decision making in the material-evaluation task of the coal program.

EXPERIMENT

The test sample used in the study was a rectangular ceramic tile (228.6 x 114.3 x 30.2 mm) known commercially as Greenlite-28. Its chemical composition and thermal properties are summarized in Table 1.

The infrared camera used in the experiment was an AGA Model 750 Thermovision System consisting of a scanning camera and an electronic picture display unit. Since this camera has not been widely used in heat -transfer research, a brief description of the camera operation is given here (the interested reader is referred

Table 1. Thermal Properties for Test Sample as Reported by Manu-
          facturer

Commercial Name	Greenlite-28
Manufacturer	A.P. Green Refractories Company
Chemical Analysis	Al_2O_3, 60-63%
	SiO_2, 33-36%
	$Na_2O + K_2O$, 1-1.5%
	Fe_2O_3, 0.5-1%
	CaO, mgO, TiO_2 -- remaining
Porosity (ASTM C20)	69-75%
Bulk Density (ASTM C437)	737-801 kg/m^3
Thermal Conductivity	3.01×10^{-3} W/cm^oC at 204^oC
	5.07×10^{-3} W/cm^oC at 1315^oC
	(Linear Variation)

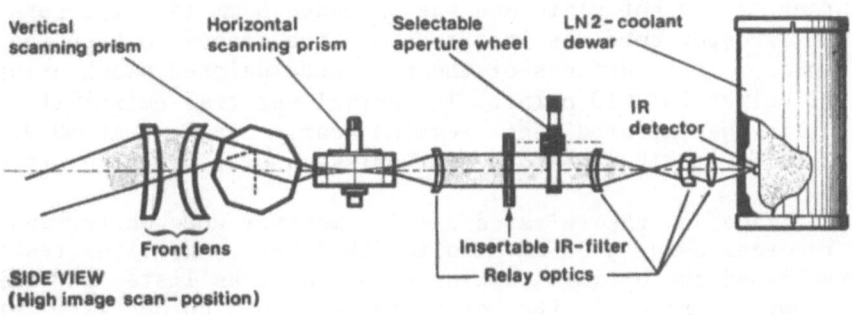

SIDE VIEW
(High image scan-position)

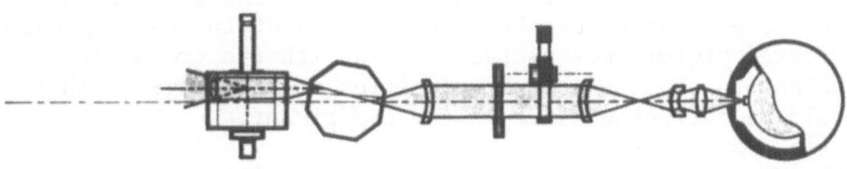

TOP VIEW
(High/right image scan-position)

Fig. 1.   Schematic Diagram of Construction of
          Infrared Camera (Courtesy of AGA)

to Ref. 10 for details). Figure 1 depicts the optical system in
the camera. The scanning optics consists of two eight-sided rotating prisms, a chopper, a train of silicon collimating lens, and
a liquid-nitrogen-cooled InSb detector. The prisms scan the field
of view at a frequency of 25 pictures/s. The electronics in the
display monitor are synchronized with the positions of the prisms
in the camera scanning optics in such a way that each point in
the optical field of view is transformed into a corresponding
point on the monitor screen. The intensity of the modulated beam
in the monitor is a function of the received infrared radiation.
In a normal viewing mode, a real-time infrared picture is displayed
with the tone of the warm parts lighter gray and cold darker gray.

The experimental set-up is depicted in Fig. 2. The ceramic
tile was supported in a horizontal position with the top surface
in contact with a hot plate. The bottom surface of the tile was
facing a front surface mirror that was inclined 45^0 toward both
the camera and the tile surface. The bottom surface temperature
of the tile was continuously monitored with the tile held in a
postiion such that the convective heat dissipation was minimized.
To compensate for the lateral heat transmission in the tile due to
edge effects, a narrow groove was cut on its bottom surface. A
nichrome heating wire was bent into a circular loop that follows
the contour of the hot plate and was embedded 6 mm into the surface. An asbestos cord was inserted into the groove to keep the
wire in place. All surfaces of the tile were painted black using
3M Nextel Velvet 101-C10 paint. The normal spectral emissivity of
the paint in the infrared range remains nearly constant at $\sim$0.974[11]
which is sufficiently high to permit omission of reflection effects.

The images on the infrared display monitor were photograph-
ically recorded using a 35-mm Minolta SLR camera. Previous tests
had established the optimum working conditions, as listed in Table
2. The power supplies to the hot plate and the nichrome wire were
adjusted separately until the infrared picture showed a uniform
gray tone over the view field, see Fig. 3(a). Nevertheless, a
white circular arc appeared in the lower left and right corners
of the picture. These revealed the location of the heating wire.
This infrared picture was useful to establish the ground-level
energies and to show these irregularities in the image so that
they could be corrected later in the experiment.

The strength of the infrared scanning method lies in its
capability to scan a large field of view. This also brings about
a large body of data output, which forbids conventional methods of
analysis. To gather this large body of data, an image-processing
technique was used. The infrared data at each picture element
were digitized to 64 gray levels in a 120 x 64 array. In this
data array, any nonuniformity in the photographic film tended to
become apparent. This noise effect was reduced by taking an

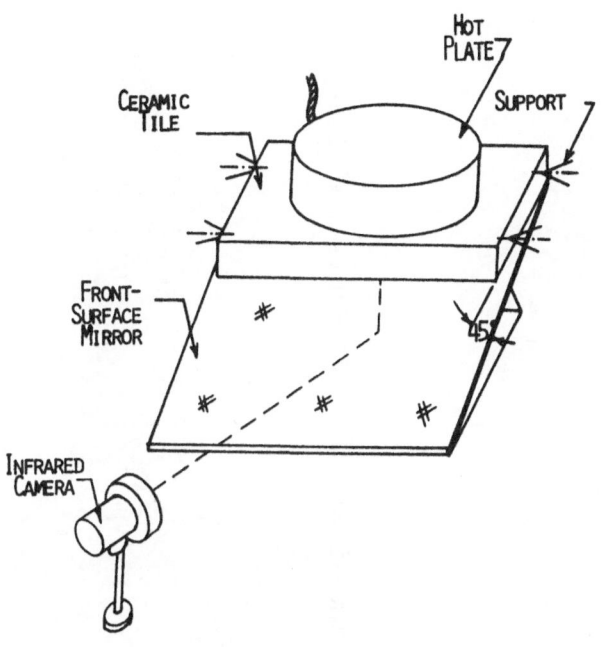

Fig. 2.   Schematic Diagram of
           Experimental Set-up

Table 2. Optimum Camera Control Settings

AGA 750 Thermovision		35 mm Minolta SLR Camera	
Spot Wobbling	On	Shutter	1 s
Field	4	f stop	8 (for normal mode operation)
Contrast	Minimum		2.8 (for inverted mode operation)
Photo Preset	Off	Focus	Intentional slight out of focus to blur raster lines
Temperature Range	10	Film	Kodak Panatomic-X (ASA 32)

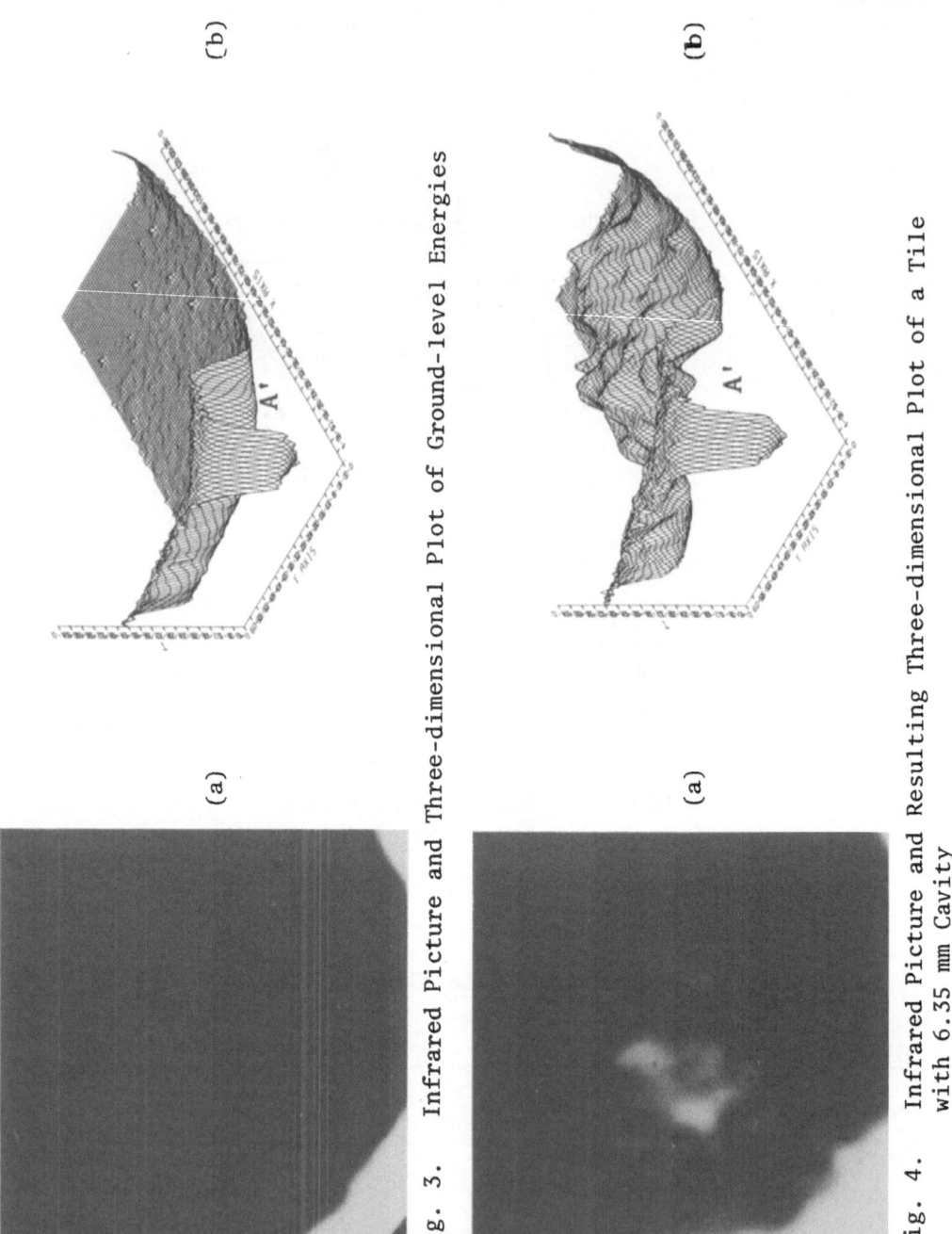

Fig. 3.   Infrared Picture and Three-dimensional Plot of Ground-level Energies

Fig. 4.   Infrared Picture and Resulting Three-dimensional Plot of a Tile
          with 6.35 mm Cavity

average of nine scans of the image processor. This was followed
by another data-averaging process to remove the raster lines in
the infrared image. The data at each picture element were averaged
with those of the eight nearest neighbors to calculate a weighted
average for that point. To visualize the effect of this noise removal routine, the resultant data were plotted using Argonne's
DISSPLA computer graphics, see Fig. 3(b). For the shades of gray
in the darker region in Fig. 3(a), the energy levels have a span
between 2 and 6. The data are well behaved in the sense that only
a few isolated spikes are found in a sea of ripples. To facilitate
the establishment of a one-to-one correspondence in these figures,
alphabets have been added.

Following this proparation work, a cavity 12.7 mm in diameter
and 6.35 mm deep was drilled on the back side of the title, and
the infrared picture was again photographed, digitized, and graphed
as shown in Fig. 4. Because of the presence of the cavity, a nonuniform surface temperature distribution resulted, which is demonstrated more vividly in the three-dimensional picture.

Finally, the cavity was deepened to 14.29 mm and the same
procedures repeated. This yielded another set of data as shown
in Fig. 5.

DATA ANALYSIS

To study the net effect due to material defects, the data in
Figs. 4(b) and 5(b) were adjusted by subtracting out the ground-
level energies given in Fig. 3(b). This procedure also removed
the irrelevant signals that resulted from the presence of the nichrome heater. The adjusted data sets were subsequently entered
into the computer memory and used for analysis.

Pattern recognition was the method used for quantitative prediction of material defects in the present paper. According to
the method, the simplest technique one could use is the membership-
roster recognition concept.[12] A large number of thermal patterns,
corresponding to various degrees of material defects, can be read
into the computer and stored in its memory. When a new pattern is
exposed to the computer, this pattern is compared with the stored
patterns. Hence, the method is essentially a template matching
process. It is in essence simple and yet will work satisfactorily
if the pattern samples are nearly perfect and free from distortions.
For the complicated patterns normally encountered in heat-transfer
work, the computer storage can pose a serious limitation if a large
number of patterns are to be dealt with. Hence, this recognition
scheme was not pursued. A more favorable common-property recognition concept was employed.

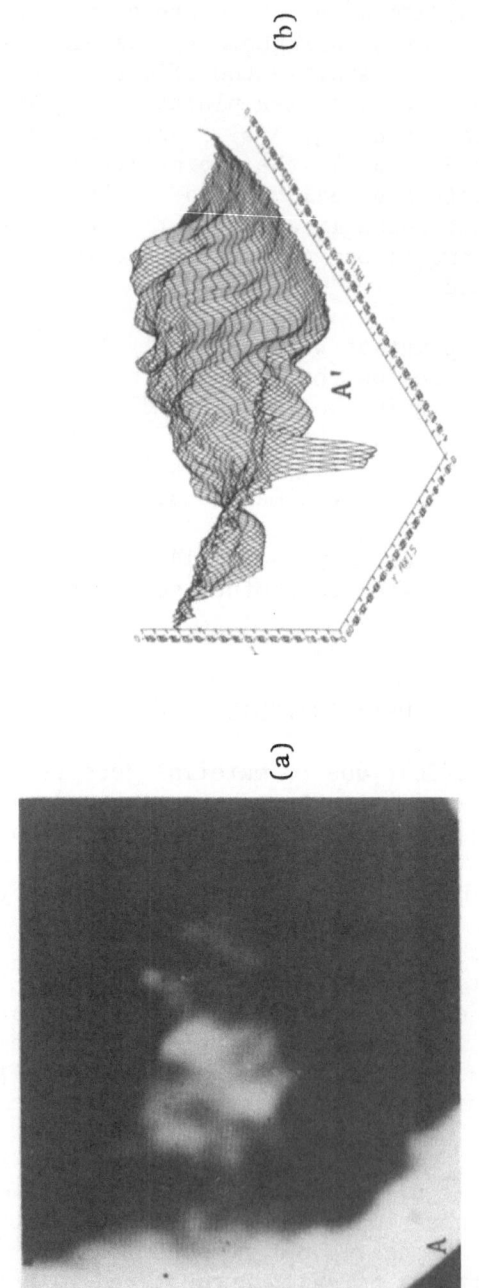

(a)

(b)

Fig. 5.   Infrared Picture and Resulting Three-dimensional Plot of a Tile
          with 14.3 mm Cavity

The common-property recognition concept is based on the existence of distinctive features or attributes in the patterns to be classified.[12] Instead of the entire patterns, only those features of interest are extracted and then prestored and coded in the computer memory. This concept is superior to the previous membership-roster concept in several ways. The storage requirements for pattern features are much less demanding than those for entire patterns. Secondly, significant pattern variations are intolerable in the membership-roster scheme but are acceptable in this common-property concept. In essence, the previous template matching is replaced by feature matching, which is less restrictive to detailed point-by-point similarity as called for in the previous scheme. An interpolation or extrapolation can often be made in the feature matching to extend its usefulness.

The features selected for comparison in this study consist of the following: (1) the area enclosed inside isotherm contours, (2) the volume contained underneath the energy surfaces, (3) the division of the volume in item (2) by the area in item (1), and (4) the slopes of the above curves. Figure 6 shows these feature plots. The abscissas in these figures represent the temperature axis. To visualize these features more vividly from a geometrical point of view, it is similar to passing a section plane of constant T through the three-dimensional surfaces in Figs. 4(b) and 5(b). The features extracted become functions of the elevation of this plane. It should be noted that the volume contained underneath the energy surface is actually related to the radiant energy emitted by the surface. This is a result of the surface being painted black, and the total intensity emitted by the surface at each picture element is proportional to $(\sigma T^4/\pi)$, where σ is the Stefan-Boltzmann constant and T is the surface temperature. For the same reason, the energy is arbitrarily interchanged with temperature in labeling the axes in the plots. Their distinction should nevertheless be noted.

DISCUSSION

Four of the six feature plots shown in Fig. 6 have two curves. The curve marked with squares indicates data extracted from Fig.5(b). The dashed curves come from Fig. 4(b), which are transposed here for comparison. Obviously, a deepening of the cavity shifts the feature curves upward. The integrated volume curves in Fig. 6(b) are particularly attractive in that both curves are smooth and amendable to curve fitting. This permits the use of the displacement of these curves to identify the flaw. Note the logarithmic ordinate in this plot. The distance between these curves is thus of sufficient magnitude to predict the cavity size with acceptable resolution.

AREA ENCLOSED INSIDE ISOTHERM LOOP

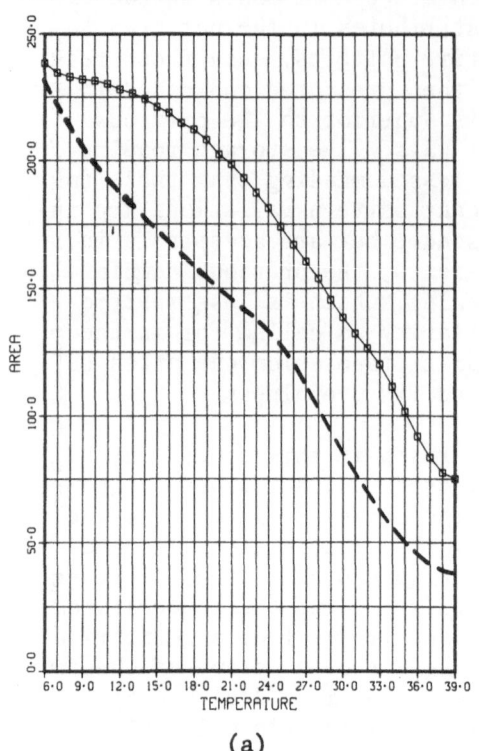

(a)

INTEGRATED VOLUME

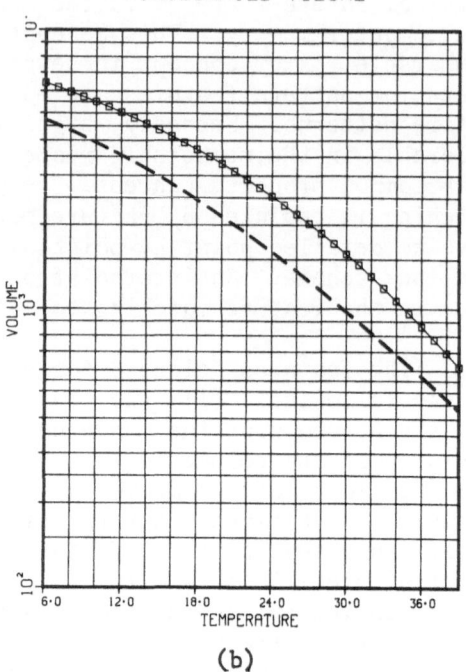

(b)

MEAN TEMPERATURE

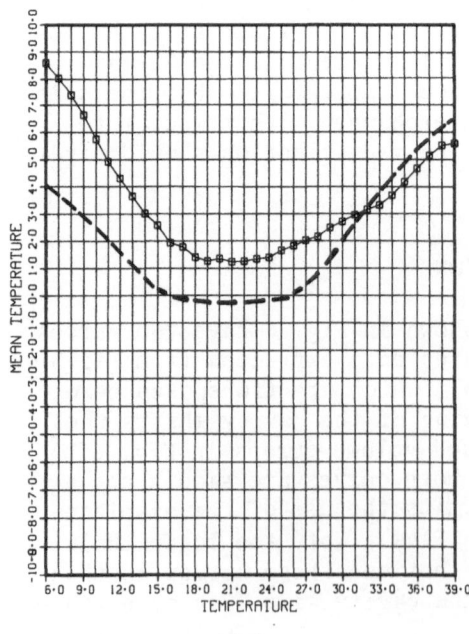

(c)

CHANGE IN ENCLOSED AREA

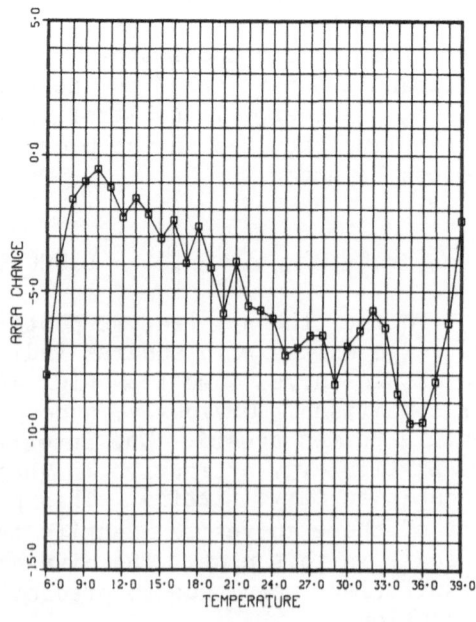

(d)

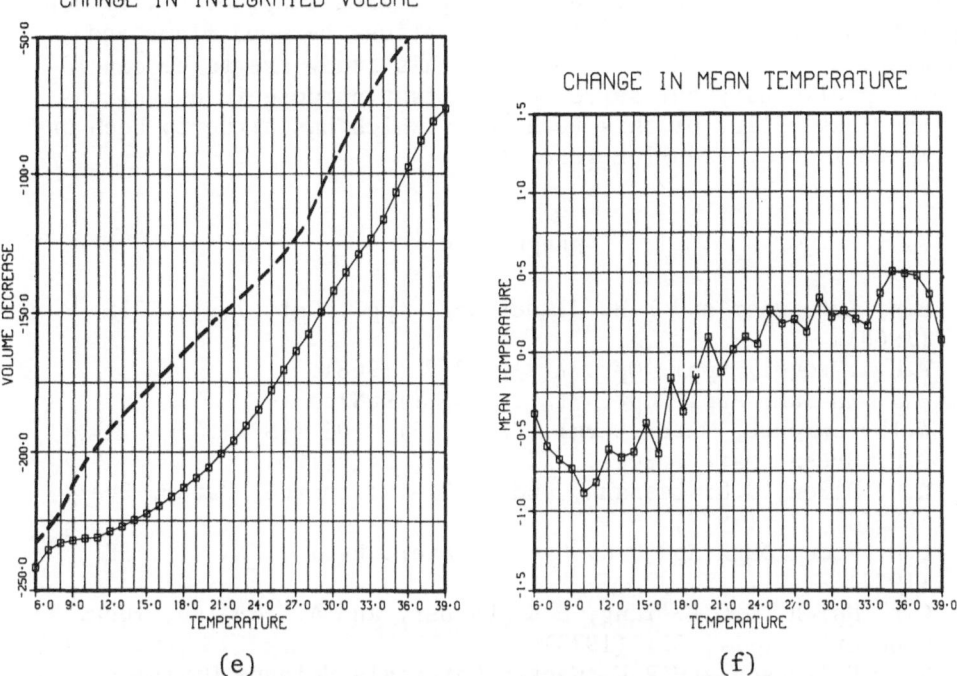

Fig. 6. Presentation of feature plots using computer graphics with
adjusted data from Figs. 4 and 5. Somewhat less regular are
the curves in Figs. 6(a),(c), and (e). An intersection point
due to the crossover of curves is observed in Fig. 6(c). In
Figs. 6(a) and (e), the curves tend to converge at lower
temperature readings. These represent regions of poor resolution, which should be avoided in the application. Aside
from these minor discrepancies, broad regions still exist
where curves in Figs. 6(a), (c), and (e) are useful. They
can be used together with Fig. 6(b) to upgrade the certainty
of flaw detection. On the other hand, the curves in Figs.
6(d) and (f) are quite irregular, signifying these features
are useless for our purpose.

CONCLUDING REMARKS

It should be noted that the number of features extracted in
the present study is not exhaustive in the sense that other features can always be found and extracted and used to support the
pattern-recognition resolution. In practice, the number of possible features often exceeds the number that can be used. The selection of useful features is limited by statistical considerations
and equipment limitations, among others.

The tests reported in the present paper are limited to one type
of defect--cylindrical cavities. For broad application, addition-

al tests need to be conducted covering a wide range of cavity configurations in order to develop correlations between flaws and
features. The limited tests made in this study are to be considered as preliminary but serve an important purpose in establishing the feasibility of this technique, which has hitherto been un-
attempted in heat-transfer research.

REFERENCES

1. E.W. Kutzscher and K.H. Zimmermann, Appl. Opt. $\underline{7}$(9), 1715
    (1968).
2. P.E.J. Vogel, Appl. Opt. $\underline{7}$(9), 1739 (1968).
3. E.J. Kubiak, Appl. Opt. $\underline{7}$(9), 1743 (1968).
4. W.T. Lawrence, Mater. Eval., $\underline{XXIX}$ (5), 105 (1971).
5. D.R. Green, Mater. Eval., $\underline{XXIX}$ (11), 241 (1961).
6. R.D. Dixon, G.D. Lassahn, and A. DiGiallonardo, Mater. Eval.,
    XXX (4), 73 (1972).
7. L.D. McCullough and D.R. Green, Mater. Eval., $\underline{XXX}$ (4), 88
    (1972).
8. C.K. Hsieh, M.C.K. Yang, E.A. Farber, and A. Jorolan, Thermal
    Conductivity $\underline{14}$, 521 (1975).
9. W.A. Ellingson and R.W. Weeks, Materials Science Division Coal
    Technology Ninth Quarterly Report, Argonne National Laboratory,
    1977.
10. W. H. Meyfarth, Eng. Digest, $\underline{18}$(1), 1969.
11. Private communication with J. C. Richmond, NBS.
12. J. T. Tou and R. C. Gonzalez, Pattern Recognition Principles,
    Addison-Wesley, 1974.

PREDICTION OF PRECOOLING HEAT TRANSFER IN CITRUS FRUITS

M.G. Raviv and I.J. Kopelman

Technion - Israel Institute of Technology

Dept. of Food Engineering & Biotechnology, Haifa, Israel

INTRODUCTION

Precooling of fresh fruits and vegetables is an effective tool in extending the shelf-life of the fresh produce. This area becomes quite important in Israel especially due to the growing interest in containerization of citrus boxes for overseas shipping.

The importance of precooling is evident especially in the peak of season where there is need to regulate and to balance fruit picking and packaging on one side and overseas shipping on the other.

Analytical and systematical approach to precooling of fresh produce is quite limited (5) and most of the available data is basically of empirical nature (1).

Using analytical approach for precooling could be a powerful tool in predicting cooling rates and cooling pattern as function of product properties and system characteristics.

This investigation was undertaken with aims of:

(a) determining the mode and the rate of heat transfer during the precooling and cold storage of citrus fruits in boxes.
(b) outlining a heat transfer simulation technique so that prediction of precooling of citrus fruit in boxes and shipping pallets can be made.

THEORETICAL CONSIDERATIONS

Taking a spherical shape to approximate the citrus fruit, the conduction heat transfer (with zero internal heat generated) within the sphere can be expressed by the following partial differential equation:

$$\frac{\partial u}{\partial t} = \alpha \left(\frac{2}{r} \frac{\partial u}{\partial r} + \frac{\partial^2 u}{\partial r^2} \right)$$

where: u – is a dimensionless temperature ($u = \frac{T-T_0}{T_1-T_0}$)

T_0 – body temperature at $t = 0$

T_1 – media temperature

t – time

α – thermal diffusivity ($\alpha = \frac{k}{\rho C_p}$)

k – thermal conductivity

ρC_p – heat capacitance

r – spherical polar coordinate.

The exact analytical solution for the above differential equation depends upon the initial and boundary values of specific physical system. However such exact analytical solutions are available only for fairly simple physical situations. For example the analytical expression:

$$\frac{2(\sin\beta_i - \beta_i \cos\beta_i)}{\beta_i - \sin\beta_i \cos\beta_i} \times \frac{\sin(\beta_i \frac{r}{k})}{\beta_i \frac{r}{k}} \exp\left(- \frac{\beta_i \alpha t}{k^2}\right)$$

$$N\beta_i = 1 - \beta_i \cot\beta_i \qquad N\beta_i = \frac{hR}{k}$$

is the exact solution for the following initial and boundary conditions:

1. $T = T_1$ at $t = 0$ for all r

2. $\frac{\partial T}{\partial t} = 0$ at $r = 0$ for all t

3. $k\frac{\partial T}{\partial r} = h(T-T_1)$ at $r = R$ for all t

These conditions describe a case whereby a constant and temperature independent thermal properties spherical object, having an initially uniform temperature (condition #1), is cooled down by being exposed to constant media temperature having a constant uniform convective heat transfer coefficient at the surface (condition #3). Condition #2 is due to symmetry. In many cases such conditions present an oversimplified description of the physical situation. Cases typical to precooling such as fluctuating or programmed media temperature; non uniform initial body temperature; substantially different thermal properties within the body - i.e., peel and mesocarp, etc. - pose such difficult conditions that no analytical exact solution is available.

To solve such cases we have used the Crank Nicolson finite difference technique (4). In essence, the technique solves numerically the differential equations by approximating the derivatives using Taylor expansion series.

The direction vs time plane (r vs t) is described by an equal space mesh grid (Figure 1) where the i and j notation stand for direction and time axes, respectively. The method is unconditionally stable at any mode of division, and will provide a solution for all cases once the initial (line j = 0) and the boundary (line i = n) conditions are known.

Employing such technique, the set of n + 1 equations describing a convective boundary conduction heat transfer for a sphere (subdivided arbitrarily into 10 intervals, i = 0,1,2,...10) is given below:

- one equation consists of i = 0,1
- 9 equations consist of i = 0-10
- one equation consists of i = 9,10

(1) $\quad -3\dfrac{\Delta t}{(\Delta r)^2}\, U_{i,j+1} + (1 + 3\dfrac{\Delta t}{(\Delta r)^2})U_{o,j+1} = 3\dfrac{\Delta t}{(\Delta r)^2}\, U_{i,j} +$

$\qquad (1 - 3\dfrac{\Delta t}{(\Delta r)^2})U_{o,j}$

(2-10) $\quad -\dfrac{\Delta t}{(\Delta r)^2}(1 - \dfrac{1}{i})U_{i-1,j+1} +(2 + 2\dfrac{\Delta t}{(\Delta r)^2})U_{i,j+1} -$

$\qquad \dfrac{\Delta t}{(\Delta r)^2}\,(1 - \dfrac{1}{i})U_{i+1,j+1} = \dfrac{\Delta t}{(\Delta r)^2}(1 - \dfrac{1}{i})U_{i-1,j} +$

$\qquad (2 - 2\dfrac{\Delta t}{(\Delta r)^2})U_{i,j} + \dfrac{\Delta t}{(\Delta r)^2}(1 + \dfrac{1}{i})U_{i+1,j}$

(11) $\quad 2\dfrac{\Delta t}{(\Delta r)^2}\, U_{9,j+1} + (2 + 2\dfrac{\Delta t}{(\Delta r)^2} + 0.22\dfrac{\Delta t}{(\Delta r)^2}\, h)U_{10,j+1} =$

$\qquad 2\dfrac{\Delta t}{(\Delta r)^2}\, U_{9,j} + (2 - 2\dfrac{\Delta t}{(\Delta r)^2} - 0.22\dfrac{\Delta t}{(\Delta r)^2}h)U_{10,j}$

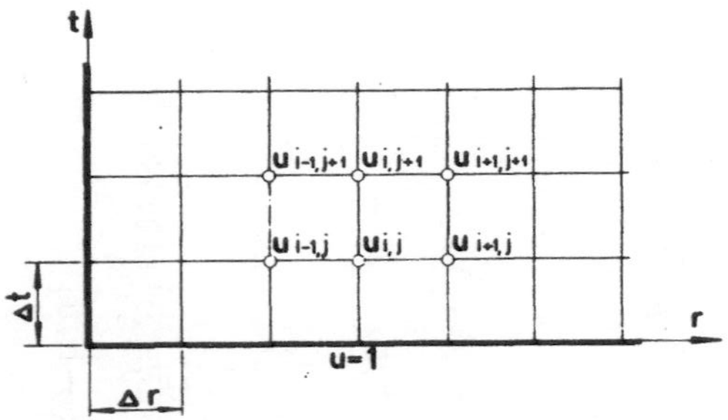

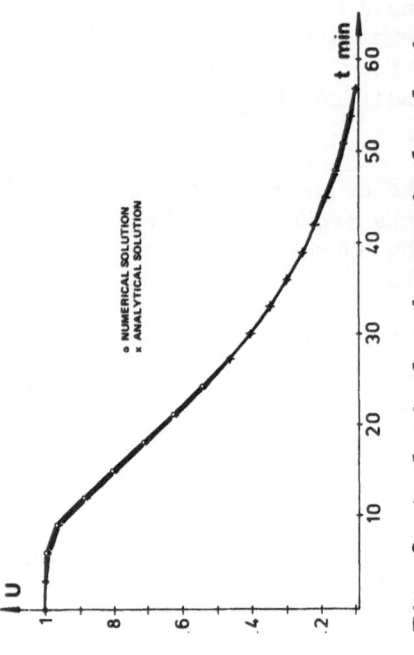

Fig. 1. Schematic diagram of the mesh grid.

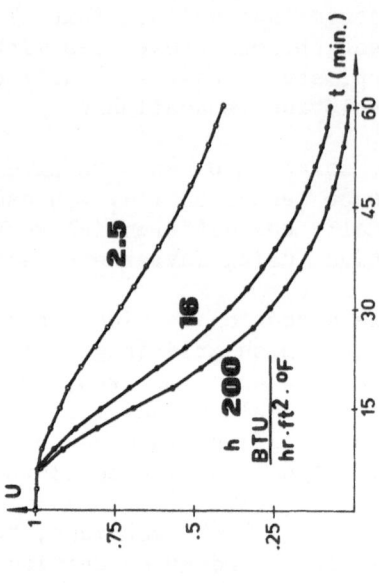

Fig. 2. Analytical and numerical solution
at the center of a spherical body.

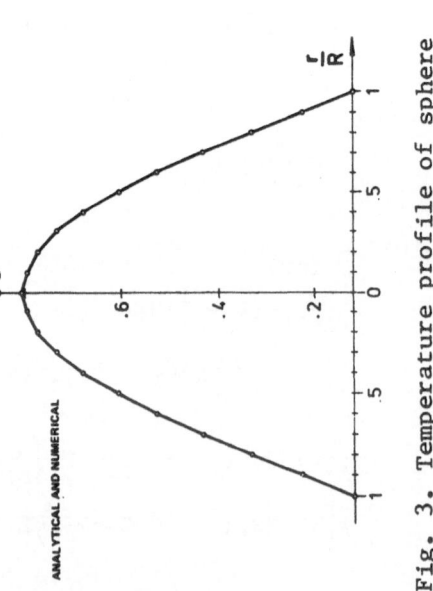

Fig. 3. Temperature profile of sphere
during cooling. Time=30 minutes.

Fig. 4. Cooling rates in air and water.

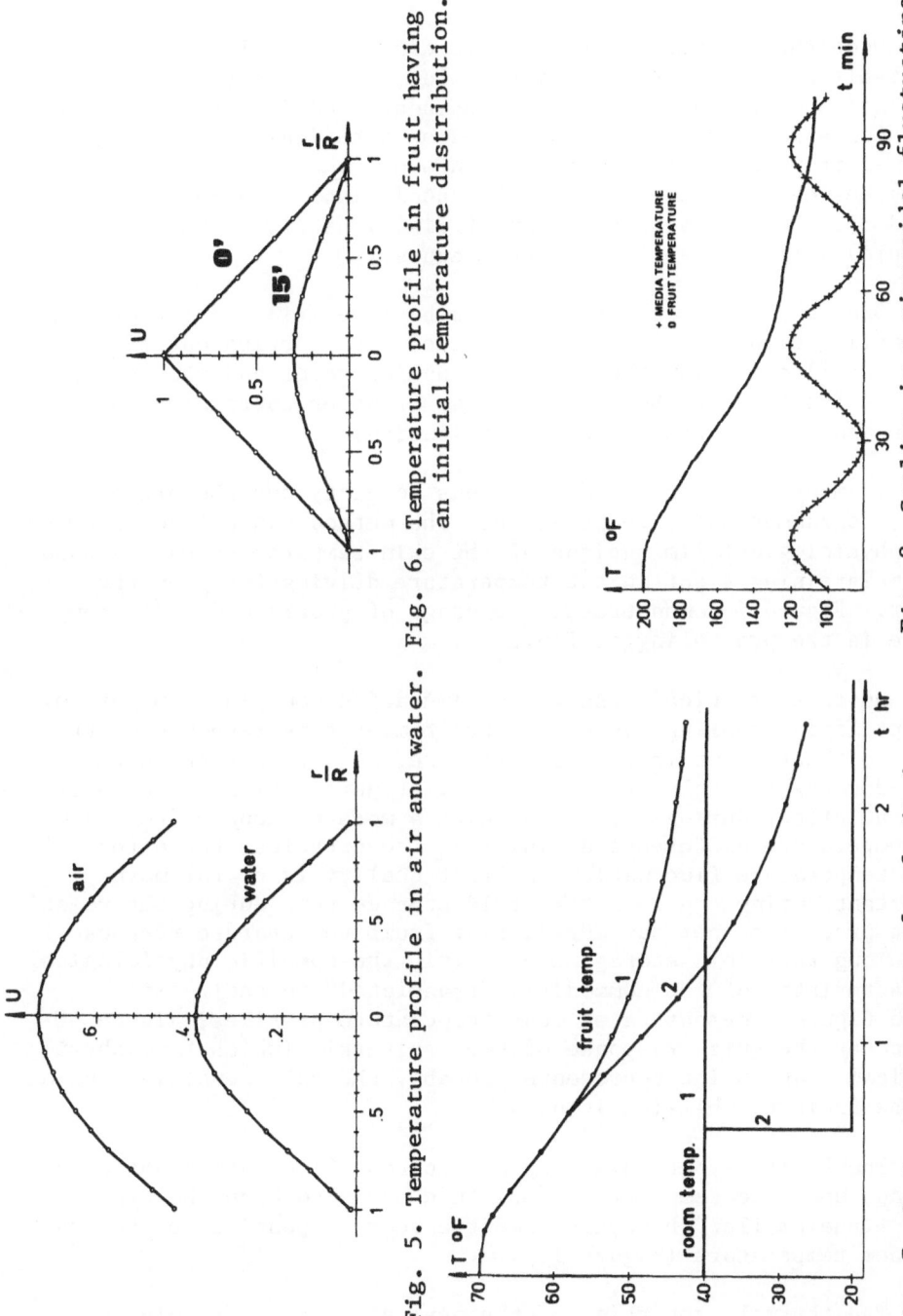

Fig. 5. Temperature profile in air and water.

Fig. 6. Temperature profile in fruit having an initial temperature distribution.

Fig. 7. Center temperature of a fruit in a programmed media temperature.

Fig. 8. Cooling in a sinusoidal fluctuating media temperature.

These n + 1 equations with n + 1 unknowns are solved simultaneously
to yield the 11 dimensionless temperatures.

Once the accuracy of the finite difference solution compares
favorably with a known analytical solution, it can be used for
simulation of any different given boundary and initial conditions.
Such comparison (done for 0.1 ft radius spherical body having the
thermal properties of an apple) points out the excellent agreement
(±1%) between the numerical and the analytical solutions (Figures
2 and 3). Increasing the number of divisions (10 in the above
example) will improve further the accuracy.

Once the numerical solution has been verified, a set of fruit
precooling simulations in air and water were carried out (Figures
4, 5 and 6) outlining the relative cooling rates and the temperature as function of the external heat transfer coefficient of the
media and initial temperature distribution.

In many cases it is advantageous to carry out the precooling
in a programmed media temperature. The method can balance between
the physiological limitations of the skin temperature on one hand
and maintaining a sufficient temperature driving force on the
other. Figure 7 demonstrates the usage of programmed media temperature in the precooling of fruit.

Another practical case where simulation can play a major role
in optimizing cooling systems is the temperature response of the
body to fluctuation in room temperature, occurring both in the
pull-down or the cold storage period. Figure 8 clearly demonstrates
such an effect during precooling with a more pronounced response
as temperature difference diminished. Nevertheless the effect of
room temperature fluctuation on fruit quality is by far more
important during the long term cold storage than during the relative
short precooling period. Predicting fruit temperature response in
such long term cold storage can be (via the specific physiological
characteristic of the commodity) "translated" to shelf-life.
While Figure 9 demonstrates body temperature profiles, Figure 10
indicates the quick response of the fruit skin (which from physiological stand point represents probably the most sensitive region
in the fruit to chilling injury).

The fruit temperature response to room fluctuation not only
depends upon the temperature amplitude but also upon the cycle
time – the smaller the cycle time the less responsive is the fruit
to room temperature (Figure 11).

The thermal properties of the peel plays a major role in
determining the cooling rates of citrus. Citrus peel, contains
large amount of air and its thermal conductivity is substantially

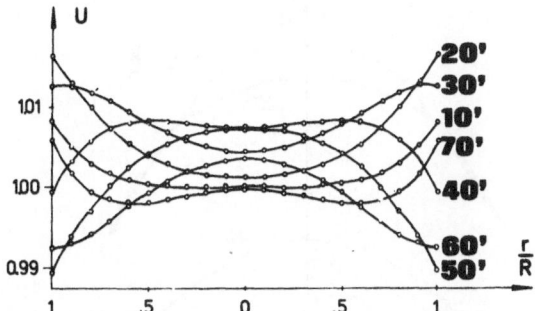

Fig. 9. Fruit temperature exposed to sinusoidal fluctuating media temperature during storage.

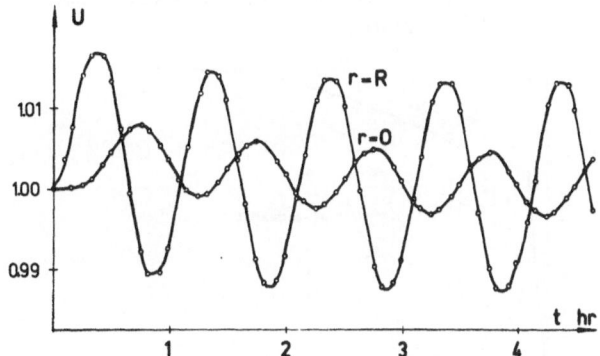

Fig. 10. Surface and center body temperature
during storage in sinusoidal fluctuating media temperature.

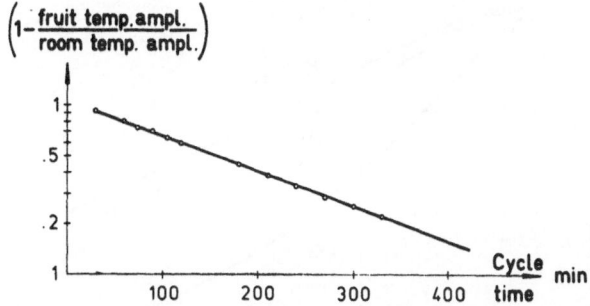

Fig. 11. The dependency of fruit temperature
amplitude with cycle time.

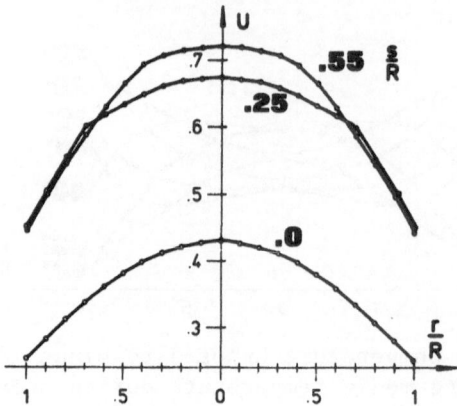

Fig. 12. Temperature profile in fruits as
         related to peel thickness. Time=
         60 minutes.

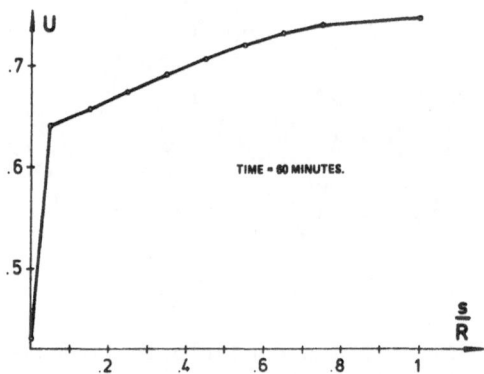

Fig. 13. Center fruit temperature as related
         to peel thickness.

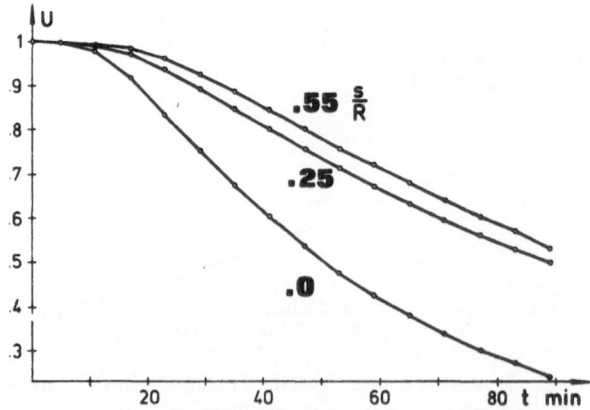

Fig. 14. Cooling pattern as related to
         peel thickness.

smaller than that of the mesocarp, thus affecting cooling rates (Figure 12), and temperature profiles (Figure 13). Citrus fruits vary substantially with respect to peel thickness and its thermal properties. It can be seen (Figure 14) that the effect of peel cannot be neglected and should always be taken into account when cooling rates are calculated.

In conclusion the present study suggests the finite difference technique for simulation of precooling of fruits under practical conditions such as fluctuating and programmed media temperature; non uniform initial body temperature; multiple thermal properties body, etc. The simulation is also useful in other aspects - such as optimizing of cooling systems; and matching physiological requirements of the produce with cooling rates, to mention some.

Further work in using the technique in determining the heat transfer and cooling rates of citrus fruits in boxes in containers is being presently carried out.

Nomenclature: R - body radius 0.1 ft

k - 0.2 btu/hr °F ft^2/ft

ρ - 50 lb/ft^3

C_p - 1.0 btu/lb°F

α - 0.004 ft^2/hr

α_{peel} - 0.002 ft^2/hr

REFERENCES

1. Bennett, A.H. and Yost, G.E. "Temperature response of Florida citrus to forced air precooling", ASHRAE Journal, April 1966.

2. Gibbon, J.M. "Some observation of temperature and cooling rates of vegetables in commercial cold stone", J. Agric. Engin. Res. (1972), 17, 332-337.

3. Pflug, I.J. and Kopelman, I.J. "Correlating and predicting transient heat transfer rates in food products", Supplement to I.I.R. Bulletin 1966-2.

4. Smith, G.D. "Numerical solution of partial differential equations", Oxford University Press, London 1965.

5. Van Beek, G. "Heat transfer through layers of agriculatural products of near spherical shape", Ann 1974-3 to the Bull. I.I.R. pp. 183-188.

INTERPRETATION OF OPTICAL INTERFEROGRAMS USING A NEW

THEORY FOR REFRACTIVE INDEX - TEMPERATURE DEPENDENCE*

Walter P. Schimmel, Jr.

Fluid Mechanics & Heat Transfer Division 1261
Sandia Laboratories
P.O. Box 5800
Albuquerque, NM   87115

INTRODUCTION

A severe limitation in the application of quantitative interferometry to heat transfer problems is a lack of necessary optical property information. To extract heat transfer coefficients from interferometric records, it is necessary to know the relationship between refractive index and temperature. In the case of gases, one can determine a closed form relationship via fundamental physical laws subject to some simplifying assumptions. For liquids, however, no such simple relationship has been observed.

The purpose of the present work is to present a relationship derived from first principles in which the first temperature derivative of the refractive index is determined from physical and optical properties. Results of the analysis will be compared with experimentally measured values for some common liquids.

ANALYSIS

An interferogram or pictorial distribution of phase difference can be related to physical property changes in one of the interferometer legs via the idealized equation of interferometry,

---

*This work was supported by the Department of Energy.

$$S(x,y) = \frac{\ell}{\lambda}\left[n_o - n(x,y)\right] \quad . \tag{1}$$

The term $S(x,y)$ can be interpreted as the "fringe" distribution, ℓ is the length of the test section, λ is the wavelength of the light source producing the interference, $n(x,y)$ is the distribution of refractive index within the test section and n_o is the refractive index in the interferometer reference arm. Note that two-dimensional distributions are assumed for both $n(x,y)$ and $S(x,y)$ with the third dimension being along the light source beam. If a refractive index change, Δn is defined,

$$\Delta n = n_o - n(x,y) \quad , \tag{2}$$

a Taylor's series expansion can be made in terms of the temperature,

$$\Delta n = \frac{dn}{dT} \Delta T + \frac{d^2n}{dT^2} (\Delta T)^2 + \ldots \quad . \tag{3}$$

If one notes that the higher order terms are small, there results,

$$\Delta n \simeq \frac{dn}{dT} \Delta T \quad . \tag{4}$$

Rewriting Equation (1) in terms of ΔT,

$$\Delta T = \frac{\lambda}{\ell}\left(\frac{dT}{dn}\right)S(x,y) \quad . \tag{5}$$

The point of the above is that the temperature distribution can be related to the fringe distribution via the refractive index gradient (dn/dT) provided it is constant over the temperature interval considered.

For a gas, the relationship can be determined in closed form (Reference 1). The refractive index is related to temperature via the Gladstone-Dale equation,

$$n = \rho G + 1 \quad , \tag{6}$$

and the ideal gas equation of state,

$$\rho = \frac{P}{RT} \quad . \qquad (7)$$

Eliminating the density, ρ, between Equations (6) and (7) results in,

$$n = \frac{GP}{RT} + 1 \quad . \qquad (8)$$

The R in Equations (7) and (8) is the so-called ideal gas constant. For a constant pressure process, the refractive index gradient becomes,

$$\frac{dn}{dT} = - \frac{GP}{RT^2} \quad , \qquad (9)$$

which is constant only for an isothermal process. If the temperature does not change greatly from the initial temperature, T_o, the assumption can be made,

$$\frac{dn}{dT} \approx - \frac{GP}{RT_o^2} \simeq constant \quad . \qquad (10)$$

Now Equation (5) can be written as,

$$\Delta T = \frac{\lambda}{\ell}\left(- \frac{RT_o^2}{GP}\right)S(x,y) \quad . \qquad (11)$$

This relationship and others similar to it have been used for years to evaluate interferometric results in gases. For liquids, however, there is usually no closed form equation of state.

One must measure dn/dT experimentally and such measurements are generally available for only a limited number of liquids. We now proceed to derive an expression for a generalized liquid which will be checked with existing dn/dT values for several specific liquids.

The molar refraction, $\bar{R}$, of a substance is defined by,

$$\bar{R} = \left(\frac{n^2 - 1}{n^2 + 2}\right)\frac{M}{\rho} = constant \quad , \qquad (12)$$

where n is the refractive index, M is the molecular weight
and ρ is the density. Solving for n results in,

$$n = \left(\frac{1 + 2 \frac{\overline{R}\rho}{M}}{1 - \frac{\overline{R}\rho}{M}} \right)^{\frac{1}{2}} \quad . \tag{13}$$

The equation of state of a material is actually a
relationship between any three independent variables.
For example, the density can be expressed as a function
of pressure and temperature,

$$\rho = \rho(P,T) \quad . \tag{14}$$

From Equation (13), it is clear that the refractive
index is a function of ρ and thus, (P,T). We are interested in the temperature derivative of the refractive
index at constant pressure, $(\partial n/\partial T)_P$. The chain rule
of differentiation can be applied as follows,

$$dn(\rho) = \left(\frac{dn}{d\rho}\right) d\rho \quad , \tag{15}$$

$$\rho = \rho(P,T) \quad , \tag{16}$$

$$d\rho = \left(\frac{\partial \rho}{\partial P}\right)_T dP + \left(\frac{\partial \rho}{\partial T}\right)_P dT \quad , \tag{17}$$

$$dn(\rho) = \frac{dn}{d\rho}\left[\frac{\partial \rho}{\partial P}\bigg|_T dP + \frac{\partial \rho}{\partial T}\bigg|_P dT \right] \quad . \tag{18}$$

Now,

$$\frac{\partial n}{\partial T}\bigg|_P = \frac{dn}{d\rho}\left(\frac{\partial \rho}{\partial T}\right)_P \quad , \tag{19}$$

but,

$$\frac{\partial \rho}{\partial T}\bigg|_P = -\beta\rho \quad ,$$

where β is the coefficient of volumetric expansion. Our
final expression is thus,

$$\left.\frac{\partial n}{\partial T}\right|_P = -\rho\beta\left(\frac{dn}{d\rho}\right) \quad . \tag{20}$$

Returning now to Equation (13) and taking the derivative with respect to ρ results in,

$$\frac{dn}{d\rho} = \frac{3\overline{R}/M}{2n\left(\frac{\overline{R}\rho}{M} - 1\right)^2} \quad , \tag{21}$$

and thus,

$$\left.\frac{\partial n}{\partial T}\right|_P = -\frac{3\beta}{2n}\left[\frac{\overline{R}\rho/M}{\left(\frac{\overline{R}\rho}{M} - 1\right)^2}\right] \quad . \tag{22}$$

It is usual to have refractive index values rather than molar refraction values so Equation (22) is recast in terms of n via Equation (12),

$$\left.\frac{\partial n}{\partial T}\right|_P = -\frac{3\beta}{2n}\left[\frac{\dfrac{n^2 - 1}{n^2 + 2}}{\left(\dfrac{n^2 - 1}{n^2 + 2} - 1\right)^2}\right] \quad . \tag{23}$$

This simplifies to,

$$\left.\frac{\partial n}{\partial T}\right|_P = -\frac{\beta}{6n}(n^2 - 1)(n^2 + 2) \quad , \tag{24}$$

which is the desired expression.

Note that for a perfect gas (where n is approximately unity), Equation (24) simplifies to Equation (9) as it should. That is,

$$\beta \simeq 1/T, \quad n = 1$$

$$\left.\frac{\partial n}{\partial T}\right|_\rho \simeq \lim_{n \to 1}\left[-\frac{1/T}{6n}(n - 1)(n + 1)(n^2 + 2)\right] \quad . \tag{25}$$

Taking the limit for all but the n - 1 term,

$$\frac{\partial n}{\partial T}\bigg|_{\rho} = -\frac{1/T}{6}(n-1)(2)(3) = -(\frac{n-1}{T}) \quad , \qquad (26)$$

but recall Equations (6) and (7),

$$n - 1 = \rho G \quad , \qquad (6)$$

$$\rho = \frac{P}{RT} \quad , \qquad (7)$$

therefore,

$$\frac{\partial n}{\partial T}\bigg|_{\rho} = -(\frac{GP}{RT^2}) \quad , \qquad (27)$$

which is Equation (9) for $T = T_o$.

COMPARISON WITH PUBLISHED DATA

As mentioned in the previous section, some existing values in the literature will be compared with results from the theory. The values are all taken from Reference 2:

Liquid	β, 1/K	n	$-\frac{dn}{dT}\bigg|_{Eqn\ 24}$ x 10^4	$-\frac{dn}{dT}\bigg|_{Ref\ 2}$ x 10^4
Water	0.256	1.3314	0.934	0.985
Methyl Alcohol	1.19	1.3253	4.25	4.0
Ethyl Alcohol	1.10	1.3583	4.39	4.0
Isopropyl Alcohol	1.06	1.3726	4.42	4.5
Benzene	1.23	1.4950	7.17	6.4
Toluene	1.06	1.4901	6.11	5.55
Nitrobenzene	0.83	1.5458	5.46	4.68
Acetone	1.43	1.3542	5.63	4.31
Carbon Tetrachloride	1.22	1.4547	6.42	5.98

Note that in the case of the homologous series, methanol, ethanol, proponal, etc., the theory predicts an increase of dn/dT with molecular weight. This is an expected result which appears to be lost in the experimental measurements given in Reference 2. It may well be that the theory suffers from the values of β and n used. For example, a 1% reduction in n for benzene results in a value of -6.91×10^{-4} for dn/dT$|$P.

Regardless of this, the theory is observed to predict values of $dn/dT|P$ which are within 10% of the measured results (always on the high side). It appears likely, therefore, that Equation (24) can be used to evaluate interferograms generated in liquids for which $dn/dT|P$ is not known.

REFERENCES

1. Schimmel, W. P., Jr., "An Optical Measurements Laboratory for Determining Heat Transfer Coefficients," SAND76-0162, Sandia Laboratories, Albuquerque, NM, June 1976.

2. Hauf, W., and Grigull, U., "Optical Methods in Heat Transfer," Advances in Heat Transfer, Vol. 6, Academic Press, New York, 1970.

APPLICATION OF LASER HOLOGRAPHIC INTERFEROMETRY

TO NATURAL CONVECTION IN ENCLOSURES*

Walter P. Schimmel, Jr.
Fluid Mechanics & Heat Transfer Division 1261
Sandia Laboratories
P.O. Box 5800
Albuquerque, NM   87115

INTRODUCTION

The understanding of the natural convection processes occurring within and around a closed vessel is crucial
to many proposed "alternate" energy schemes. Solar energy
extraction, for example, is concerned with both convective
losses from the collectors and long term stability of
liquid filled thermal storage tanks. With respect to the
nuclear industry, a long list of natural convection in
enclosures exists. Transportation of spent fuel and
storage of thermally active wastes are only two of the
problems falling into this category.

Because the natural convection fields are extremely
sensitive to disturbance from probes, it is desirable to
investigate them using some sort of optical measurement
technique. The present work is concerned with such a
technique, laser holographic interferometry, and its
application to convection in enclosures.

A laboratory being used for determining heat transfer
coefficients using the technique of laser holographic
interferometry (holometry) is described.[1] Using a relationship between temperature and index of refraction,
several cases of free convection in enclosures are evaluated. Results are presented for a heated hexagonal
cylinder in an isothermal circular enclosure. In addition,

---

*This work was supported by the Department of Energy.

a heated hexagonal cylinder in an isothermal box is evaluated. These provide check cases to some numerical code development which is being done in parallel with the optical measurements laboratory effort.[2]

Example holometrograms or holometric interferograms are shown to agree with analytical predictions as to number of fringes in the convective field. For the case of the hexagonal cylinder in the circular cross-section enclosure, it is shown that convection has a significant influence upon the temperature gradient (heat transfer coefficient) when compared with thermal conduction.

DESCRIPTION OF LABORATORY AND EXPERIMENTAL APPARATUS

The components of the holometric setup are presented in schematic form in Figure 1. The entire arrangement is isolated from building vibrations by a Modern Optics V-12 stable table system with viscous mechanical dampening. Resonant frequency of the air mount system is less than 1.25 Hz. Illumination of the system is provided by a Spectra-Physics model 125-A helium-neon laser with RF excitation. The electronic shutter system is a Jodon model ES-100 used in the integrating exposure mode with the power density detector located behind the hologram plate (Figure 1). The camera used to record the resulting interferograms is a Rolleiflex SL-66, 6 cm x 6 cm, single lens reflex with a 150 mm f4.0 lens. All photographs were taken with Polaroid type 105 black and white P-N film. The negatives were then analyzed on a Gaertner toolmaker's microscope with substage illumination.

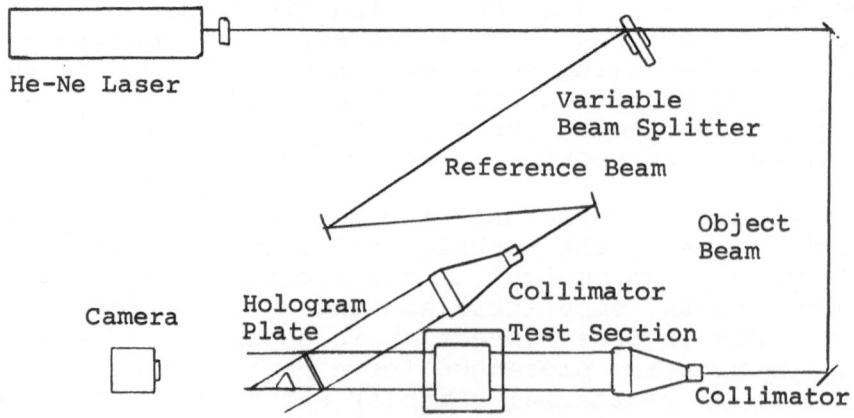

FIGURE 1. Laser Holometric Setup, Schematic

The test sections used in this preliminary study were mounted on an interferometrically stable laboratory jack. To monitor surface and ambient temperatures, a series of chromel-alumel thermocouples were read on a digital temperature indicator.

The first configuration examined in the present study was a heated horizontal right circular cylinder in an isothermal rectangular cross-section box. A transparent glass plate covered the entrance end of the box (with respect to the laser beam) and a ground glass plate formed the exit end boundary. The 2.8 cm diameter cylinder was supported by a phenolic stand which provided stability to the cylinder and acted as a thermal insulator between the cylinder and box. Thermocouples were mounted on the inside of the cylinder next to the electrical heating coil and on the outer surface between the cylinder and the stand. The heater and thermocouple leads were routed to the rear of the cylinder and then to the temperature recorder. The surrounding box was constructed of 1.9 cm thick aluminum plate. Cross hairs on the entrance and exit planes helped in the alignment process. The opening in the box was 6.4 cm x 8.9 cm and the inside was painted flat black to minimize reflection. The configuration approximated a heated reactor rod in a box-like container.

Of some direct interest to the fuel cycle of the Liquid Metal Fast Breeder Reactor is the case of a heated hexagonal cross-section cylinder in a circular cross-section container. This configuration was also investigated in the laboratory. Dimensions of the model were 2.5 cm across the flats on the hexagon and 3.7 cm inside diameter on the surrounding circular cylinder. Wall thickness of this outer cylinder was 1.3 cm. Again, the ends of the outer cylinder were covered by glass plates-- clear on the entrance and diffuse on the exit.

TEMPERATURE FIELD DETERMINATION FROM FRINGE SHIFT DATA

In order to use the interference effect to deduce temperature profiles, one must derive a relationship between index of refraction and temperature. The preceding developments can thus be used to relate temperature to fringe shift data. In the discussion to follow, only gases will be considered because the added consideration of mechanical strain in solids is not encountered in convection studies, and liquids usually do not have explicit equations of state. It should also be mentioned

that the analysis is not restricted to holographic interferometry but applies to all types of interferometry.

The index of refraction of a gas can be related to
its density under the appropriate conditions, that is

$$n = \rho G + 1 \quad . \tag{1}$$

In the range of parameters in which gases "obey" the
Gladstone-Dale equation, the ideal gas equation of state
can be used

$$P = \rho RT \quad . \tag{2}$$

Here, P is the pressure, ρ the density, T the temperature
in absolute degrees, and R the ideal gas constant for the
particular gas being considered. Eliminating ρ between
Equations (1) and (2) yields

$$n = G\left(\frac{P}{RT}\right) + 1 \quad . \tag{3}$$

In interferometric experiments, the retardation of
the optical wavefront by one wavelength corresponds to
one complete fringe shift. Thus, a shift of ε fringes
corresponds to a wave retardation of $\lambda_r \varepsilon$, where λ_r is
the wavelength of light in the reference medium used.
A general expression for fringe shift is thus,

$$\varepsilon(x,y) = \frac{L}{\lambda_o}\left[n_2(x,y) - n_1(x_1 y)\right] \quad . \tag{4}$$

The question of three-dimensionality which is unique to
holographic interferometry is avoided at the present and
only two-dimensional (x,y) fringe shifts are considered.

To apply these general results to heat transfer
experiments, Equation (3) is inserted in Equation (4),

$$\varepsilon = \frac{L}{\lambda_o}\left[\left(\frac{GP_2}{RT_2} + 1\right) - \left(\frac{GP_1}{RT_1} + 1\right)\right] \quad , \tag{5}$$

or,

$$\varepsilon \frac{GL}{\lambda_o R}\left(\frac{P_2}{T_2} - \frac{P_1}{T_2}\right) \quad . \tag{6}$$

For the care where $P_2 \approx P_1 = P$, the expression becomes

$$\varepsilon = \frac{GLP}{\lambda_o R} \left(\frac{1}{T_2} - \frac{1}{T_1} \right) . \qquad (7)$$

Because the quantity of interest is usually T_2, and because T_1 and ε are measured, Equation (7) is rewritten as,

$$\frac{T_2}{T_1} = \frac{1}{1 + \beta \varepsilon} , \qquad (8)$$

where,

$$\beta = \frac{\lambda_o R T_1}{PLG} . \qquad (9)$$

Note in Equation (9) that $T_2 > T_1$ corresponds to $\rho_2 < \rho_1$ and thus ε will be negative (β is positive for all cases).

LABORATORY RESULTS

The configuration of the heated circular cylinder in the isothermal rectangular cross-section box produced the double exposure holometrogram (holometric interferogram) presented in Figure 2. The cylinder surface temperature was determined to be 351.5K and the box temperature was 294.3K. Both measurements were made with thermocouples as described in the previous section. The quantity, β (Equation (9)), was calculated to be 0.0268. Use of the data reduction technique outlined above indicates that there should be six fringes. The photograph clearly confirms this prediction. Temperatures corresponding to the fringes are:

ε	T, K	$T - T_1$, K
cold surface	294.3	0
1	302.4	8.1
2	311.0	16.7
3	320.0	25.7
4	329.6	35.3
5	339.8	45.5
6	350.7	56.4
hot surface	351.5	57.2

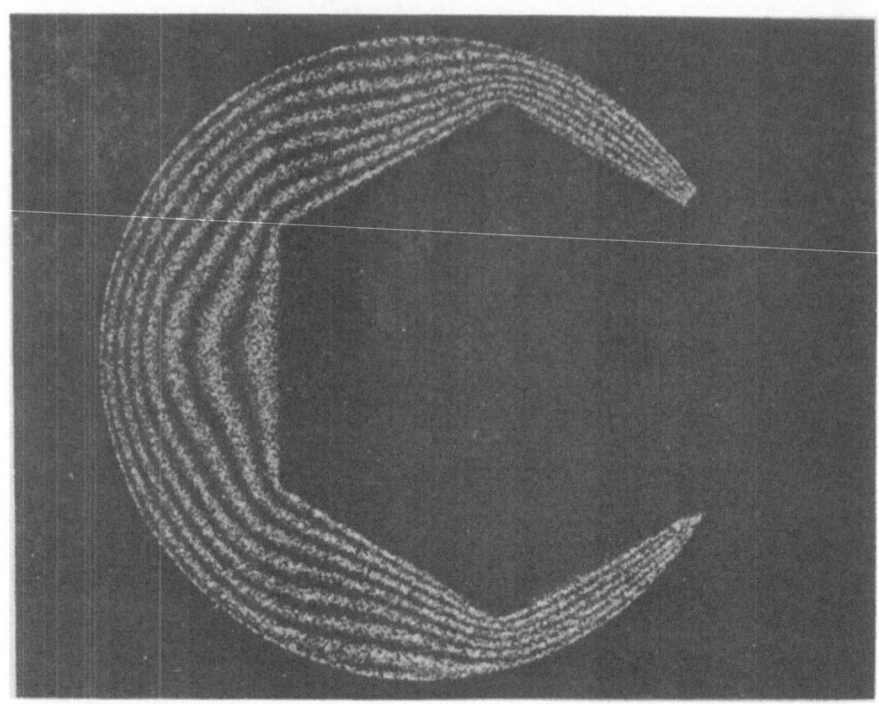

FIGURE 3
Holometrogram of Test Section 2

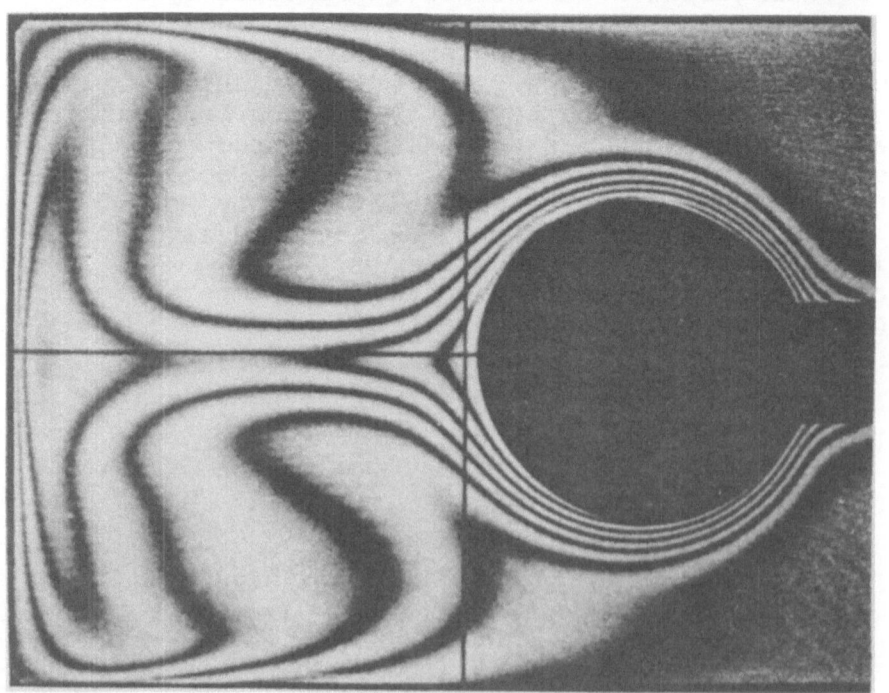

FIGURE 2
Holometrogram of Test Section 1

Because of the preliminary nature of this study, no attempt was made to deduce heat transfer coefficients for this configuration. Such a determination will be made for the next test section considered.

The heated hexagonal cross-section cylinder in the circular cylinder container was the next configuration to be examined in the laboratory. The double exposure holometrogram corresponding to this case is presented in Figure 3. The fringe shift parameter, β, was calculated to be 0.0185. Again, the predicted number of fringes agreed with the observed number in this photograph. It should be pointed out that this is not always the case. In some earlier studies, in which the length of the test section was not carefully controlled, a different number of fringes resulted. In the two cases reported in the present work, the test section length was delineated by the containing glass plates. The temperatures corresponding to the fringes are:

ε	T, K	$T - T_1$, K
cold surface	294.4	0
1	299.9	5.5
2	305.7	11.3
3	311.7	17.3
4	317.9	23.5
5	324.4	30.0
6	331.2	36.8
7	338.2	43.9
hot surface	344.4	50.0

For this case, the temperature gradients along the vertical were determined as outlined above. The gradient corresponding to pure conduction is 49.0 K/cm. At the hot surface, the gradient is 29.6 K/cm while at the cold surface it is about 116.0 K/cm. Because energy must be conserved across the interface between fluid and solid, this means that the presence of the convective field has resulted in a twofold (at least) increase to the heat flux at the top of the container when compared with conduction. Were this an actual test case rather than a demonstration model, a similar determination of the temperature profiles across the layer would be made for other locations. Because the holometrogram is a permanent record, this data reduction task could be carried out at some later time.

REFERENCES

1.   Schimmel, W. P., Jr., "An Optical Measurements
     Laboratory for Determining Heat Transfer
     Coefficients," SAND76-0162, Sandia Laboratories,
     Albuquerque, NM, 87115, June 1976.

2.   Larson, D. W., Gartling, D. K., and Schimmel, W. P.,
     Jr., "Computational and Experimental Methods for
     Enclosed Natural Convection," SAND77-0645, Sandia
     Laboratories, Albuquerque, NM, 87115, June 1977.

THERMAL CONDUCTIVITY ANALYSIS

OF A MULTICHIP MICROPACKAGE

E. A. Wilson

Honeywell Information Systems, Inc.

Phoenix, Arizona

I. INTRODUCTION

As integrated circuit speeds increase, the length of interconnecting signal runs become a significant portion of the delay
in a logic network. This forces increased packaging density with
the related problems of heat removal. One of Honeywell's answers
to increasing packaging density is the multichip micropackage
which is an alumina substrate with layers of dielectric and signal
runs screened and fired on it to form a small, high density, multilayer circuit "board". The chips are soldered to metal pads on the
surface of the micropackage and the solder joint has voids. The
solder under the chip not only holds the chip in place, but is the
prime path for heat flow from the chip. Two of the questions which
arose during the thermal analysis of the micropackaging concept
were how void free must the solder joint be in order to guarantee
an adequate thermal path (especially since the dielectric is a very
poor thermal conductor) and how much can the beam leads be depended
on as a thermal path as this void area increases?

In addition to these basic questions, a need was found for a
preliminary design aid which could be used to predict junction
temperatures as integrated circuit chips and micropackages are
designed. Since this tool would be used by other than engineers
experienced in heat transfer analysis, it has to be simple and
easy to use and preferably accessible via a terminal instead of
batch jobs.

This paper presents the investigation into the influence of
beam leads and solder voids on integrated circuit chip temperatures

and the development of the preliminary design program. The following is contained in the subsequent sections:

- A finite element for chip/micropackage thermal analysis;

- The finite element model used for the beam lead and solder
  void example problems;

- Results of beam lead and solder void example problems.

II. THE FINITE ELEMENT

The shape of the element is shown in Figure 1. This is an
"approximate" element because all heat flow is assumed to be
perpendicular to the boundaries, hence the isothermal lines in the
plan view are all squares. Such an assumption is not true in the
corners of the element, but since the finite element model in the
next section cannot exactly represent the true pattern of heat
sources (transistors and resistors) of all chip types, it is
reasonable to use an "approximate" element for an approximate
solution.

The basic variational equation which serves as a conduction

matrix generator is $[K] = [A^{-1}]^T \iiint (\{\nabla\}\lceil L \rceil)^T (\{\nabla\}[L]) \, dV \, [A^{-1}]$ (1)

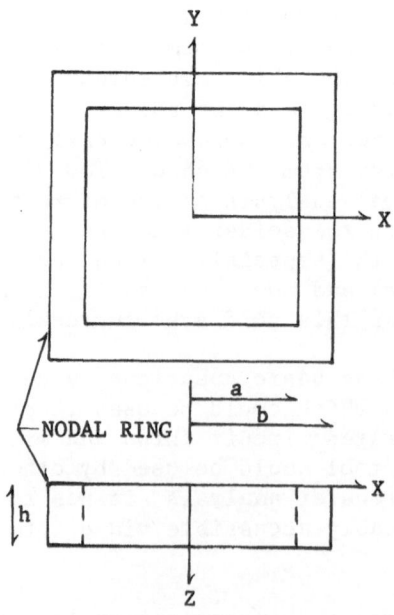

FIGURE 1

where

    [K] = Conduction Matrix
    [A] = Nodal Coordinate Matrix
    [L] = Temperature Function Vector
    {∇} = Differential Operator Vector
    dV = Differential Volume for Integration

Because of the simplifying assumption mentioned above, the problem
is reduced to just two coordinates, x and z. The nodes then are
actually nodal perimeters or square rings as shown in Figure 1.
This yields

$$[L] = [1 \ x \ z \ xz] \tag{2}$$

$$\{\nabla\} = \begin{array}{c} \frac{\partial}{\partial x} \\[6pt] \frac{\partial}{\partial z} \end{array} \tag{3}$$

$$[A] = \begin{array}{cccc} 1 & a & 0 & 0 \\ 1 & b & 0 & 0 \\ 1 & a & h & ah \\ 1 & b & h & bh \end{array} \tag{4}$$

$$dV = 8x \ dx \ dz \tag{5}$$

The integration limits are 0 to h for z and a to b for x. The
results of the matrix multiplications and integrations indicated
in equation (1) are presented below.

$$k_{11} = (b^2-a^2)[h/6 = (3b^2+a^2)/4h] + (b^3-a^3) \ (-2b/3h) \tag{6}$$

$$k_{12} = (b^2-a^2)[-h/6 - (a+b)^2/4h] + (b^3-a^3)(a+b)/3h \tag{7}$$

$$k_{13} = (b^2-a^2)[h/12 - (3b^2+a^2)/4h] + (b^3-a^3)(2b/3h) \tag{8}$$

$$k_{14} = (b^2-a^2)[-h/12 + (a+b)^2/4h] + (b^3-a^3)(-a-b)/3h \tag{9}$$

$$k_{22} = (b^2-a^2)[h/6+ (3a^2+b^2)/4h] + (b^3-a^3)(-2a/3h) \tag{10}$$

$$k_{23} = (b^2-a^2)[-h/12 + (a+b)^2/4h] + (b^3-a^3)(-a-b)/3h \tag{11}$$

$$k_{24} = (b^2-a^2)[h/12 - (3a^2+b^2)/4h] + (b^3-a^3)(2a/3h) \tag{12}$$

$$[K] = 8k/(a-b)^2 \begin{bmatrix} k_{11} & k_{12} & k_{13} & k_{14} \\ k_{12} & k_{22} & k_{23} & k_{24} \\ k_{31} & k_{32} & k_{11} & k_{12} \\ k_{14} & k_{24} & k_{12} & k_{22} \end{bmatrix} \tag{13}$$

where k is the material thermal conductivity.

Then the basic matrix equation becomes

$$[K]\{T\} = \{Q\} \tag{14}$$

where $\{Q\}$ is the vector of nodal heat inputs.

III. MODEL FOR EXAMPLE PROBLEMS

The model shown in Figure 2 is representative of the size of LSI high power density chips (.16 W/mm^2). The nodal heat inputs represent an active area (region in which the integrated circuit gates are located) of 1.75 mm by 1.75 mm. The beam leads are actually discrete pieces of copper, .025 mm thick and vary in width from .1 mm at the chip to .15 mm at the outer bonding pad. The model approximates the lead frames (up to 40 leads) with the ring elements by adjusting the thermal conductivity (k) to yield the same total resistance as the lead frame itself. Also, the dielectric elements in the region of the outer lead bond (designated as D' in Figure 2) have their thermal conductivity adjusted to account for the fact that the bonding pads are discrete and not a complete ring. The values

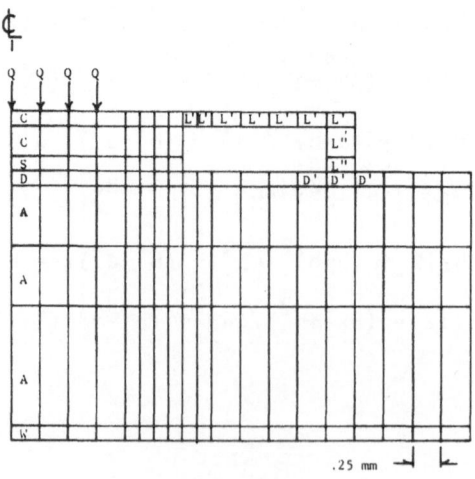

FIGURE 2

of k for the various elements are:

k (Chip)	= .1	W/(mm-$^{\circ}$C)	
k (Solder)	= .03	W/(mm-$^{\circ}$C)	
k (Dielectric)	= .0012	W/(mm-$^{\circ}$C)	
k (Dielectric)	= .0008	W/(mm-$^{\circ}$C)	
k (Alumina)	= .024	W/(mm-$^{\circ}$C)	
k (Leads)'	= .024	W/(mm-$^{\circ}$C)	(heat flow in x direction)
k (Leads)"	= .01	W/(mm-$^{\circ}$C)	(heat flow in z direction)

The row of elements along the bottom of the model (designated by W) have a thermal conductivity and thickness equivalent to the resistance between the back surface of the substrate on which the chip is mounted and the coolant inside a liquid cooled heat sink. The value of k/h used for this example is .00172 W/(mm^2-$^{\circ}$C) and corresponds to a substrate to coolant temperature drop of 8°C for a 60 Watt multichip micropackage.

The total width of the model (2x = 8mm) represents the chip mounting density of the size class of chips approximated by the chip in this model.

The effect of solder voids was investigated by setting the thermal conductivity of the solder elements to zero in a cumulative progression from the center to the edge of the chip, and also from the edge to the center. This provides two sets of data, one for continuous interior voids with a ring of solder and one for a continuous center post of solder. In addition to these progressions, individual and groups of solder elements were assigned k = 0 in order to approximate random, distributed discrete voids. These three types of voids are referred to in the next section as center void, perimeter void, and quasi random voids. The effect of beam leads for all three of the solder void patterns was easily determined by just omitting the lead elements.

IV. RESULTS OF VOID & LEAD INVESTIGATION

The curves in Figure 3 present the temperature at the center of the chip (with the power normalized to 1 Watt) relative to the minimum temperature on the bottom surface of the substrate. The variation in temperature along the bottom surface is only about 1°C and hence a simplified model which assumes an isothermal surface would not introduce a significant (greater than 10%) error due to such an assumption.

For both lead and no lead cases, the perimeter void forms an

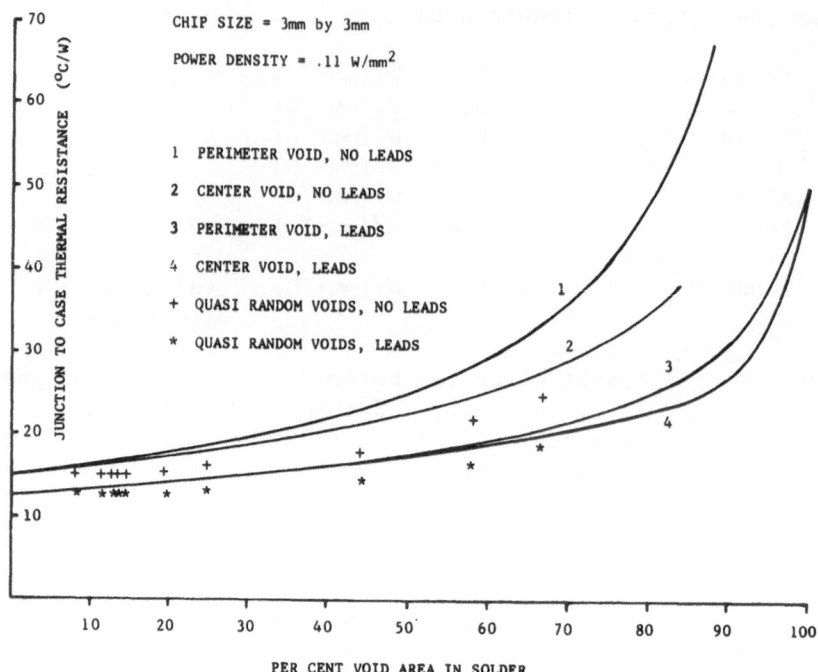

FIGURE 3

upper bond on void effects. It is important to note that for
quasi random voids below 30%, the practical result is that voids
may be ignored. The solder void pattern is relatively unimportant
when the beam leads are included in the analysis. For voids of
above 40% or 50% the effect of beam leads versus no beam leads on
the pattern becomes very important. Below 50% void area, the effect
of the beam leads is to reduce the chip temperature by 15% (0%
void area) to 25% (50% void area). It should be noted that the
assumption of 40 leads in the lead frame would probably be an upper
bound and a lesser number is more usual. The more likely range of
influence of the leads on chip temperature is 10% to 20% for the
void area range of 0% to 50%.

VI.   CONCLUSION

As a result of the work described in this paper, an X-ray
inspection was instituted in the micropackage manufacturing process
to assure that the per cent void area in the solder joint does not
exceed 25 per cent. Also, integrated circuit designs are routinely
checked with an electrical analogy program which gives a conservative
prediction of chip temperature which will still be valid up to a
40% void area when leads are included in the finite element analysis.

SESSION B   SOLIDS AT HIGH TEMPERATURE

Session Chairmen: J.G. Cook
National Research Council
Ottawa, Ont.

M.J. Laubitz
National Research Council
Ottawa, Ont.

THERMAL CONDUCTIVITY OF HEAVILY DOPED p-TYPE INDIUM ANTIMONIDE

R. D. Redin and T. Ashworth

Physics Department, South Dakota School of Mines and
Technology, Rapid City, South Dakota 57701

C. Y. Hsiung

Electrical Engineering Department, South Dakota School
of Mines and Technology, Rapid City, South Dakota 57701

INTRODUCTION

The thermal conductivity of p-type InSb is strongly influenced
by the scattering of phonons by free carriers. This was first suggested by Challis et al.[1] on the basis of measurements of the thermal conductivity of several n- and p-type samples over the temperature range of 1.2K to 4.2K. Crosby and Grenier[2] extended the measurements to dopings in excess of 10^{18} acceptors per cm^3. They successfully accounted for the observed temperature dependence in the
range 1.3K to 4.2K by modifying Ziman's[3] hole-phonon scattering relaxation time to include the effect of screening. Kosarev et al.[4]
further measured a group of n- and p-type samples from 2K to 100K.
They generalized Callaway's[5] model by using separate drift terms
for longitudinal and transverse phonons and were able to estimate
the contribution of hole-phonon scattering to both types of modes.
They used a formula for hole-phonon scattering which included the
influence of the coulomb field of the impurity ions on the scattering process. A good fit to the data was obtained in the range of
about 2K to 7K. In previous work[6] at this laboratory Holland's[7]
thermal conductivity formalism was applied to measurements made on
a heavily doped p-type InSb sample. The analysis showed that information about hole-phonon scattering could be obtained at temperatures as high as 80K.

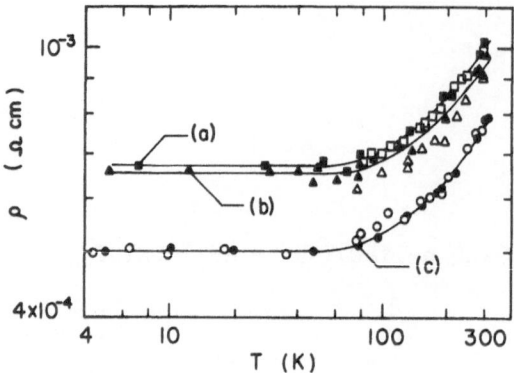

Fig. 1. Electrical resistivity
(a) Sample S5-A-31, (b) Sample S5-
A-42, (c) Sample E2-A. Solid points
were taken during the thermal conductivity measurements, open points
were taken during the Hall effect
measurements.

The principal features of the work reported in this paper are: 1.) Use of samples with doping levels greater than 10^{19}cm^{-3}; 2.) Measurements of greater accuracy and over a wider temperature range than those previously done at this laboratory; 3.) Improved sample characterization by measurements on the same samples of the Seebeck coefficient, Hall coefficient, and electrical resistivity; 4.) An attempt to provide a more rigorous theoretical treatment of three phonon processes and impurity scattering by use of a variational calculation.

MEASUREMENTS AND RESULTS

Three samples were measured. Samples S5-A-31 and S5-A-42 were single crystals with their [100] directions along the direction of heat flow in the thermal conductivity measurement. They were doped with zinc. Sample E2-A was polycrystalline (etching of the surface disclosed seven different single-crystal regions). The dopants were 50% Zn, 50% Cd, reported by the supplier[8] to be necessary to obtain the high impurity concentration. The hole concentrations obtained from Hall effect measurements, were 2.51 x 10^{19}cm^{-3}, 2.56 x 10^{19}cm^{-3}, and 3.55 x 10^{19}cm^{-3} for samples S5-A-31, S5-A-42, and E2-A respectively. The Hall coefficients of all three samples were constant from room temperature to liquid-nitrogen temperature. Hall coefficient measurements on sample E2-A were extended to liquid-helium temperature; the Hall coefficient remained constant. The three samples were, therefore, considered to be completely degenerate over the entire temperature range of the measurements.

The linear absolute method with precision guarding and temperature control[9] was used for the thermal conductivity measurements. Above 20K, the estimated random error introduced by the system was less than 2%. Below 20K, however, the uncertainty gradually increased to 8% at 5K. The uncertainty in sample geometry and the distance between thermocouples introduced an uncertainty of up to 11% in the thermal conductivity at all temperatures.

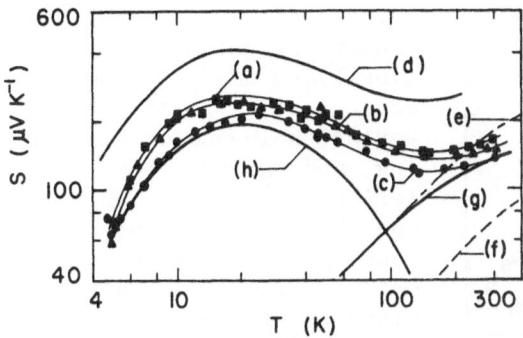

Fig. 2. Thermoelectric power. (a) Sample S5-A-31, (b) Sample S5-A-42, (c) Sample E2-A, (d) Tamarin et al., Ref. 10, $N_A = 10^{18} cm^{-3}$. Diffusion components for E2-A: (e) impurity scattering, (f) acoustic phonon scattering, (g) mixed impurity and phonon scattering. (h) Phonon–drag component for E2-A.

Electrical resistivity and thermoelectric power measurements were taken concurrently with the thermal conductivity measurements. The results for the electrical resistivity are shown in Fig. 1. Below 80K, the resistivities of the three samples were independent of temperature and decreased with increased doping. Above 80K, the resistivity increased with increasing temperature due to the additional hole scattering by phonons.

In Fig. 2, the measured thermoelectric power, curves (a), (b), and (c), show a temperature dependence similar to that of curve (d) from Tamarin et al.[10] The rise in thermoelectric power at low temperatures clearly exhibits the drag effect of phonons on carriers even though the effect is not as strong as for purer samples. For sample E2-A curves (e), (f), and (g) show the estimated diffusion thermoelectric power due to impurity scattering, acoustic phonon scattering, and the mixed impurity and phonon scattering respectively. Curve (h) gives the phonon–drag component for E2-A, which was obtained from the difference of the measured thermoelectric power and the estimated diffusion component curve (g). The phonon–drag component shows approximately the same temperature dependence as the thermal conductivity below 20K.

The measured thermal conductivities are shown as points in Fig. 3. The corresponding lines give the estimated lattice thermal conductivity in which a correction for the electronic thermal conductivity is included. At the high temperature end of these curves, the results agree within 10% with Holland,[11] and with Bush and Steigmeier.[12] At the low temperature end, the results are consistent with those of Crosby and Grenier. The electronic contribution to the thermal conductivity was estimated from the electrical resistivity by using the Wiedemann–Franz law. The charge carriers contributed 5.5% to the heat conduction at 300K and 1% at 100K.

For quantitative calculations of the thermal conductivity, we first applied Holland's[7] model. This model separates the contri-

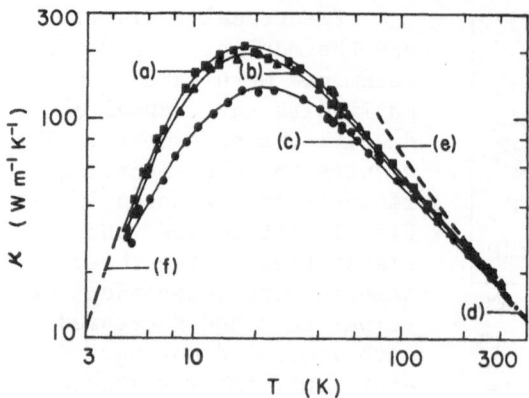

Fig. 3. Thermal conductivity. (a) Sample S5-A-31, (b) Sample S5-A-42, (c) Sample E2-A, (d) Bush and Steigmeier, Ref. 12, $N_A = 1.6 \times 10^{16}$ cm^{-3}, (e) Holland, Ref. 11, $N_D = 7 \times 10^{13}$ cm^{-3}, (f) Crosby and Grenier, Ref. 2, $N_A = 7.6 \times 10^{13}$cm^{-3}.

butions of longitudinal and transverse phonons. Relaxation times due to boundary scattering, impurity scattering, hole-phonon scattering, and phonon-phonon scattering were used. The phonon-phonon scattering terms contain three unknown parameters. These were determined by fitting the equation to Holland's[11] data for pure InSb. The values found for the parameters were identical to those reported by Bhandari and Verma.[13]

For the application to impure InSb the phonon-phonon scattering parameters were held at the same values as for the pure material. The boundary scattering and the impurity scattering terms were changed to correspond to the actual sample.

Two important adjustments were necessary to produce a good fit to the data. First the impurity scattering term had to be increased above that predicted from the known impurity content (3 times higher for sample E2-A). This effect has been frequently noted by other workers. Second, it was necessary to include hole-phonon scattering of transverse modes. With spherical electron energy surfaces only longitudinal modes are affected. This has been verified for n-type InSb by Moneval[14] in heat-pulse experiments. The scattering of transverse modes in p-type InSb is apparently due to the non-spherical hole band. We find $E_{12}/E_{11} = 0.4$, where E_{12} is the shear deformation potential and E_{11} the longitudinal deformation potential. Kosarev found 0.2 for this ratio. We used -1.5 eV for E_{11}. By adjusting the deformation potentials and adding electron screening or Kosarev's coulomb field effect one can get better fits to the data. However, one quickly has so many adjustable parameters that no definite conclusions can be reached.

We have also tried the variational method using the isotropic continuum model of Hamilton and Parrott.[15] Three-phonon processes are treated more rigorously and the parameters are the second-and third- order elastic constants which are available in the liter-

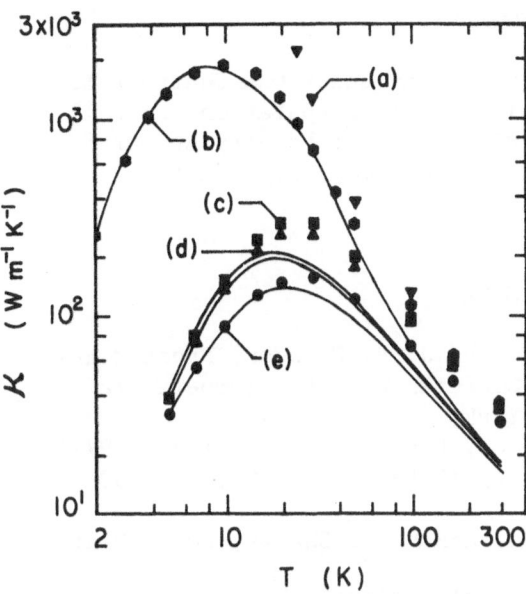

Fig. 4. Computed thermal conductivity of InSb with Hamilton and Parrott's model with modified elastic constants. Points are computed values and lines are experimental values. (a) three-phonon processes only, (b) Holland, Ref. 11, (c) Sample S5-A-31, (d) Sample S5-A-42, (c) Sample E2-A.

ature.[16] We added the Ziman hole-phonon interaction to their model. For pure InSb the theory over-estimates the thermal conductivity by a factor of 10 at 300K. Hamilton and Parrott applied their model to germanium and found better agreement although the theoretical values were still high at high temperatures. The lack of agreement could be due to assumptions made in the model, particularly the neglect of phonon dispersion, or to the presence of some additional type of scattering mechanism.

By changing the assignment of elastic constants we were able to get a better fit. The expansion of the elastic energy in terms of the nonlinear Lagrangian strains involves the five constants A, B, C, λ, μ. Numerical values for these constants are obtained by relating them to averaged values of the experimentally determined second- and third- order elastic constants. Based on the theory of Thurston and Brugger[17] the correct assignment should be $A = 4 \bar{C}_{456}$, $B = \bar{C}_{144}$, $C = \frac{1}{2}\bar{C}_{123}$, $\lambda = \bar{C}_{12}$, and $\mu = \bar{C}_{44}$. We found that if the assignments of A and B were changed to $A' = A-3\mu$ and $B' = B-\frac{1}{2}\lambda$ we obtained the fit shown in Fig. 4. Furthermore this assignment also produced very good fits for germanium, silicon, and GaAs. The basis for the modified assignment of constants was the idea that the experimental constants were based on an expansion in terms of the linear part of the strain. Since this is apparently not true, this was apparently a fortuitous choice of constants which compensated for other deficiencies in the theory. We also find by this method that impurity scattering is higher than that predicted from known impurities. Also it is necessary to include scattering of transverse modes in the hole-phonon interaction. We find that 80 to 95% of the thermal conductivity is due to transverse modes.

ACKNOWLEDGMENTS

A portion of this work was supported under ONR Contract No.
N00014-68-A-0160 (Project Themis). Sample preparation and Hall
Effect measurements were made during a Summer Faculty Research
Appointment (R. D. R.) at the Ames Laboratory, Iowa State
University, Ames, Iowa.

REFERENCES

1. L. J. Challis, J.D.N. Cheeke and J. B. Harness, Proceedings
    of the Ninth International Conference on Low Temperature
    Physics, Plenum Press, New York, (1976) p.1145.
2. C. R. Crosby and C. G. Grenier, Phys. Rev. B $\underline{4}$, 1258 (1971).
3. J. M. Ziman, Phil. Mag. $\underline{1}$, 191 (1956); Phil. Mag. $\underline{2}$, 292
    (1957).
4. V. V. Kosarev, P. V. Tamarin, and S. S. Shalyt, Phys. Stat.
    Sol. b, $\underline{44}$, 525 (1971).
5. C. Callaway, Phys. Rev. $\underline{113}$, 1049 (1959).
6. R. G. Morris, R. D. Redin, and E. J. Stephans, Bull. Am. Phys.
    Soc., Series II, $\underline{12}$, 574 (1967).
7. M. G. Holland, Phys. Rev. $\underline{132}$, 2461 (1963).
8. Exotic Materials, Inc., Santa Ana, Calif. (Samples S5-A-31
    and S5-A-42 were cut from material supplied by Semitronics,
    Inc., Winchester, Mass.).
9. T. Ashworth, L. R. Johnson, C. Y. Hsiung, and M.M. Kreitman,
    Cryogenics $\underline{13}$, 34 (1973).
10. P. V. Tamarin, S. S. Shalyt, I. G. Lang, and S. T. Pavlov,
    Sov. Phys. - Solid State $\underline{14}$, 47 (1972).
11. M. G. Holland, Phys. Rev. $\underline{134}$, A471 (1964).
12. G. Busch and E. Steigmeier, Helv. Phys. Acta, $\underline{34}$, 1 (1961).
13. C. M. Bhandari and G. S. Verma, Phys. Rev. $\underline{140}$, A2101 (1965).
14. J. P. Maneval and D. Huet, Phys. Letters, $\underline{48A}$, 463 (1974).
15. R. A. Hamilton and J. E. Parrott, Phys. Rev. $\underline{178}$, 1284 (1969).
16. G. R. Barsch, J. Appl. Phys. $\underline{39}$, 3780 (1968).
17. R. N. Thurston and K. Brugger, Phys. Rev. $\underline{133}$, A1604 (1964).

THE THERMAL CONDUCTIVITY OF THE GROUP V SEMIMETALS

J-P. Issi and J. Heremans

Université Catholique de Louvain - Laboratoire PCES

1, Place Croix du Sud, B-1348 Louvain-la-Neuve, Belgique

The relative contribution of the various low temperature thermal conductivity mechanisms are analyzed in the three group V semimetals; bismuth, antimony and arsenic.

INTRODUCTION

Many experimental data on the thermal conductivity of the group V semimetals have recently been reported. It is aimed here to review them briefly showing how the rhombohedral crystal structure and the resulting electronic band structure singularize the group V semimetals from the less exotic isotropic insulators or metallic conductors. Fig.1 illustrates a typical behaviour of the thermal conductivity of bismuth, antimony and arsenic below room temperature. Also, it may be seen from Table I that most low temperature heat transport mechanisms contribute to the thermal conductivity of these semimetals.

The rhombohedral structure, which may be considered as a slightly distorted cubic one, is responsible for the small overlap of the conduction and valence bands. Fig.2 shows a schematic representation of this band structure together with some relevant band parameters for the three group V semimetals. In momentum space the electron and hole Fermi surfaces consist of ellipsoids in the case of Bi and Sb and have a rather more complicated shape in the case of As [1]. As another result of the rhombohedral structure, the transport properties of the group V semimetals are highly anisotropic. The zero-field thermal conductivity κ has two components, one along the trigonal axis; $\kappa_{//}$, and the other in the

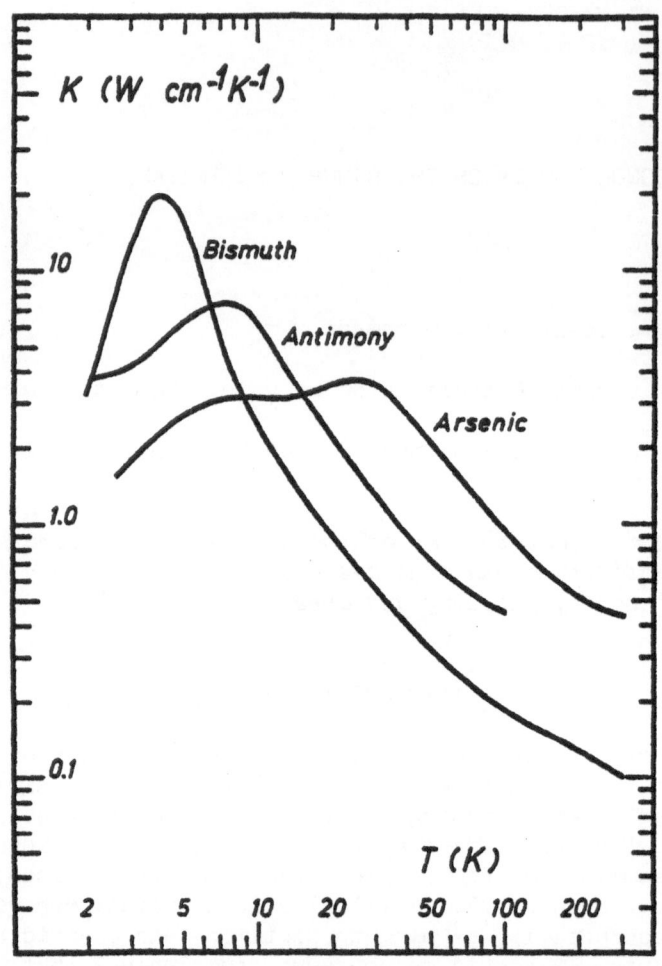

Fig.1 *The thermal conductivity of the group V semimetals Bi, Sb and As versus temperature. The three maxima which are those of κ_L lie around 4, 10 and 30 K for Bi, Sb and As respectively; the Debye temperatures being 120, 200 and 280 K respectively. Note that in the higher temperature range where κ_E dominates in Sb and As, and is important in Bi, the thermal conductivity increases with increasing density of carriers as expected.*

trigonal plane; $\kappa_\perp$. In arsenic, only $\kappa_\perp$ has been measured [2], while for bismuth both $\kappa_{//}$ and $\kappa_\perp$ have been reported for the whole temperature range below room temperature [3][4]. However, for bismuth, we have represented in Fig.1 and 3 only $\kappa_\perp$, which is approximately equal to 1.4 $\kappa_{//}$ at 100K and 1.6 $\kappa_{//}$ at 300K [3].

THE ELECTRONIC THERMAL CONDUCTIVITY

In bismuth [5][6] and antimony [7] single crystals the electronic component of the thermal conductivity, κ_E has been separated from the lattice one, κ_L, by applying a high magnetic field. The experiments were performed up to 150K for Bi [6] and 100K for Sb [7]. In As [2], the two contributions have been tentatively separated (Fig.4). In the case of Bi the results confirmed the earlier findings of White and Woods on polycrystalline material [8], i.e; that, in the range 2-20K, κ is purely lattice and that increasing the temperature enhances the relative contribution of κ_E. The results for single crystals are shown in Fig.3 and indicate that, in the binary direction, κ_E amounts to 14% of the total thermal conductivity at 35K and 43% at 150K. Note that Gallo and co-workers [3], using a simple model, computed a value of $\sim$60% at 150K.

TABLE I

HEAT TRANSPORT MECHANISMS BELOW 300 K			
	LATTICE	*ELECTRONS HOLES*	*BIPOLAR*
INSULATORS	All T	-	-
SEMICONDUCTORS	All T	Usually weak	-
METALS	-	All T	-
BISMUTH	2 - 20 K	below 1K, above 20K	above 50 K
ANTIMONY	Low T	All T	Weak
ARSENIC	Low T	All T	-

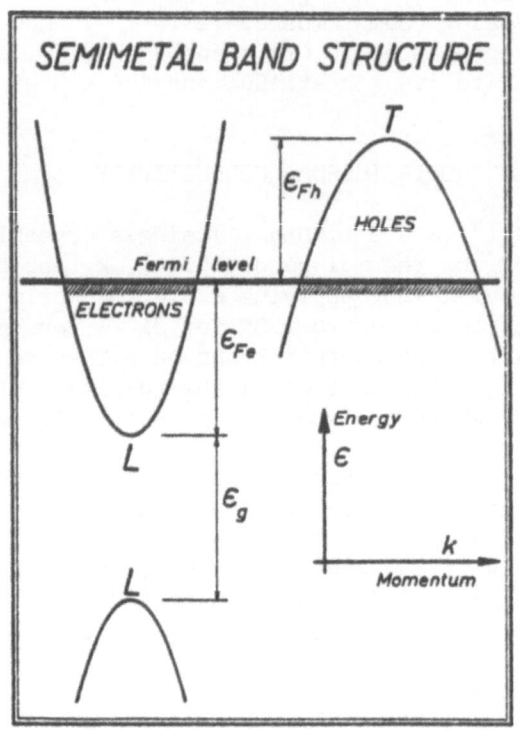

	Bi	Sb	As
Fermi energy of electrons ε_{Fe} (eV)	0.0276	0.096	0.01905
Fermi energy of holes ε_{Fh} (eV)	0.0109	0.104	0.177
Electron and hole density (cm^{-3})	2.75×10^{17}	5.5×10^{19}	2.2×10^{20}

Fig.2 *Schematic representation of the electronic band structure and band parameters of the group V semimetals at 0 K. Note that, while for Sb and As the band parameters are almost temperature insensitive, in bismuth the carrier density varies by an order of magnitude from 0 K to room temperature. The Fermi energies in Bi are also temperature dependent, as well as the direct energy gap ε_g and the effective masses.*

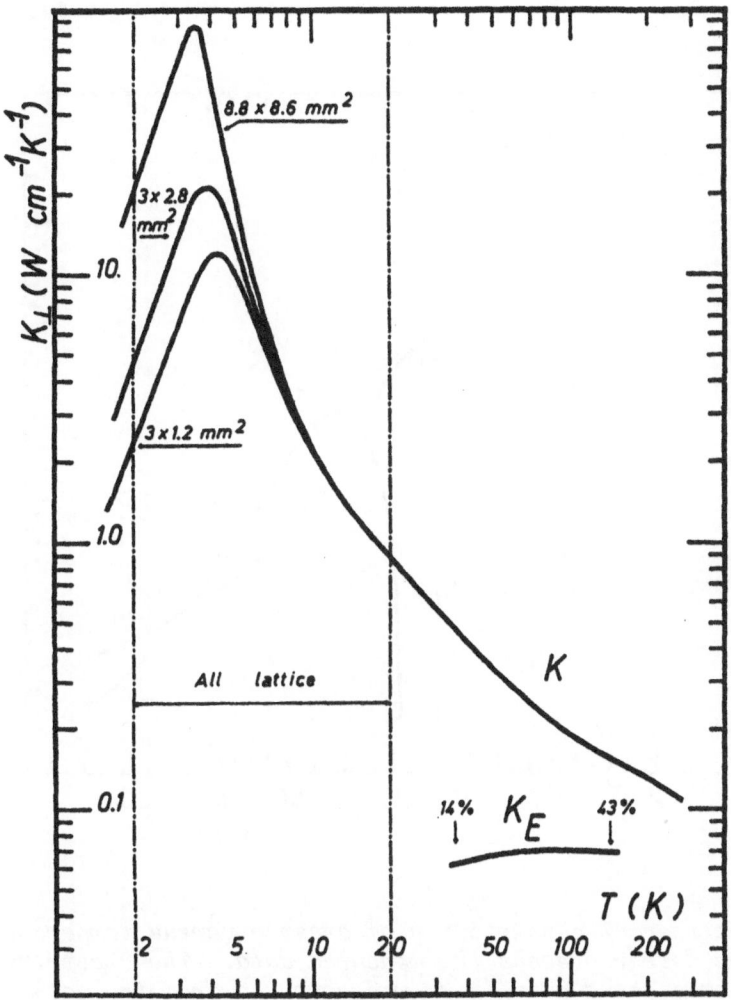

<u>Fig.3</u> *The thermal conductivity of three samples of bismuth of*
different cross sections([10])([11]). In the range from 2 to
20 K, the lattice thermal conductivity is the sole mechanism
and the dielectric size effect is clearly demonstrated.
The total electronic thermal conductivity κ_E at higher
temperature as determined by Uher and Goldsmid ([6]) is
represented too. It amounts 14% of the total thermal
conductivity at 35 K and 43% at 150 K.

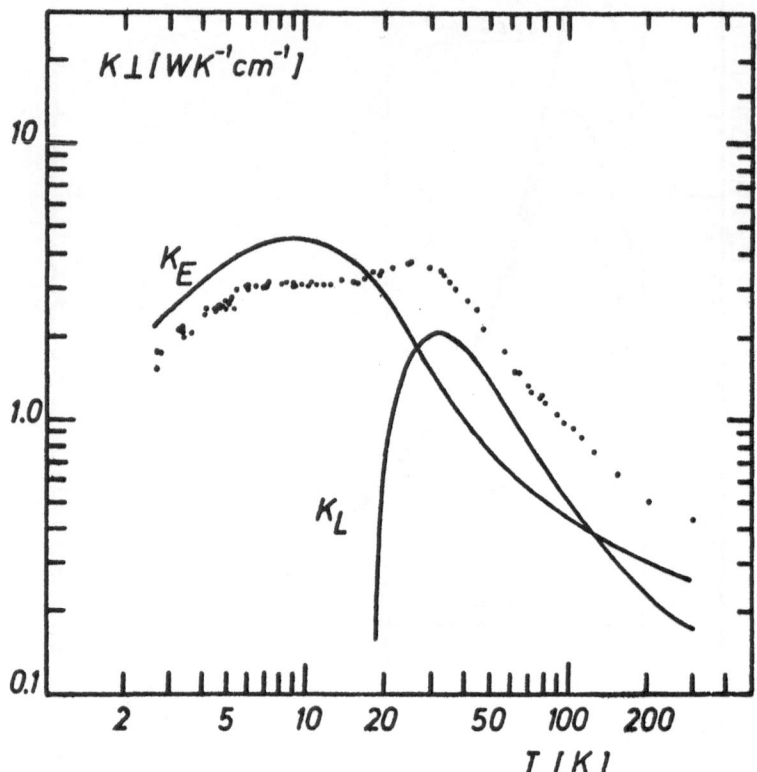

<u>Fig.4</u> *The thermal conductivity of arsenic versus temperature* (^{2}).
*The dots represent the measured data. The electronic
thermal conductivity* κ_E *was computed from the data of the
electrical conductivity by means of the Wiedemann-Franz
law. At higher temperatures, a small correction was
brought to account for the slight departure of the
charge carriers system from total degeneracy (relation 1').
The lattice thermal conductivity* κ_L *was obtained by
substracting* κ_E *from the measured total conductivity.*
κ_L *has a maximum around 30 K.*

In semimetals, electrons and holes may participate independently to heat transport, or in pairs, as a bipolar contribution.
The Wiedemann-Franz law is applicable to the partial electronic
thermal conductivities κ_i, with a suitable adjustement of the
Lorenz number to account for an eventual partial degeneracy of the
group i of the charge carriers considered

$$L_i = (\frac{k}{q})^2 \cdot \gamma (\xi_i, r) \qquad \qquad \ldots(1)$$

where k is the Boltzmann constant, q the electronic charge and
$\gamma (\xi_i, r)$ a function of the reduced Fermi energy $\xi_i = \varepsilon_i/kT$ and
of the scattering parameter r in the relaxation time approximation.
The bipolar term is given by

$$\kappa_{eh} = T \frac{\sigma_e \cdot \sigma_h}{\sigma_e + \sigma_h} (S_h - S_e)^2 \qquad \qquad \ldots(2)$$

where the σ_i are the partial electrical conductivities, and the
S_i the partial thermoelectric powers; the indices e and h refer
to electrons and holes respectively. The total electronic
conductivity is then

$$\kappa_E = \kappa_e + \kappa_h + \kappa_{eh} \qquad \qquad \ldots(3)$$

In the case of bismuth the partial degeneracy of the carriers
above 50K and the non parabolicity [1] as well as the extreme
sensitivity of the band structure to temperature [9] lead to
complicated expressions of γ. For arsenic, if it is assumed that
the relaxation time τ takes the form $\tau = \tau_0 \, \varepsilon^{-1/2}$ (pure acoustic
lattice scattering), expression (1) reduces to [2]

$$L_i = \frac{1}{3} (\frac{k}{q})^2 \left[\pi^2 - \frac{\pi^4}{3} \cdot (\frac{kT}{\varepsilon_i})^2 \right] \qquad \qquad \ldots(1')$$

and the bipolar term is found to be negligible (less than 1%)
below room temperature. The second term in the last brackets is
a small correction which accounts for a slight departure from full
degeneracy of the charge carriers in the higher temperature range.
In bismuth this departure is so large that, even in a parabolic
rigid band model, we cannot accept a second order approximation
but must use the complete analytical formulae. In the residual
resistivity range of arsenic, where the electrical resistivity is
temperature insensitive, and the carrier system is fully degenerate,
one would have naively expected to find the Sommerfeld value for
the Lorenz number, L_0. However, a value of 0.75 L_0 is experimentally found [2] (Fig.4). At higher temperatures, because of the

onset of intervalley scattering, which is essentially inelastic, a
deviation from L_0 is expected to be found for the three group V
semimetals.

THE LATTICE THERMAL CONDUCTIVITY

    In the range 2-20K, the observed thermal conductivity of Bi
exhibits the typical dielectric behaviour, as may be seen from
Fig.3. It starts with an almost T^3 dependence at low temperatures
which reflects the temperature variation of the lattice specific
heat c_v when the mean free path l is constant and equal - within
a numerical factor of the order of unity - to a transverse
dimension of the sample, as predicted by the Debye relation

$$\kappa_L \sim c_v.v.l \qquad\qquad ...(4)$$

the velocity of sound v being almost temperature insensitive.
The power of T in this region is not strictly 3, but varies
according to the authors from 2.6 to 3.1. Fig.3 shows the low
temperature size dependence of the lattice thermal conductivity
of bismuth in the binary direction ([10])([11]). Kopylov and Mezhov-
Deglin ([12]) investigated also in detail this size-effect on samples
which axes were inclined with respect to the principal crystallo-
graphic directions. However, at intermediate temperatures, and up
to 40K, a smaller but detectable size-effect has been recently
observed ([13]), evidencing a sizeable contribution of the low energy
phonons to the total κ_L. This confirmed one of the early predictions of Herring ([14]), and was consistent with the expected q^{-3}
dependence of the relaxation time of low energy phonons on wave
number q in the single mode approximation. This q^{-3} dependence,
which characterizes the rhombohedral structure, has dramatic
consequences when compared to the less pronounced q^{-2} dependence
of cubic materials : it enhances to an appreciable extent the
relative contribution of the low q - large l phonons with respect
to the large q - low l thermal phonons.

    The behaviour of antimony is less spectacular, in the sense
that we do not have regions where the dielectric type behaviour
is the sole mechanism. Here, the carrier density is sufficiently
high at all temperatures to maintain κ_E comparable to κ_L and a
high magnetic field is needed to separate the two contributions at
any temperature. Besides, at low temperatures, the high carrier
density is mainly responsible for phonon scattering and the typical
phonon-electron scattering T^2 variation of the lattice thermal
conductivity is expected below the maximum. The maximum is less
pronounced than for the case of bismuth and the higher temperature
T^{-1} variation characteristic of phonon-phonon U-processes is
evidenced by White and Woods' measurements ([8]). More recent

measurements by Blewer and Zebouni ([15]) extended this work on single crystals at lower temperatures, and a pure T^2 law was found in the region 0.4 - 1.4K.

The interpretation of the data for arsenic is, for the time being, speculative. From the results of Fig.4 one may however state that in the lowest and highest temperature range κ_L is negligible and that it presents a maximum around 30K, where it is comparable to κ_E. Above the maximum the decrease in κ_L is probably due to phonon-phonon umklapp processes and below this temperature to phonon-electron scattering.

It is worth noting that, in addition to the fundamental interest which stimulates lattice thermal conductivity studies, a comprehensive work on rhombohedral materials may help the search for more efficient thermoelectric materials. The lattice thermal conductivity is the only thermoelectric parameter which one can control almost independently of the others i.e. the thermopower, the electrical conductivity and the electronic thermal conductivity. Further, it has been recently shown ([10]) that pure bismuth has the highest thermoelectric figure of merit ever recorded in the liquid helium range, and is the best candidate for thermoelectric refrigeration below 4K.

CONCLUSIONS

The experimental results obtained on the group V semimetals below 300K reflect what could be expected from the competition of phonons and charge carriers in a system where the latter are varying. The comparison is made, either from one material to another at a given temperature, or in the same material (bismuth) as a function of temperature. Arsenic, which has the highest electronic population at all temperatures, exhibits an almost metallic behaviour, since κ_E is predominant except for a small temperature range where κ_L is of comparable magnitude. At the other end of the scale, because of it low carrier density, bismuth shows a pure dielectric behaviour in the range 2-20K, while, at higher temperature, κ_E and κ_L are comparable. Antimony behaves in a way intermediate between that of arsenic and bismuth.

ACKNOWLEDGEMENTS

The authors are indebted to Professor G.A. Saunders for enriching collaboration and to Drs C. Uher and W.P. Pratt Jr. for kindly showing and discussing their experimental results prior to publication. They are also grateful to Dr. O.P. Hansen for critical reading of the manuscript.

Note added in proof

   Preliminary results of experiments carried out at ultralow temperatures by C. Uher and W.P. Pratt Jr., at Michigan State University, extend beautifully the curves reported above. At the present stage, their work seem to confirm what is expected from the competition of κ_E *and* κ_L *at these temperatures.*

REFERENCES

1. See for example M.S. DRESSELHAUS J. Phys. Chem. Solids 32, suppl.1, 3 (1971)

2. J. HEREMANS, J-P. ISSI, A.A.M. RASHID and G.A. SAUNDERS J. Phys. C : Solid State Phys., (in press)

3. C.F. GALLO, B.S. CHANDRASEKHAR and P.H. SUTTER J. Applied Phys., 34, 144 (1963)

4. M.E. KUZNETSOV, V.S. OSKOTSKII, V.I. POLSHIN and S.S. SHALYT, Sov. Phys. JETP, 30, 607 (1970)

5. I.Ya. KORENBLIT, M.E. KUZNETSOV, V.M. MUZHDABA and S.S. SHALYT, Sov. Phys. JETP, 30, 1009 (1970)

6. C. UHER and H.J. GOLDSMID, Phys. Stat. Sol. (b) 65, 765 (1974)

7. N.A. REDKO, M.S. BRESLER and S.S. SHALYT, Sov. Phys. - Solid State, 11, 2435 (1970)

8. G.K. WHITE and S.B. WOODS, Phil. Mag. 3, 342 (1958)

9. M.P. VECCHI and M.S. DRESSELHAUS, Phys. Rev. B10, 771 (1974)

10. J. BOXUS and J-P. ISSI, J. Phys. C : Solid State Phys. 10, L397 (1977)

11. J-P. ISSI and J.H. MANGEZ, Phys. Rev. B6, 4429 (1972)

12. V.N. KOPYLOV and L.P. MEZHOV-DEGLIN, Sov. Phys. - JETP, 38, 357 (1974)

13. J-P. ISSI, J-P. MICHENAUD and J. HEREMANS, Proceedings of the Fourteenth International Conference on Thermal Conductivity, Storrs, Connecticut (Plenum, New York, 1976) p.127

14. C. HERRING, Phys. Rev. 96, 1163 (1954)

15. R.S. BLEWER and N.H. ZEBOUNI, Phys. Letters, 23, 297 (1966)

THE TRANSPORT PROPERTIES OF THE CUBIC ALKALINE EARTHS Ca, Sr AND Ba

J.G. Cook and M.J. Laubitz

National Research Council of Canada

Ottawa, Canada K1A 0S1

The electrical resistivity (ρ), thermoelectric power (S) and thermal conductivity (κ) of two Sr samples and two Ba samples have been determined from 30K to 300K. Large DMR were observed. The estimated transport properties for ideally pure Sr and Ba indicate that these elements, like Ca, show large deviations from the Bloch-Gruneisen form for $\rho(T)$ at all temperatures, large and positive diffusion thermopowers with a negative phonon-drag contribution, and large deviations from the Wiedemann-Franz relationship (DWFR). In this respects, they are much more like the transition metals than the monovalent metals.

In the second, analytical, portion of the paper we study the DWFR in some detail. First, the effect of lattice conduction is estimated, and found to be large. Then, a function X(E) of the electron energy, closely related to the conventional conductivity function $\sigma(E)$, is estimated from the ρ and S data now available for Ca, Sr and Ba above 300K, and used to compute S and the Lorenz function for elastic electron-phonon scattering below 300K. Comparison with the experimental data indicates that the energy dependence of the electron parameters is responsible for the electronic DWFR, and effects the diffusion thermoelectric power. Such "band effects" may also be seen in the thermal resistivity due to inelastic scattering in at least Sr. Regrettably, we are not able to explain the observed DMR.

THE ETTINGSHAUSEN-NERNST COEFFICIENTS AND PHYSICAL PROPERTIES OF NICKEL AND ALUMEL*

J. P. Moore and R. S. Graves

Oak Ridge National Laboratory

Oak Ridge, TN   37830

Although the emf from Chromel-Alumel thermocouples is normally considered to be relatively insensitive to magnetic fields, a large error can occur due to the Ettingshausen-Nernst effect. This effect is largest below the Curie transition temperature where the Ettingshausen-Nernst coefficient, Q, for Alumel is large. A technique is described for directly measuring Q by employing an electromagnet in conjunction with an apparatus normally used for measuring thermal conductivity, electrical resistivity, and Seebeck coefficient.

As shown in Fig. 1, the Q of Alumel increases sharply with increasing temperature and then drops by a factor of 7 at the Curie temperature (423 K). The Q results are a factor of 600 above previous literature values for Alumel and this assists in the interpretation of thermometry errors noted during a simulated reactor test. The physical property data are discussed and compared to data on other nickel-base alloys.

---

*Research sponsored by the Department of Energy under contract with the Union Carbide Corporation.

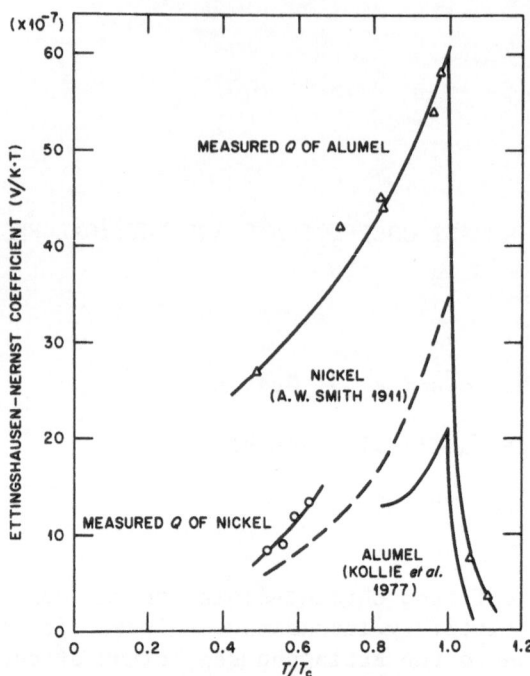

Fig. 1. Ettingshausen–Nernst Coefficients of Nickel and Alumel (Ni + 3% Mn + 2% Al + 1% Si + 0.5% Co + ...) Versus Reduced Temperatures. T. G. Kollie, R. L. Anderson, J. L. Horton, and M. J. Roberts; Rev. Sci. Inst. 48(5), 501 (May 1977). A. W. Smith, Phys. Rev. 33, 295 (1911).

THERMAL CONDUCTIVITY AND THERMAL DIFFUSIVITY OF SELECTED CARBON STEELS, CHROMIUM STEELS, NICKEL STEELS, AND STAINLESS STEELS

C. Y. Ho, M. W. Ackerman, R. H. Bogaard, T. K. Chu, and P. D. Desai

Center for Information and Numerical Data Analysis and Synthesis
Purdue University, W. Lafayette, Indiana   47906

The available data and information on the thermal conductivity and the thermal diffusivity of selected carbon steels, chromium steels, nickel steels, and stainless steels have been searched, compiled, and evaluated. Through the analysis and synthesis of the fragmentary data and information, preliminary recommended values have been generated over the temperature range from room temperature to about 1150 K, which are presented in this paper in graphical form. Estimated values have been synthesized for those steels and temperature ranges for which no data are available.

The results show that the largest spread of the thermal conductivity versus temperature curves of the various steels within each group occurs at room temperature and the curves converge at the high end of the temperature range covered. The thermal conductivity curves of the stainless steels increase with temperature over the entire temperature range while those of the other three groups of steels generally decrease over most of the temperature range with some showing a minimum near 1050 K.

The thermal diffusivity versus temperature curves of the carbon steels, the chromium steels, and the nickel steels show sharp dips at the magnetic transition, corresponding to the spikes in the specific heat curves, while the thermal diffusivity curves of the stainless steels increase smoothly over the entire temperature range.

THERMAL CONDUCTIVITY AND ELECTRICAL RESISTIVITY OF EIGHT

SELECTED AISI STAINLESS STEELS

T. K. Chu and C. Y. Ho

Center for Information and Numerical Data Analysis
and Synthesis
Purdue University, W. Lafayette, Indiana 47906

ABSTRACT - The thermal conductivity and the electrical resistivity
of eight technologically and commercially important AISI stainless
steels were studied. The eight selected stainless steels were AISI
303, 304 (including 304L), 316, 317, 321, 347, 410, and 430. Experimental data on these properties of the stainless steels were
exhaustively searched, systematically compiled, and critically evaluated. The evaluated data were further analyzed with regard to
factors that affect the properties such as the amount and nature of
the various alloying elements, the metallurgical structure of the
steels, heat treatment, cold working, etc. As a result, recommended
values were generated for both the solid and molten states covering
the temperature range from 1 K to above the melting point. Estimated
values were synthesized for those stainless steels and temperature
ranges for which no data were available. As an aid to the data
analysis and synthesis, the properties of other non-AISI designated
stainless steels with similar compositions and structures were also
studied. These included British, German, and Russian stainless steels
and other American stainless steels.

I.   INTRODUCTION

The primary objective of this study was to generate recommended
engineering design data for the thermal conductivity and the electrical resistivity of six of the 300 series and two of the 400 series of AISI stainless steels, which were selected because of their
current technological and commercial importance.

The eight selected AISI stainless steels and their chemical
composition ranges and limits are given in Table 1.

Table 1. Chemical Composition Ranges and Limits of Eight Selected AISI Stainless Steels

AISI No.	Chemical Composition Ranges and Limits, percent								
	C	Mn Max.	P Max.	S Max.	Si Max.	Cr	Ni	Mo	Other Elements
303	0.15 Max.	2.00	0.20	0.15 Min.	1.00	17.00/ 19.00	8.00/ 10.00	0.60* Max.	
304	0.08 Max.	2.00	0.045	0.030	1.00	18.00/ 20.00	8.00/ 10.50		
316	0.08 Max.	2.00	0.045	0.030	1.00	16.00/ 18.00	10.00/ 14.00	2.00/ 3.00	
317	0.08 Max.	2.00	0.045	0.030	1.00	18.00/ 20.00	11.00/ 15.00	3.00/ 4.00	
321	0.08 Max.	2.00	0.045	0.030	1.00	17.00/ 19.00	9.00/ 12.00		Ti 5 x C Min.
347	0.08 Max.	2.00	0.045	0.030	1.00	17.00/ 19.00	9.00/ 13.00		Nb-Ta 10 x C Min.
410	0.15 Max.	1.00	0.040	0.030	1.00	11.50/ 13.50			
430	0.12 Max.	1.00	0.040	0.030	1.00	16.00/ 18.00			

* Optional

First of all, general background information on the electrical resistivity and the thermal conductivity of stainless steels is given. In particular, the effects of chemical composition, metallurgical structure, heat treatment, cold working, nuclear irradiation, and porosity on these two properties of stainless steels are discussed. Since the recommended values are generated through a process in which available data are exhaustively searched, systematically compiled, critically evaluated, and then analyzed and synthesized, the methodology of data evaluation is then outlined.

The recommended values for the electrical resistivity and the thermal conductivity of the eight selected stainless steels are presented in both tabular and graphical forms. The values given for seven of the eight stainless steels cover the temperature range from 1 K to above the melting point into the molten state and for one steel cover from 1 K to the melting point. The recommended values are for the quench-annealed state in the cases of the austenitic 300-series stainless steels. For stainless steel 410, the values are for the oil-quenched state. In the case of stainless steel 430, they are for the air-cooled, annealed state. It is noted that values above 1500 K are extrapolated, since the majority of the measurements are below this temperature. Unless stated otherwise, the uncertainty in the values are of the order of ±5%. In the majority of cases the recommended values in the tables are given beyond the physically significant figures, which is merely for retaining the smoothness of the tabulated values and should not be interpreted as indicative of the degree of accuracy of the values.

There is a discussion text for each stainless steel, in which the particular feature of the said steel is discussed, the available data and information on the electrical resistivity and the thermal conductivity are reviewed, and the considerations involved in arriving at the recommendations are discussed. Whenever appropriate, the effects of heat treatment and cold working on the recommended values are discussed individually in the text. Finally, a general discussion of the thermal conductivity and the electrical resistivity of the eight stainless steels is given.

II. GENERAL BACKGROUND

In practical applications stainless steels are chosen mainly for their most important property - corrosion resistance. Other service requirements and properties may then become important after a stainless steel with the desired corrosion resistance has been chosen. In general, the chemical composition, the heat treatment, and the cold-work state are the major factors that determine the various properties of the steels. These will be discussed briefly below, with emphasis on their effects on the electrical resistivity and thermal conductivity.

The corrosion-resistant property of stainless steels is achieved by the addition of chromium, in excess of 12%, to iron. Thus chromium is the major alloying element in all stainless steels. In the 300 series stainless steels, the chromium content is about 18%. Nickel, which is usually present at about 8-12% in the 300 series stainless steels, serves to stabilize the austenite. The element next in abundance is manganese (≤2%), which also tends to stabilize the austenite. Manganese in the sulfide form also improves the hot workability and the machinability. Silicon usually appears as trace impurity (≤1%), though it might improve the corrosion resistance to a limited extent. Other elements are added to acquire certain desired properties for special service requirements. The 400 series stainless steels differ from the 300 series by the complete absence of nickel. Their structures also differ from that of the austenite, and may be ferritic or martensitic depending on the chromium concentration.

In an alloy significant contribution to the electrical resistivity comes from the solute atoms. In dilute binary alloys the electrical resistivity is directly proportional to the solute concentration. In concentrated binary alloys and in complicated multiple alloys such as stainless steels, this direct proportionality no longer holds. However, estimates can still be made by considering their relative abundance and their proximity to the host (i.e., iron in ferrous alloys such as steels) in the periodic table. In the 300 series stainless steels at room temperature, chromium contributes the most to the electrical resistivity. Measurements on chromium steels [1] show this contribution to be about 50%. Contribution from nickel is about 18%, as estimated from binary Fe + Ni alloys [2]. About 15% is due to the host iron matrix. The remainder of the electrical resistivity can then be attributed to the other impurities, probably equally divided between silicon and manganese. Even though manganese is usually more abundant than silicon in stainless steels, its effect on the electrical resistivity is relatively smaller because of its closeness to iron in the periodic table. For the 400 series stainless steel, the major contribution (~65%) comes again from the chromium atoms. The host iron matrix contributes about 20% and the other alloying elements the rest of the resistivity. These relative contributions are temperature dependent. At elevated temperatures 90% of the electrical resistivity comes from the host iron matrix.

In metals and alloys, thermal energy is transported by electrons and by lattice waves (atomic vibrations). In the 400 series stainless steels at room temperature, for example, each of these carries about equal amount of thermal energy. At higher temperatures electrons are more effective in transporting thermal energy. As the temperature lowers, lattice waves become more and more effective, and usually give the largest relative contribution to the total

thermal conductivity of about 50 K. At liquid helium temperatures
(<4 K), these two contributions are again about equal.

To a first approximation, the electronic thermal conductivity
is inversely proportional to the electrical resistivity. The relative effects of the various alloying elements on the electronic
thermal conductivity can therefore be estimated from their relative
contributions to the electrical resistivity. The behavior of the
lattice component is more complicated. In general, the lattice
thermal conductivity decreases with increasing amount of alloying
elements and with the dissimilarity between the alloying element
atom and the host iron atom.

Impurity effect on the lattice thermal conductivity is more
prominent at low temperatures (<250 K). At elevated temperatures
the lattice thermal conductivity is severely limited by the interaction between the lattice waves, and impurity effect is relatively
unimportant. In principle, the lattice thermal conductivity can be
calculated given the impurity content and the crystal structure.
However the calculation is often not exact because the manner in
which the host matrix is altered by an impurity is not known.

The electrical resistivity and thermal conductivity are also
affected by different heat treatments. In the 300 series stainless
steels, heat treatments are very limited. In order to maintain
these steels in the metastable austenitic condition, they are in-
variably quench-annealed from about 1300 K (1900°F) and perhaps
stress-relieved by heating up to ~700 K (800°F). Stainless steel
430 is ferritic at all temperatures and its properties are therefore also not changed by heat treatment.

Only stainless steel 410 among those studied is subject to
different heat treatments. This steel is martensitic when oil-
quenched or air-cooled from ≥1300 K (1850°F) and is ferritic when
in the annealed condition after slow cooling from 1050 K (1450°F).
Results on chromium steels [1] indicate that the annealed state
should have a higher electrical resistivity (by about 12%) at room
temperature and correspondingly a lower thermal conductivity.

Cold-worked metals contain dislocations, which affects both
the thermal conductivity and the electrical resistivity. The effect
on the latter, however, is quite negligible since these steels have
high electrical resistivities (~70 x 10^{-8} Ωm at room temperature).
Even if assuming a saturation density of 1 x 10^{16} m^{-2}, the electrical
resistivity due to dislocations is estimated to be of the order of
2 x 10^{-8} Ωm. Dislocations have a much larger effect on the lattice
component of the thermal conductivity. This will be discussed later
individually for each of the stainless steels.

Cold-working may also produce metallurgical transformations in steels. In the austenitic stainless steels there is a tendency to form martensite. The degree of transformation depends on the deformation and on the type of steel. Measurements on low Ni-Cr alloy steel and on Mn steels [3] indicates that the martensitic state is lower in electrical resistivity (and hence higher in thermal conductivity) than the austenitic state by approximately a factor of 1.5. Reduction in electrical resistivity upon cold drawing has been observed in 302 and 304 stainless steels [4]. To a first approximation, a 50/50 martensite/austenite would have a decrease in electrical resistivity of about 20% from that of the austenitic.

Special service conditions may also change the electrical resistivity and the thermal conductivity. One notable instance is the nuclear reactor application. Experimental investigation on a 347 stainless steel [5] has shown that there is an increase in electrical resistivity and a decrease in thermal conductivity after being irradiated by a neutron fluence of 3.3×10^{17} cm^{-2} (neutron energy >1 Mev). The increase in electrical resistivity is 0.7% and 0.3% at 77 K and 297 K respectively, and the decrease in thermal conductivity is 0.9% and 0.1% at the same temperatures. These changes are, however, completely recovered after annealing at 463 K. Thus, it appears that neutron irradiation would not change significantly the electrical and thermal transport properties of bulk material, especially at elevated temperatures. Prolonged nuclear reactor service may produce another problem. The bulk material may become porous. In that case, the electrical resistivity and thermal conductivity can be estimated from values for the bulk material and the porosity.

The effect of porosity has been investigated for 304L stainless steel sintered from powders [6]. The results indicate that both the thermal conductivity and the electrical conductivity (reciprocal of the resistivity) decrease with porosity. The exact manner in which these properties vary with porosity is complicated and may depend on other factors such as the kind of material and the process by which the material is fabricated. For practical purposes, one can estimate properties for the porous material for porosity less than 10% by the equation [7]

$$k = k_b \frac{1 - P}{1 + 0.5 \, P}$$

where k and k_b are the conductivities (thermal or electrical) of the porous material and of the bulk material, respectively, and P is the fractional porosity.

III. METHODOLOGY OF DATA EVALUATION

Due to the difficulties of accurately measuring the properties of materials and of adequately characterizing the test specimens, especially solids, the property data available from the scientific and technical literature are often conflicting, diverging widely, and subject to large uncertainty. Indiscriminate use of literature data for engineering design calculations without knowing their reliability is dangerous and may cause inefficiency or product failure, which at times can be disastrous. Therefore, it is imperative to evaluate critically the validity, reliability, and accuracy of the literature data and related information, to resolve and reconcile the disagreements in conflicting data, and to synthesize often fragmentary data in order to generate a full range of internally consistent recommended values.

Considering the thermal conductivity, for example, in the critical evaluation of the validity and reliability of a particular set of experimental data, the temperature dependence of the data is examined and any unusual dependence or anomaly is carefully investigated. The experimental technique is reviewed to see whether the actual boundary conditions in the measurement agreed with those assumed in the theory and whether all the stray heat flows and losses were prevented or minimized and accounted for. The reduction of data is examined to see whether all the necessary corrections were appropriately applied. The estimation of experimental inaccuracies is checked to ensure that all the possible sources of error, particularly systematic error, were considered by the author(s). Experimental data could be judged to be reliable only if all sources of systematic error were eliminated or minimized and accounted for. Major sources of systematic error include unsuitable experimental method, poor experimental technique, poor instrumentation and poor sensitivity of measuring devices, sensors, or circuits, specimen and/or thermocouple contamination, unaccounted for stray heat flows, incorrect form factor, and, perhaps most important, the mismatch between actual experimental boundary conditions and those assumed in the theoretical model used to define and derive the value of the property. These and other possible sources of errors are carefully considered in critical evaluation and analysis of experimental data. The uncertainty of a set of data depends, however, not only on the estimated inaccuracy of the data, but also in the inadequacy of characterization of the material for which the data are reported.

In many cases, however, research papers do not contain adequate information for a data evaluator to perform a truly critical evaluation. In these cases, some other considerations may have to be used for data evaluation. For instance, if several authors' data agree with one another and, more importantly, these were obtained by using different experimental methods, these data are likely to be reliable. However, if the data were observed by using the same

experimental method, even though they all agree, the reliability of
the data is still subject to questioning, because they may all suffer from a common, but unknown source of error. Secondly, if the
same apparatus has been used for measurements of other materials
and the results are reliable, and if the result of measurement on
the new material is in the same range, the result for the new material is likely to be reliable. However, if the information given by
the author is entirely inadequate to make any value judgment, the
data assessment becomes subjective. At times judgments may be based
upon factors and considerations such as the purpose and motivation
for the measurement, general knowledge of the experimenter, his past
performance, the reputation of his laboratory, etc.

In the process of critical evaluation of experimental data
outlined above, unreliable and erroneous data are uncovered and
eliminated from further consideration. The remaining evaluated
data are then subjected to further analysis and correlation in regard to the various factors that affect the property under study.
In the cases where available data are scarce, estimated values are
synthesized by theoretical calculations such as calculating the
electronic thermal conductivity from the electrical resistivity and
by semiempirical techniques such as intercomparing the property
values of various stainless steels (both domestic and foreign) of
similar chemical compositions and metallurgical structures accounting for the various affecting factors. By using theoretical relationships, several properties of the same material can be cross-
correlated for checking the consistency of data or for data estimation. For example, the thermal conductivity and specific heat can
be correlated with the thermal diffusivity.

IV. ELECTRICAL RESISTIVITY AND THERMAL CONDUCTIVITY OF
SELECTED AISI STAINLESS STEELS

1. AISI 303 Stainless Steel

This variation of the basic 18-8 austenitic stainless steel
contains additional amounts of sulfur and phosphorous. The high
sulfur content makes it more machinable. Another variation, 303 Se,
contains some additional selenium for the same purpose. The sulfur
may also improve the ductility of the steel. These elements are
present in such small amounts (≤0.20%), that the electrical and
thermal properties of these steels are almost identical to those of
the more common type, AISI 304, though their mechanical properties
are different.

There are four sets of experimental data available for the
electrical resistivity [8,9,10,11], but only one is for room temperature and above [8]. These results indicate that, to within

measurement errors, the electrical resistivity of AISI 303 is the same as that of AISI 304. Measurements at high temperatures on foreign steels of compositions similar to AISI 303 [12,13] also substantiate this conclusion. Therefore the recommended values for the electrical resistivity of AISI 303 are taken to be the same as those for AISI 304.

Nine sets of experimental data are available for the thermal conductivity of AISI 303, covering a wide temperature range [8,9,14, 15,16,17]. These results show that the thermal conductivity of AISI 303 is also not significantly differed from that of AISI 304, and the recommended values are taken to be the same as those for AISI 304.

The recommended values for both the electrical resistivity and the thermal conductivity are tabulated in Table 2 and shown in Figure 1. For the effects of cold working on these properties, see the discussion on AISI 304 stainless steel.

2. AISI 304 and 304L Stainless Steels

AISI 304 is a low-carbon member of the 18-8 type austenitic stainless steel, with a slightly higher chromium content for improved corrosion resistance. This steel is susceptible to intergranular corrosion in the temperature range 700-1150 K (800-1600°F) due to carbide precipitation. AISI 304L is a still lower carbon version of AISI 304. The reduced carbon content improved the resistance to intergranular corrosion.

There are twenty-six sets of data available for the electrical resistivity of AISI 304, eighteen of which deal with the change of resistivity upon drawing [4]. In addition, there are ten sets of data on the lower carbon version, AISI 304L. Evaluation of these measurements leads to the conclusion that the small difference in carbon content between AISI 304L and AISI 304 is not significant enough to warrant separate recommendations. The recommended values therefore are based on measurements on both of these two stainless steels, such as those of Tye et al. []8,19], Tye [6], Clark et al. [10], Feith et al. [20], and of Stutius and Dillinger [21].

Twenty sets of data are available for the thermal conductivity of AISI 304 and seven sets for AISI 304L. Again, these results show that the thermal conductivity of these two steels are virtually the same. The recommended values are based on the results of Feith et al. [20], Tye et al. [18,19], Powell [22], Taylor et al. [23], Ewing et al. [24], Deverall [25], Powers [26], Brown and Bergles [27], Tye [6], and Stutius and Dillinger [21].

Table 2. Electrical Resistivity and Thermal Conductivity of
AISI 303, 304, 304L, and 316 Stainless Steels[†]

Temperature (K)	Electrical Resistivity (10^{-8} Ω m)		Thermal Conductivity (W m^{-1} K^{-1})	
	AISI 303, 304, and 304L	AISI 316	AISI 303, 304, and 304L	AISI 316
1	48.5	54.4	0.069	0.056
4	48.5	54.4	0.305	0.248
10	48.5	54.4	0.875	0.713
20	48.3	54.4	1.99	1.63
30	48.2	54.4	3.27	2.69
40	48.3	54.6	4.65	3.87
50	48.5	55.1	6.02	5.01
60	48.9	55.8	7.11	6.03
70	49.5	56.7	7.98	6.88
80	50.3	57.5	8.69	7.55
90	51.2	58.5	9.26	8.10
100	52.2	59.5	9.75	8.54
150	57.5	64.5	11.55	10.29
200	62.7	69.1	12.89	11.49
250	67.5	73.6	13.90	12.51
273	69.6	75.4	14.39	12.97
300	71.9	77.7	14.89	13.44
350	76.0	81.5	15.79	14.32
400	79.8	85.2	16.61	15.16
500	86.8	91.7	18.28	16.80
600	93.3	97.7	19.77	18.36
700	99.2	103.1	21.21	19.87
800	104.3	108.0	22.59	21.39
900	108.6	112.1	23.99	22.79
1000	112.5	115.7	25.33	24.16
1100	115.8	118.9	26.58	25.46
1200	118.7	121.7	27.81	26.74
1300	121.4	124.3	29.18	28.02
1400	124.0	126.8	30.34	29.32
1500	126.4	129.2	31.55	30.61
1600	128.7	131.3	32.70	31.86
1644	–	132.2*	–	32.41*
1672	130.4*	152.0*	33.53*	26.90*
1700	–	152.6	–	27.24
1727	150.0*		28.15*	
1800	151.7		28.99	

* Melting Range.
[†] The uncertainty in the recommended values is of the order of ±5%.
At or above melting, the uncertainty is of the order of ±15%.

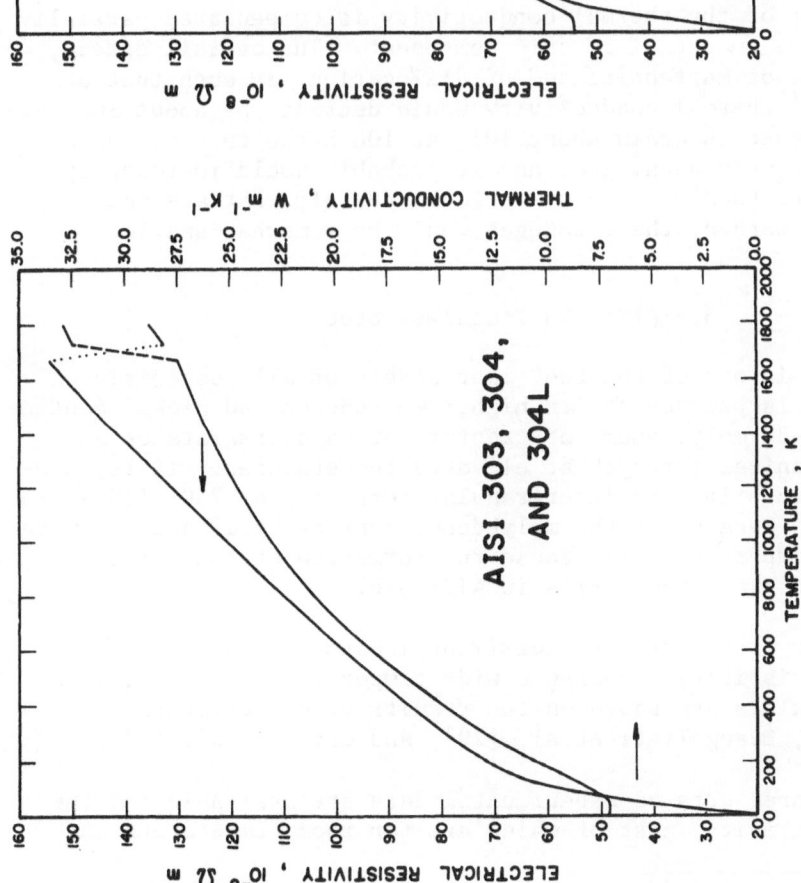

Figure 2. Electrical Resistivity and Thermal
Conductivity of AISI 316 Stainless
Steel.

Figure 1. Electrical Resistivity and Thermal
Conductivity of AISI 303, 304, and
304L Stainless Steels.

The recommended electrical resistivity and thermal conductivity values for AISI 304 and 304L are tabulated in Table 2 and shown in Figure 1, which are also for AISI 303.

AISI 304 has a tendency to undergo martensitic transformation upon cold working: the amount of martensite is probably less than 20% at 80% cold-work reduction. The existence of martensite decreases the electrical resistivity and thus increases the electronic component of the thermal conductivity. At 20% transformation to martensite, the electrical resistivity may decrease by about 7%. The martensitic transformation temperature of austenitic stainless steels is believed to be slightly lower than room temperature. Hence this change may diminish as the temperature is raised, since the martensite may partially revert back to the austenitic phase.

Severe cold working also produce dislocations which decrease the lattice component of the thermal conductivity at low temperatures. The percentage decrease depends on the temperature: for a saturation dislocation density of 10^{16} m^{-2}, the decrease is about 20% at 10 K; about 17% at 50 K, and about 7% at 100 K. The decrease in the lattice component of the thermal conductivity is compensated partially by the increase in the electronic component. The overall effect, including that of martensite and of dislocation, is such that at 10 K the total thermal conductivity would decrease by about 10%, at 50 K the decrease is again about 10%, at 100 K the thermal conductivity would remain unchanged, and it probably would increase by about 3% around room temperature. For materials that are not as severely cold worked, these changes would be somewhat smaller.

3. AISI 316 Stainless Steel

AISI 316 is one of the two* most stable of all austenitic stainless steels because of its higher molybdenum and nickel content. It contains 2-3% molybdenum for greater corrosion resistance and improved mechanical strength at elevated temperatures. It is, however, also susceptible to intergranular corrosion at 750-1150 K (900-1600°F). Because of the molybdenum content, prolonged service at elevated temperatures may cause the formation of the brittle sigma phase at grain boundaries in AISI 316.

There are eight sets of experimental data available for the electrical resistivity covering a wide temperature range. The recommended values are based on the results of measurements of Matolich [28], Evangelisti et al. [29], and Clark et al. [10].

Twenty-three sets of experimental data are available for the thermal conductivity, most of which are for room temperature and

---

* The other is AISI 317.

above. The recommended values are based on the results of Lucks et al. [30], Matolich [28], Watson and Flynn [31], Feith et al. [32], Fieldhouse et al. [33], Williams and Blum [15], Iacobelli et al. [34], Evangelisti et al. [29], and Foster et al. [35]. Since there are no data on the thermal conductivity at low temperatures, the recommended values at those temperatures are obtained by comparison with other 300 series stainless steels, taking into account variations in composition.

The recommended values for the electrical resistivity and thermal conductivity of AISI 316 are tabulated also in Table 2 and shown in Figure 2.

Upon cold-working, this steel displays very little tendency for martensitic transformation. Thus, the electrical resistivity is not expected to change by cold working. The thermal conductivity may be decreased by about 10% at temperatures below 50 K and by about 5% at 100 K at a saturation dislocation density of 10^{16} m^{-2}. This decrease is probably negligible at room temperature and above.

4. AISI 317 Stainless Steel

AISI 317, because of its high molybdenum and nickel contents is probably the most stable of all austenitic stainless steels. Its properties are almost identical with those of AISI 316.

No experimental data are available for the electrical resistivity and the thermal conductivity of this steel. The recommended values are derived from those for AISI 316 with adjustments according to the variation in composition. These are tabulated in Table 3 and shown in Figure 3.

The effects of cold work on these properties of this steel are most likely to be the same as those for AISI 316.

5. AISI 321 Stainless Steel

AISI 321 is a titanium-stabilized version of the 18-8 austenitic stainless steel. It is mostly used in welding applications where post-welding annealing is impractical and where intergranular corrosion in the temperature range 700-1150 K (800-1600°F) would be high for the other austenitic steels.

There are two sets of experimental data available for the electrical resistivity of this steel, both of which are for low temperatures [10,36]. However, there are a number of data sets available for foreign stainless steels of similar compositions,

Table 3. Electrical Resistivity and Thermal Conductivity of
          AISI 317, 321, and 347 Stainless Steels[†]

Temperature (K)	Electrical Resistivity (10^{-8} Ω m)		Thermal Conductivity (W m^{-1} K^{-1})	
	AISI 317	AISI 321 (solid) and 347	AISI 317	AISI 321 (solid) and 347
1	56.8	52.4	0.051	0.062
4	56.8	52.4	0.226	0.275
10	56.8	52.4	0.645	0.796
20	56.8	52.3	1.47	1.83
30	56.8	52.2	2.43	3.01
40	57.0	52.3	3.49	4.14
50	57.5	52.6	4.60	5.60
60	58.3	53.1	5.61	6.67
70	59.2	53.7	6.42	7.51
80	60.2	54.6	7.10	8.19
90	61.1	55.6	7.67	8.72
100	62.1	56.7	8.19	9.16
150	66.8	61.7	9.80	10.95
200	71.5	66.6	10.96	11.96
250	75.8	71.1	11.97	13.23
273	77.7	73.0	12.41	13.66
300	79.9	75.2	12.91	14.16
350	83.6	79.0	13.81	15.03
400	87.1	82.7	14.66	15.86
500	93.6	89.3	16.31	17.47
600	99.4	95.4	17.90	19.01
700	104.8	100.9	19.45	20.53
800	109.4	106.0	20.98	22.02
900	113.5	110.3	22.48	23.44
1000	116.9	113.9	23.86	24.78
1100	120.0	117.1	25.19	26.10
1200	122.8	120.0	26.50	27.37
1300	125.4	122.7	27.84	28.61
1400	128.9	125.3	29.15	29.88
1500	130.2	127.8	30.45	31.13
1600	132.5	130.0	31.72	32.34
1644	133.5*	–	32.28*	–
1672	153.5*	131.6*	26.63*	33.21*
1700	154.1	–	26.97	–
1727		151.3*		27.90*
1800		152.9		28.78

* Melting Range.
† The uncertainty in the recommended values is of the order of
  ±5%. At or above melting, the uncertainty is of the order of
  ±15%.

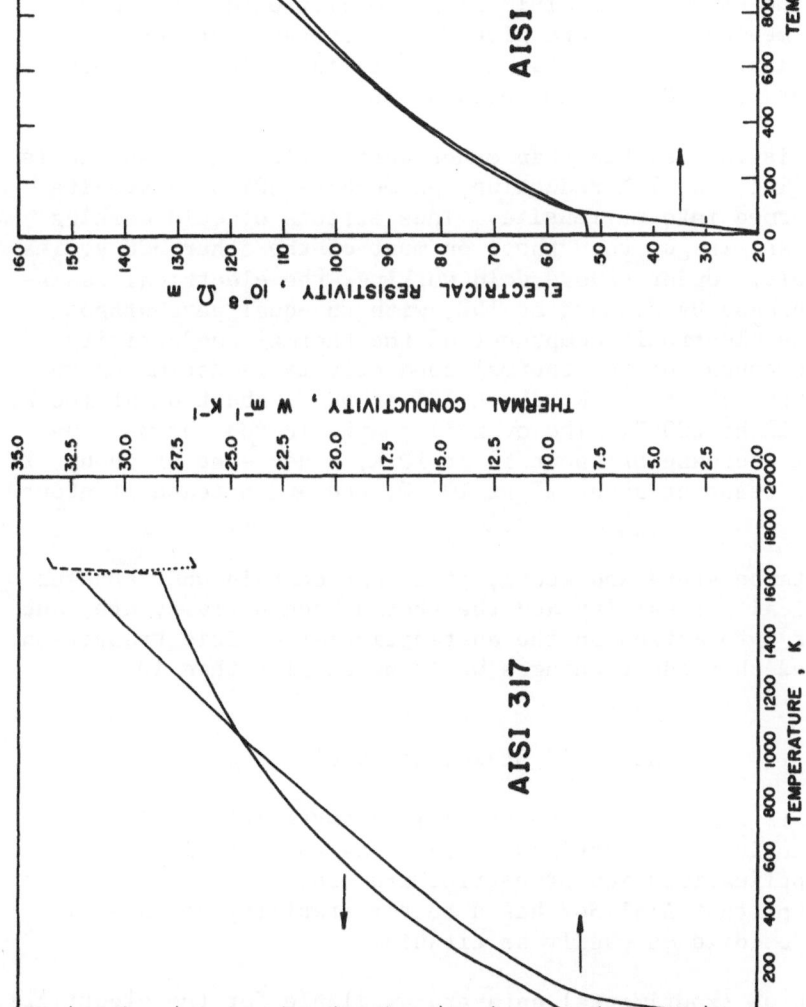

Figure 4. Electrical Resistivity and Thermal
Conductivity of AISI 321 (solid
only) and 347 Stainless Steels.

Figure 3. Electrical Resistivity and Thermal
Conductivity of AISI 317 Stainless
Steel.

covering temperatures up to about 600 K [37,38]. The recommended
values below 600 K are based on the data for the above references.
Above 600 K they are derived from the electrical resistivity of
other austenitic stainless steels with adjustments according to
variations in composition.

Two sets of experimental data are available for the thermal
conductivity, both for temperatures below 159 K [14,21]. For low
temperatures the recommended values are based on these results.
For room temperature and above, results on other austenitic steels
are well as on foreign steels of similar compositions [39,40,41,42]
are used as basis for deriving values for AISI 321.

Because the titanium in AISI 321 is readily oxidizable upon
melting, recommended values are given only for temperatures below
the melting point. These are tabulated also in Table 3 and shown
in Figure 4 for AISI 321 in the solid state only.

AISI 321 is less stable than other austenitic stainless steels
upon cold working: at 75% reduction, as much as 50% of austenite
may be transformed into martensite. Thus effects of cold working
on this steel are larger than those on most of the other 300 series
stainless steels. Under severe cold working, the electrical resistivity may decrease by as much as 15%, with an equal percentage
increase in the electronic component of the thermal conductivity.
The lattice component of the thermal conductivity is estimated to
decrease by about 30% at 10 K, about 17% at 50 K, about 6% at 100 K,
and less than 1% at 200 K. The overall change in the thermal conductivity is a decrease of about 5% at 10 K, a decrease of about 3%
at 50 K, an increase of about 2% at 100 K, and an increase of about
7% at 200 K.

At room temperature and above, it is not certain what changes
in the electrical resistivity and the thermal conductivity are, due
to the lack of information on the austenitic-martensitic transition.
It is estimated that these changes would be smaller than 10%.

6. AISI 347 Stainless Steel

AISI 347 is another carbide-stabilized austenitic stainless
steel, with niobium and sometimes tantalum as the stabilizing elements. Its applications and properties are essentially the same as
AISI 321, except that AISI 347 has a better stability because niobium does not oxidize as easily as titanium.

Four sets of experimental data are available for the electrical
resistivity of this steel [8,9,10,43]. There are two additional
experimental data sets on foreign stainless steels of similar compositions for higher temperatures [37,44].

These measurements indicate that the electrical resistivity of
AISI 347 is the same as that of AISI 321 to within measurement
errors.

Thirteen sets of experimental data are available for the thermal
conductivity of this steel [8,9,30,33,43,45,46,47]. For temperatures higher than 1000 K, these data sets [8,30,33] give values too
low compared with those for other 300 series stainless steels and
are inconsistent with composition variations. The recommended thermal conductivity values are therefore based on the data for a stainless steel of similar composition [37] and on intercomparison with
the data for other austenitic stainless steels, especially stainless
steel 321. It comes to the conclusion that the small variation in
composition between AISI 321 and 347 is not significant to justify
separate recommendations. Consequently, the same recommended values
for the electrical resistivity and thermal conductivity are generated for AISI 347 and 321 as tabulated in Table 3 and shown in Figure 4. It is important to note that the values for the molten state
are for AISI 347 only.

Besides electrical and thermal properties, the stability of
stainless steel 347 under cold work is also very similar to that of
stainless steel 321. Thus the effect of cold working on stainless
steel 347 is expected to be the same as that on stainless steel 321.

7. AISI 410 Stainless Steel

AISI 410, because of its low chromium content and of the
absence of nickel, is austenitic at elevated temperatures and ferritic or martensitic at room temperature. It is usually produced
in the fully annealed, martensitic condition: being heated to the
austenitic range and then slow-cooled in air. In this way, it is
hardened by the formation of martensite within the ferrite matrix.
It is susceptible to temper embrittlement in the temperature range
700-800 K (750-950°F); services at these temperatures are therefore
to be avoided. Because of its low chromium content, AISI 410 is
inferior in corrosion resistance than the austenitic stainless
steels. It is a low cost, general-purpose stainless steel of wide
commercial applications.

Five sets of data are available for the electrical resistivity
of this steel. Most of these are compilations for value at room
temperature. The only set that contains original data is by Clark
et al. [10] and is for room temperature and below. However, there
are some data on foreign stainless steels of similar compositions
[48,49,50]. The recommended values below room temperature are derived from the result of Clark et al. [10]; for higher temperatures,
they are based on the results on foreign stainless steels with minor
adjustments according to composition variation.

There are two experimental data sets for the thermal conductivity
of this steel [14,45], both for temperatures lower than room temperature. The recommended values for low temperatures are derived
from these measurements. For high temperatures, results on foreign
stainless steels of similar compositions [49,50,51] are analyzed and
the recommended values are based on these results. There are no
data for temperatures above 1000 K, for which the recommended values are extrapolations from values at lower temperatures.

The recommended values for the electrical resistivity and
thermal conductivity of AISI 410 are tabulated in Table 4 and shown
in Figure 5.

Because AISI 410 is heat-treatable, the electrical resistivity
and the thermal conductivity of this steel depend somewhat on its
annealing condition. The uncertainty in the recommended values
are therefore higher than that for the austenitic stainless steels,
and is estimated to be ±12% for all temperatures. This includes
errors in the estimation procedure as well as the lack of exact
information on the metallurgical structure of the steel under different annealing conditions.

Since there is no significant metallurgical transformation in
stainless steel 410 upon cold working, the electrical resistivity
is not expected to change significantly. The thermal conductivity
may be affected through the lattice component, which is more sensitive to defects, especially dislocations. It is estimated that for
severely cold worked material, the thermal conductivity may decrease
by 3% at 200 K, 10% at 100 K, and 20% below 50 K.

8. AISI 430 Stainless Steel

Because of its high chromium content, AISI 430 is ferritic at
all temperatures. It is not hardenable by heat treatment, even
though it may develop some austenitic structure on heating. The
amount of austenite is probably not sufficient to alter significantly its properties. Similarly, no significant martensite is
formed on cooling. It is susceptible to the embrittlement at 750 K
(885°F), and therefore is usually annealed above this temperature.

There are five data sets available for the electrical resistivity
of this steel. Only one [10] contains original results and is for
temperatures below 300 K. In addition, there are two data sets on
foreign stainless steels of similar compositions for higher temperatures [13]. The recommended values are based on these results.

Five sets of experimental data are available for the thermal
conductivity of this steel above room temperature [52,53,54]. Three
of these are values from different experimental runs, and show a

Table 4. Electrical Resistivity and Thermal Conductivity of
AISI 410 and 430 Stainless Steels[†]

Temperature (K)	Electrical Resistivity (10^{-8} Ω m)		Thermal Conductivity (W m^{-1} K^{-1})	
	AISI 410	AISI 430	AISI 410	AISI 430
1	38.2	39.4	0.09	0.089
4	38.2	39.4	0.43	0.40
10	38.2	39.4	1.32	1.17
20	38.2	39.4	3.40	2.78
30	38.2	39.4	6.14	4.64
40	38.3	39.5	8.97	6.73
50	38.4	39.7	11.64	8.76
60	38.7	40.2	14.03	10.65
70	39.1	40.7	16.00	11.20
80	39.6	41.2	17.73	13.58
90	40.1	41.9	19.28	14.77
100	40.7	42.6	20.42	15.76
150	44.6	47.2	23.69	19.08
200	49.3	51.8	24.98	20.55
250	53.8	56.6	25.78	21.42
273	55.9	58.7	26.06	21.68
300	58.3	61.1	26.33	21.86
350	62.8	65.3	26.71	22.11
400	67.3	69.4	26.99	22.32
500	76.3	77.9	27.31	22.66
600	85.0	86.8	27.46	22.98
700	93.2	95.4	27.50	23.10
800	100.4	103.4	27.30	23.62
900	106.4	110.7	26.63	23.93
1000	111.1	115.8	26.04	24.47
1100	114.0	117.9	26.16	25.25
1200	116.3	119.7	26.98	26.23
1300	118.6	121.5	28.04	27.35
1400	120.9	123.4	29.20	28.55
1500	123.2	125.4	30.36	29.80
1600	125.6	127.3	31.52	31.05
1750	–	130.2*	–	32.93*
1755	129.2*	–	33.32*	–
1775	–	149.7*	–	28.98*
1805	148.5*	150.3	29.70*	29.35
1900	150.7		30.81	

* Melting Range.
† The uncertainty in the recommended values is of the order of
±12% for AISI 410 and ±7% below 1200 K and ±12% at or above
1200 K for AISI 430. At or above melting, the uncertainty is
of the order of ±15%.

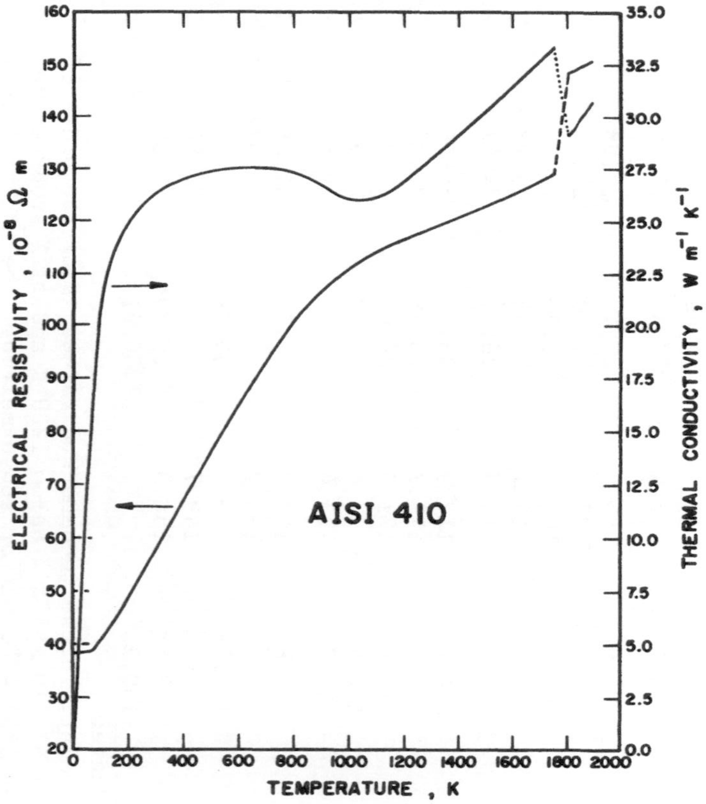

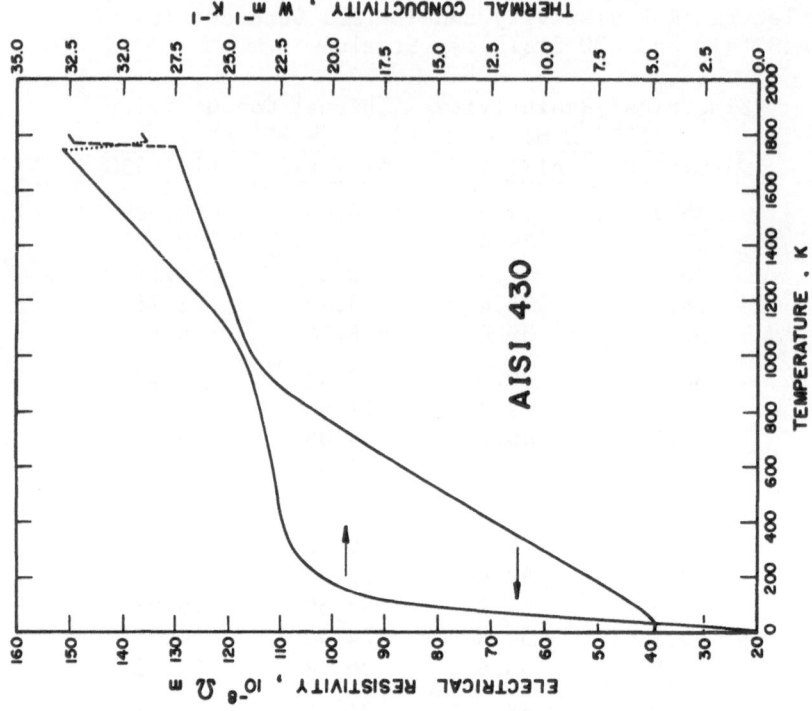

Figure 6. Electrical Resistivity and Thermal
Conductivity of AISI 430 Stainless
Steel.

Figure 5. Electrical Resistivity and Thermal
Conductivity of AISI 410 Stainless
Steel.

temperature dependence that is too strong compared with the other two sets, by Silverman [52] and by Moeller and Wilson [53]. The recommended values are based on the latter results, which cover slightly different temperature ranges and are in agreement in the region where their measurement temperatures overlap. For low temperatures, the recommended values are extrapolated, with consideration given to the temperature variations of its electrical resistivity and of the thermal conductivity of other stainless steels.

The recommended values for the electrical resistivity and thermal conductivity of AISI 430 are tabulated also in Table 4 and shown in Figure 6.

As in the case for AISI 410, there is no metallurgical transformation in AISI 430 upon cold working. The electrical resistivity is therefore not significantly altered. The decrease in thermal conductivity due to dislocations generated by severe cold working is estimated to be 2% at 200 K, 8% at 100 K, 16% at 50 K, and 20% at liquid helium temperatures.

V. DISCUSSIONS AND CONCLUSIONS

Results of the present study of the thermal and electrical properties of 6 stainless steels of the 300 series and 2 stainless steels of the 400 series indicate that the thermal and electrical properties of stainless steels of the same series are very similar.

The 300 series austenitic stainless steels are variations of the 18-8 type: they contain as major alloying elements about 18% chromium and about 8% nickel. Their crystal structure is the metastable face-centered cubic. The thermal conductivity and the electrical resistivity of the 300 series stainless steels are very similar both in magnitude and in temperature variation. The basic types, such as AISI 303 and 304, show the highest thermal conductivity and the lowest electrical resistivity. The titanium-stabilized AISI 321 and niobium-stabilized AISI 347 are next in order. AISI 316 and 317, with 2 to 4% molybdenum, show the lowest thermal conductivity and the highest electrical resistivity. These variations are consistent with their compositions.

The major difference in composition between the 400 series and the 300 series stainless steels is the absence of nickel in the former. AISI 410 contains, as major alloying element, about 12% chromium. It is produced in the martensitic state by quenching from above the austenizing temperature. The most interesting feature is the relatively high thermal conductivity of this steel at ordinary temperatures. A rough estimate showed that the lattice thermal conductivity of this stainless steel is about 2.5 times that of the austenitic stainless steel at room temperature. It is

believed that the major reasons for this behavior are the lower
alloying content, the lower values of the Grüneisen parameter (by
a factor of about 1.4, as evidenced by the lower thermal expansion
coefficient [55]), and the higher phonon velocity [56] of the steel
compared with that of the austenitic steels. The first factor reduces the electrical resistivity and also the magnitude of the impurity scattering of the lattice waves. The second factor reduces
the strength of the Umklapp thermal resistance. The last factor
affects both the impurity and the Umklapp thermal resistance, even
though to smaller extents than the first two factors. The minimum
in the thermal conductivity of AISI 410 at around 1100 K is probably
related to solute reactions (e.g., the precipitation of carbides,
the partial transformation to ferrite, etc. [57,58]).

The thermal and the electrical conductivity of AISI 430 are
lower than those of AISI 410. The Grüneisen parameter and the
phonon velocity of AISI 430 are not significantly different [2,3]
from those of AISI 410, even though AISI 430 is ferritic in structure. The lower conductivities in AISI 430 are interpreted as mainly
due to its higher chromium content (about 17%).

Finally, even though the electrical resistivities of AISI 410
and 430 are lower than those of the austenitic stainless steels at
room temperature, they have a higher temperature coefficient, so
that at around 1000 K their values become higher. This behavior
is most likely to be magnetic in origin.

As evidenced by the available experimental data on both the
electrical resistivity and the thermal conductivity of the 8 stainless steels reviewed and discussed in this work, it is clear that
for some of the stainless steels the available data are very scarce
and serious gaps exist for the temperature dependence for all of
them. The resulting recommended values that cover the full range
of temperature up to above the melting point go far beyond the very
limited experimental data.

VI. REFERENCES

1. Matsushita, T., Sci. Repts. Tohoku Impl. Univ., 1, 9, 243-50
    (1920).

2. Schwerer, F.C. and Cuddy, L.J., Phys. Rev. B, 6, 1575-86 (1970).

3. McReynolds, A.W., J. Appl. Phys., 17, 823-32 (1946).

4. Storchheim, S., Trans. AIME, Journal of Metals, 3(12), 1181-3
    (1951).

5. Palmer, E.E., Bell, J.R., Crabtree, R.D., Wattier, J.R., Springer, H.L., and Thomas, E.A., U.S. National Aeronautics and Space Administration Rept., NASA CR-116 519, Vol. III, 520 pp. (1969). [AD 883 344]

6. Tye, R.P., U.S. National Aeronautics and Space Administration Rept., NASA CR-72710, 22 pp. (1970).

7. Flynn, D.R., in Mechanical and Thermal Properties of Ceramics (Wachtman, J.B., Jr., Editor), U.S. National Bureau of Standards, NBS Spec. Publ. 303, 63-123 (1968).

8. Hogan, C.L., "The Thermal Conductivity of Metals at High Temperature," Lehigh University, Ph.D. Dissertation, 43 pp. (1950); Hogan, C.L. and Sawyer, R.B., J. Appl. Phys., 23, 177-80 (1952).

9. Zimmerman, J.E., "Heat Conduction in Alloys and Semiconductors at Low Temperatures," Carnegie Institute of Technology, Sc.D. Thesis, 54 pp. (1951).

10. Clark, A.F., Childs, G.E., and Wallace, G.H., Cryogenics, 10, 295-305 (1970).

11. Vierstra, A. and Butman, R.C., MIT Lincoln Lab. Tech. Rept. No. 257, 34 pp. (1962).

12. Perova, V.I. and Knoroz, L.I., Tr. Tsentral. Nauch-Issledovatel Inst. Tekhnol. I Mashinostroen, 93, 149-60 (1960).

13. Bungardt, K. and Spyra, W., Arch. Eisenhuettenw., 36(4), 257-67 (1965).

14. Androulakis, J.G. and Kosson, R.L., in Thermal Conductivity - Proceedings of the Seventh Confernece (Flynn, D.R. and Peavy, B.A., Jr., Editors), NBS Spec. Publ. 302, 337-48 (1968).

15. Williams, D.R. and Blum, H.A., in Thermal Conductivity - Proceedings of the Seventh Conference (Flynn, D.R. and Peavy, B.A., Jr., Editors), NBS Spec. Publ. 302, 349-54 (1968).

16. Brown, R.M., "Development of an Inductively Heated High Temperature Thermal Conductivity Device," University of Illinois, M.S. Thesis, 63 pp. (1962).

17. Clausing, A.M., "Thermal Contact Resistance in a Vacuum Environment," University of Illinois, Ph.D. Thesis, 156 pp. (1963).

18. Tye, R.P., Hayden, R.W., and Spinney, S.C., High Temp.-High Pressures, 4, 503-11 (1972).

19. Tye, R.P., Hayden, R.W., and Spinney, S.C., in Advances in Cryogenic Engineering (Timmerhaus, K.D., Reed, R.P., and Clark, A.F., Editors), Vol. 22, 136-44 (1977).

20. Feith, A.D., Hein, R.A., Johnstone, C.P., and Flagella, P.N., USAEC Rept. GEMP-643, 13 pp. (1968).

21. Stutius, W. and Dillinger, J.R., J. Appl. Phys., 44(6), 2887-8 (1973).

22. Powell, R.W., Proc. Roy. Soc. (London), 48, 381-92 (1936).

23. Taylor, R.E., Powell, R.W., Nalbantyan, M., and Davis, F., U.S. Air Force Rept. AFML-TR-68-227, 74 pp. (1968).

24. Ewing, C.T., Seebold, R.E., Grand, J.A., and Miller, R.R., J. Phys. Chem., 59, 524-8 (1955).

25. Deverall, J.E., USAEC Rept. LA-2269, 62 pp. (1958).

26. Powers, R.W., in Liquid Hydrogen Propellant for Aircraft and Rockets - Ohio State Univ. Research Foundation Twenty-Fifth Bi-Monthly Progress Report on U.S. Air Force Contract No. W33-038-ac-14794(16243), 3 (1949).

27. Brown, W.T., Jr. and Bergles, A.E., in Proceedings of the Fourth Symposium on Thermophysical Properties (Moszynski, J.R., Editor), ASME, 184-8 (1968).

28. Matolich, J., Jr., U.S. National Aeronautics and Space Administration Rept. NASA CR-54151, 25 pp. (1965).

29. Evangelisti, R., Isacchini, F., and Premoli, A., Energia Nucleare, 12, 266-72 (1965).

30. Lucks, C.F., Thompson, H.B., Smith, A.R., Curry, F.P., Deem, H.W., and Bing, G.F., U.S. Air Force Rept. TR-6145 (Part I), 7-16 (1951). [AD 117 715]

31. Watson, T.W. and Flynn, D.R., Trans. Met. Soc. AIME, 242, 844-6 (1968).

32. Feith, A.D., Conway, J.B., and Flagella, P.N., USAEC Rept. GEMP-1008, 26-49 (1968).

33. Fieldhouse, I.B., Hedge, J.C., and Lang, J.I., U.S. Air Force Rept. WADC TR 58-274, 79 pp. (1958). [AD 150 954]

34. Iacobelli, R. and Moretti, S., Rev. Hautes Temper. Refract., 3, 215-28 (1966).

35. Foster, E.L., Nelson, S.G., Hildebrand, W.J., Griesenauer, N.W., Moak, D.P., and Allen, J.M., BMI Interim Engr. Progr. Rept. No. 7, 62 pp. (1968). [AD 830 414]

36. Berman, R., Phil. Mag., 42, 642-50 (1951).

37. Powell, R.W. and Tye, R.P., Int. J. Heat Mass Transfer, 10(5), 581-96 (1967).

38. Babakov, A.A., Chobanyan, A.A., Petrova, M.P., and Avrukh, E.L., Metal Sci. Heat Treat. Metals (USSR), 9, 726-8 (1971).

39. Shelton, S.M. and Swanger, W.H., Trans. Am. Soc. Steel Treating, 21, 1061-70 (1933); Shelton, S.M., NBS J. Res., 12, 441-50 (1934).

40. Banaev, A.M. and Chekhovskoi, V.Ya., High Temp. (USSR), 3(1), 47-52 (1965).

41. Ermolaev, B.I., Metalloved. Term. Obrab. Metal., 10, 51 (1971).

42. Egorov, B.N. and Kondratenkov, V.I., High Temp. (USSR), 6(5), 861-4 (1968).

43. Hust, J.G., NBS Rept. 9772, 30 pp. (1970).

44. British Iron and Steel Res. Assoc., "Physical Constants of Some Commercial Steels at Elevated Temperatures," Butterworths, 12-4 (1953).

45. Powers, R.W., Ziegler, J.B., and Johnston, H.L., Ohio State Univ. Cryogenic Lab. Tech. Rept. 264-6, 17 pp. (1951).

46. Vianey, L.R., USAEC Rept. NP-1989, 6 pp. (1951). [AD 140 931]

47. di Novi, R.A., J. Phys. E, 1, 379-82 (1968).

48. Tye, R.P., Engineer, 221(5761), 698-71 (1966).

49. Krzhizhanovskii, R.Ye., Metal Sci. Heat Treat. Metals (USSR), 2(74), 77-8 (1962).

50. Griffiths, E., Powell, R.W., and Hickman, M.J., Iron and Steel Inst., Special Rept. No. 24, 215-68 (1939).

51. Timrot, D.L., Zh. Tekh. Fiz., 5(6), 1011-36 (1935).

52. Silverman, L., J. Metals, 5, 631-2 (1953).

53. Moeller, C.E. and Wilson, D.R., Midwest Research Institute Rept. MRI-2059-E, 95 pp. (1960).

54. Francis, R.K., Brown, R., McNamara, E.P., and Tinklepaugh,
     J.R., U.S. Air Force Rept. WADC TR 58-6001, 69 pp. (1958).
     [AD 205 549]

55. Touloukian, Y.S., Kirby, R.K., Taylor, R.E., and Desai, P.D.,
     Thermal Expansion - Metallic Elements and Alloys, Vol. 12 of
     Thermophysical Properties of Matter - The TPRC Data Series,
     IFI/Plenum, New York, NY, 1440 pp. (1975).

56. Fink, K., Richter, F., Lotter, U., and Schrecke, K., Thyssen-
     forschung, 2, 65-80 (1970).

57. American Society for Metals, Source Book on Stainless Steels,
     American Society for Metals, Metals Park, OH, 408 pp. (1976).

58. Brick, R.M., Gordon, R.B., and Phillips, A., Structure and
     Properties of Alloys, McGraw-Hill, New York, NY, 503 pp. (1965).

ACKNOWLEDGMENT

This work was supported by the American Iron and Steel
Institute (AISI), Washington, D.C., under AISI Project Number
67-371, and under the technical direction of AISI Panel on Physical, Electrical and Magnetic Properties of AISI Subcommittee on
Product Properties.

The extensive literature search, acquisition, and documentary
activities essential to this work were supported by the Defense
Logistics Agency (DLA), U.S. Department of Defense, Alexandria, VA.

SESSION C   REACTOR MATERIALS

Session Chairmen: D.L. McElroy
                   Oak Ridge National Lab.
                   Oak Ridge, Tenn.

                   J.P. Moore
                   Oak Ridge National Lab.
                   Oak Ridge, Tenn.

THERMOPHYSICAL PROPERTIES: SOME EXPERIENCES IN RESEARCH, DEVELOPMENT AND APPLICATIONS

Paul Wagner

Los Alamos Scientific Laboratory
University of California
Los Alamos, NM 87545   USA

and

Russell U. Acton

Sandia Laboratories
Albuquerque, NM 87115   USA

Introduction

Over the years the Conferences on thermal expansion and thermal conductivity have been properly preoccupied with discussions of the theory and measurement of thermophysical properties. Actually, some of the uses to which this information has been put are of considerable interest in themselves, and some of the seemingly routine requests for thermophysical property information by workers in the applications area have led to complex, and sometimes basic research projects. Just for example, during the years 1955 to 1973 the Los Alamos Scientific Laboratory designed, built, and tested thirteen graphite-core, gas-cooled nuclear reactors with the stated purpose of developing a nuclear rocket. The principle of operation was simple, the graphite core was heated and the energy transferred to the flowing gas propellant. Since thermal gradients were generated in the graphite core, stresses were generated. In general,

$$\text{stress generated by thermal gradients} = f(\alpha, E, Q, \lambda, X_i, \nu),$$

where α is the thermal expansion coefficient, λ the thermal conductivity, and E, Q, X_i, and ν are Young's modulus, heat flux, dimensions, and Poisson's ratio respectively. Since energy generation rates were so enormous[1] (a reactor core about the size of a 55 gallon oil drum could generate energy more rapidly than Hoover Dam), thermal gradients were very high and resultant stresses were very high. For proper core

design, heat-up, control and test operation it was essential
that λ and α be known as functions of temperature (to 2500°C)
as accurately as possible. In addition to using this information as an analytical tool, maintenance of controlled
values of λ and α in fuel rods dictated that we use these
thermophysical properties for production control criteria.
The anisotropic nature of the graphites used added an
additional complication to the measurements and techniques
had to be devised [2-4] to overcome the influence of these
directional characteristics.

This paper describes just a few of the paths that were
followed as a result of involvement in some of the programs
at the Los Alamos Scientific Laboratory and at Sandia
Laboratories.

The Nuclear Rocket Program

In addition to the material discussed in the Introduciton,
there was an intimate coupling of the thermal expansion
characteristics of the graphite core and the reactor control
in the nuclear rocket engine. As the temperature was increased the graphite and uranium expanded, this decreased
the overall core density, and therefore the ability of the
core to moderate neutrons (and maintain the required excess
reactivity). This in turn meant that as the reactor operating
temperature increased the control drums in the reactor had
to allow more neutrons to reflect back into the core. The
rule of thumb used was that the decrease in the density of
the graphite caused by heating to 2500°C decreased the
excess reactivity that was used for control by 15-20%.
Since the control margin wasn't all that large to start
with, a very rigorous testing and production control schedule
designed to monitor the CTE was instituted at LASL and at
contract fuel rod fabrication centers. The most elaborate of
these was built at the Westinghouse Astro Nuclear Laboratory.[5]
This facility could make research grade measurements simultaneously on ten fuel samples from room temperature to 2500°C.
Both graphite and uranium-containing graphite fuel rods were
tested in this CTE apparatus as a routine portion of the
fabrication program.

At LASL, we actually monitored the overall expansion of
full length (137 cm) fuel rods to ensure that the sampling
procedure for CTE samples was indeed representative of the
entire rod. This non-destructive test was done by measuring
the change in length produced when a quartz-encased fuel
rod was immersed in an isothermal oil bath at 150°C. The
expanding rod moved a push rod which contacted a dial gauge.
These measurements were very sensitive and proved to be very
accurate when performed carefully. For our purposes, this
was the ultimate demonstration that measurements made on

small, appropriately sampled CTE specimens yielded results
with the required accuracy.

Graphite Research

As an adjunct to the nuclear rocket program, LASL funded a
research and development effort whose goal was to study the
structure of graphite and to understand and fabricate pre-
dictable and reproducible graphites. One of the most
important data inputs in the evaluation of experimental
materials proved to be results obtained from measurements
of linear coefficient of thermal expansion (CTE). The
reasons for this are three-fold. First, the measurement
of CTE is by its nature an averaging type measurement which
is independent of specimen shapes and is not affected by
fabrication or machining flaws. Second, by virtue of its
structure, graphite is highly anisotropic and a rapid and
convenient measure of artifact anisotropy may be obtained
from CTE measurements. Third, the absolute value of the
CTE must be considered for practical applications and uses
of the graphites (see the previous section). Crystalline
graphite has a layered structure with a C-C distance of
2.46 A and an interlayer spacing of 3.35 A. This large interlayer spacing is manifested in a much larger CTE than the
in-plane ("a") value; for example at about 800°C the "c"
direction CTE is about $8 \times 10^{-6}/^\circ$C while the "a" direction
CTE is approximately zero. Furthermore, between 25°C and
about 400°C the in-plane expansion is actually negative.
In a polycrystalline graphite the thermal expansion properties
will depend on the spatial arrangement of the crystallites
which make up the material or, turning this around, measurements of CTE yield important information on the distribution
of crystallite "c" axes. Of utmost importance is the actual
value of the CTE in graphite since its low value contributes
to the thermal stress resistance of this material. On the
other hand, a low value of the CTE in graphite is a dis-
advantage when used with metals or with metal carbide systems
since the latter have relatively high CTE's and the mismatch
may cause severe mechanical problems. The CTE of graphite
can vary widely with raw materials and manufacturing procedures
and means of tailoring the CTE of the final product improved rapidly as our understanding of the structure of
polycrystalline graphites increased.

Another spin-off from the graphite research program was
the research done on boron-doped, artificial, polycrystalline
graphites.[6,7] The original intent was to tailor the
electrical properties of the graphite by adding small
amounts of boron. Indeed our first measurements showed an
increase in the electrical conductivity of about 35% upon
addition of 0.3 percent B. X-ray diffraction studies on
these materials showed a marked decrease in the d_{002} (i.e.

interlayer) spacing with increased B content. We concluded
(as had others) that at low B concentration studied, the
boron atoms occupied substitutional sites in the lattice and
as such would be expected to enhance the electrical conductivity in the in-plane direction.

We reasoned that the observed decrease in d_{002} was a reflection of an increase in the interlayer ("c" direction)
bonding forces and should affect the thermal expansion
characteristics. Since the out-of-plane vibrational modes
are more readily excited than the in-plane modes, the
thermal expansion in this, the "c" direction is much greater
than it is in the direction parallel to the layer planes.
Because of this large anisotropy in the polycrystalline
system, the measurement of gross thermal expansion (especially
at low temperatures) in any particular direction will, to a
large extent, be a measure of the "c" direction component.
From the d_{002} data, we anticipated a decrease in the "c"
direction component. We then chose samples of identical
anisotropy (so that the crystalline effects would not be
masked by anisotropic effects) and measured α from 23° to
645°C. We did indeed find a monotonic decrease in the average
CTE of about 20% with B content (up to 0.8% B). This was
further confirmed by noting an increase in Young's modulus
with increased boron content up to the solubility limit.

Since we noted an increase in the electrical conductivity
(σ) of the graphite with increased boron content and since
there has been a long-standing challenge to the experimenter
to demonstrate unambigiously, the fractional value of lattice
thermal conductivity component in graphite,[8] we also made
measurements (at 23°C) of thermal conductivity.   Since

$$\lambda = \lambda_{electronic} + \lambda_{lattice}$$

and by the Weidmann-Franz law

$$\lambda = L\sigma T + \lambda_{lattice}$$

we felt we could extrapolate σ back to zero and get the
lattice component. Unfortunately λ <u>decreased</u> with increased B content. Apparently the B atoms act as phonon
scatterers and $\lambda_{lattice}$ decreases more rapidly than
$\lambda_{electronic}$ increases as the boron atom concentration increases; however, the magnitude of the decrease in λ exceeds that predicted by simple scattering theory[7,9] and a
truly satisfactory explanation of this phenomenon has
never been published.

The SNAP 19/Pioneer Heat Source Insulators

LASL was asked to look at the CTE and thermal cycling
behavior of the three nested, pyrolytic graphite tubes
which comprised the SNAP 19/Pioneer heat source insulator.
The source of concern was that the insulators would break
down in some manner upon thermal cycling and fail the
reliability tests. As it turned out,[10] the cylindrical
insulators distorted and also grew in length with repeated thermal cycling between room temperature and 2500°C.
X-ray diffraction measurements made before and after heating showed an increase in the crystallite height, a decrease
in the interlayer spacing, and an increase in the orientation
factor. All this indicated an overall increase in the
crystallinity of the pyrolytic graphite upon heating. It
was found that high temperature annealing the pyrolytic
graphite in a constrained configuration and remachining
produced a thermally stable insulator.

Sandia Graphites

During the years1969 to 1975, Sandia Laboratories designed,
built and flew a number of high performance re-entry bodies
using graphite nose tips, heat shields and heat shield
insulators. The kinetic energy of a re-entering body is
transformed into heat, with surface temperatures approaching
the sublimation point. Thermal gradients are thus generated
in the re-entry bodies which produced stresses, as was discussed earlier.

Thermal properties govern more than thermal stresses within
the RVs. As the graphite heat shield heats up and expands,
the payload slips forward, changing the center of gravity
and consequently, the flight characteristics of the vehicle.
Before impact, however, the heat shield begins to cool, put-
ting the insulation in compression. The insulation strength
must be able to support the shield during ambient and heat-
up conditions, but be crushable during the shrinkage phase
in order not to fracture the heat shield. The graphite
materials development program associated with these re-entry
bodies was so closely coupled to the thermophysics laboratory
that often property data was awaited in order to select the
next change in a process variable. This close coupling
necessitated the development of a properties laboratory
that could generate data on large numbers of samples
to high temperatures in short times. The thermal diffusivity
apparatus that was designed as a part of that program is
believed to still be unique--it is capable of measuring the
thermal diffusivity of 20 samples from room temperature to
2500 K in an eight-hour day. The apparatus features computer
data acquisition and reduction.

Thermal expansion measurements on these graphites were made
in dilatometers. We experimented with multi-sample dilatometers but settled on single dilatometers with multiple
furnaces. Expansion measurements were made at 20C per
minute. Comparison tests at 2 and 5°C per minute showed no
significant differences. At the end of a test, the hot
furnace was removed from the dilatometer. Without the
furnace, the sample tube and pushrod very quickly cooled
to room temperature. A second "cold" furnace was placed
on the dilatometer. In this fashion, three samples could
be run on each dilatometer in an eight-hour day.

In conjunction with LASL, Sandia has developed a graphite
nuclear fuel for advanced pulsed reactors. The same types
of problems occur as was discussed earlier.

The thermal properties of the SNAP 19/Pioneer heat source
insulators were also investigated by Sandia. The interesting
aspect of this work was in the accident analysis. The
thermal expansion of the close fitting sleeves of pyrolytic
graphite cause the heat transfer rate to be different depending on the direction of heat flow. The heat flow is outward under normal operating conditions, but could be inward
under accident or planetary re-entry conditions. The insulator thus acts somewhat as a thermal diode.

Energy Related Projects

As is the case with many other laboratories, much of our
work in the last three or four years has been on energy
related projects. At some point in any energy program,
the production, movement or storage of heat must be considered, thus the need for thermophysical properties.
Most of the properties data in these areas is generated
for use in mathematical models which attempt to describe
the entire system and/or process.

LMFBR

The liquid metal fast breeder reactor is a prime candidate
to meet the world's future energy needs. Safety considerations
are the major stumbling block of this energy program. We are
currently engaged in studying the case wherein the pressure
vessel is breached and molten sodium, stainless steel and
fuel come in contact with the concrete of the containment
vessel. A knowledge of the thermophysical properties of
the various components is essential in modeling these
 reactions, penetration rates, etc.

Waste Isolation

Isolation is the terminology now being used in conjunction
with radioactive wastes. The word disposal implies no
further accountability or control, whereas isolation means
just the opposite. Should nuclear reactors become the
energy source of the immediate future, then waste isolation
becomes a major problem.

Salt beds are being given prime consideration as radioactive
waste isolation sites for a number of reasons:

1. Salt beds are tectonically stable,

2. No ground water problems (salt is water soluble. These
    beds have been around for about two million years. If
    there was any water around, they wouldn't be.),

3. Salt transports about twice as much heat as would granite.
    Waste storage containers would thus be cooler,

4. Salt is plastic. Cracks and fissures tend to heal. Items
    placed in salt become imbedded with time, and

5. The mining and maintenance of a waste isolation
    facility would be considerably cheaper in salt
    than in granite.

Thermophysical properties are necessary in the analysis
of the transfer of heat from the waste to the salt
surroundings.

Insitu Coal Gasification

The burning of a coal seam, in place in the ground, to
produce a fuel gas, offers large energy savings and would
allow the full utilization of high sulfur coals. The
thermal modeling of this process is under investigation.
Several burns have been successfully conducted at Hanna,
Wyoming, by the Laramie Energy Center with Sandia Lab's
assistance in the areas of instrumentation and data
acquisition and analysis. Down hole instrumentation was
thermally coupled to the surrounding strata through the
use of matching grouts. The thermal properties of the
coal seams, over and under burden and grouts were necessary
to select the proper grout. The instrumentation was thermally
coupled so that the temperature field and burn front would
be distorted as little as possible.

Summary

Accurate measurements of thermophysical properties, understanding exactly any use requirements, and understanding
the behavior of the material are all potential ingredients
for in-depth research and development projects. The work
reported here is but a brief description of some of the
many LASL and Sandia projects and their chain-branching
manifestations. Unfortunately there is a wealth of information on projects that have been pursued in the past
25 years of the general type described here which are
described in detail in internal memoranda or informal
reports and almost certainly will not ever be published in
the open literature.

References

1. Spence, R. W., Sci. Technol. No. 43 59 (1965).

2. Wagner, P., and Dauelsberg, L. B., Carbon 5 271 (1967).

3. Wagner, P., and Dauelsberg, L. B., Carbon 6 373 (1968).

4. Wagner, P., Brit. J. Appl. Phys., 2 293 (1969).

5. Gaol, P. S., Thermal Expansion of Solids Symposium, Sept.
    18-20, 1968 (Gaithersburg, MD).

6. Wagner, P., and Dickinson, J. M., Carbon 8 313 (1970).

7. Kelly, B. T., and Tobin, D., Whittaker, A., and Wagner,
    P., Carbon 9 447 (1971).

8. Klemens, P. G., 4th Conference on Thermal Conductivity,
    October 13-16, 1964 (San Francisco, CA).

9. Klemens, P. G., private communcation.

10. Wagner, P. unpublished work.

EX-REACTOR DETERMINATION OF THERMAL GAP CONDUCTANCE

BETWEEN URANIUM DIOXIDE: ZIRCALOY-4 INTERFACES*

J. E. Garnier and S. Begej

Battelle, Pacific Northwest Laboratories

Richland, Washington

ABSTRACT

A study of thermal gap conductance between UO_2 and Zircaloy-4 has been initiated utilizing a unique transient pulse technique: Modified Pulse Design (MPD)--a technique employing a heat pulse (laser) and signal detector to monitor the energy transmitted through UO_2-Zircaloy samples both in contact and separated by a gap. Initial experiments have been conducted as a function of temperature (to 873°K), gas composition (He, He:Ar, Ar), and gap width 6.0 x 10^{-4} and 2.18 x 10^{-3}cm). The thermal gap conductance is dependent upon these variables, and comparison with calculated results using existing models for the temperature jump distance is made. A concept involving direct energy transfer across narrow gaps by individual gas molecules is also considered.

INTRODUCTION

Heat, as generated in nuclear fuel pellets, must be transferred across the interface between the fuel and the surrounding cladding through solid:solid contacts and/or the gas phase before entering the coolant. To ensure that the normal and off-normal conditions affecting heat transfer through the cladding are properly considered in reactor design and safety analysis, a thorough understanding of the heat transfer process between the fuel pellet and surrounding cladding is necessary [1]. The goal of this study is to extend measurements of thermal interface conductance to higher temperatures and gas pressures than obtained with previous ex-reactor investigations [2-4] and to

---

*This paper is based on work conducted under a contract with the Nuclear Regulatory Commission, Reactor Safety Research, Fuel Behavior Branch.

extend measurements of interface conductance to actual gap conditions.
Extension of these measurements would allow a further refinement of
computer programs such as KAPL-3300[5], GAPCON-THERMAL[6,7], FRAP-S[8]
and LAMBUS[9] which predict the effects of gap and contact conductance
in oxide fuel elements.

The purpose of this paper is to describe the Modified Pulse Design (MPD) technique for measuring thermal interfacial conductances
and to discuss gap conductance values obtained to date with those predicted using various analytical models for the temperature jump
distance.

Mathematical Model*

It has been shown that for parallel slab geometries used in the
flash method, the lateral heat losses from the samples may be neglected[10,11] leaving a one-dimensional heat conduction problem to
be solved. The thermal properties are assumed constant over the
transient temperature range. The heat losses from the front and back
faces are generally not negligible and are included in this analysis.
With the model geometry shown in Figure 1, the governing equations
for the temperature of the two samples are:

$$1/\alpha_i \frac{\partial T_i}{\partial t} = \frac{\partial^2 T_i}{\partial X_i^2} \qquad i = 1,2$$

where α_i, T_i and X_i are the thermal diffusivities, temperatures, and
spacial coordinates of the UO_2 and Zircaloy specimens. Applying the
Laplace transform to the associated boundary conditions shown in
Figure 1, results in partial differential equations which can be
solved (assuming that the heat pulse $w(t)$ is a delta function) to
yield an analytic expression for the back face transient temperature.
The analytic expression for temperature versus time is found using
the inversion theorem:

$$T_2(t,0) = \frac{1}{2\pi i} \lim_{L \to \infty} \int_{\sigma-iL}^{\sigma+iL} e^{tp} \overline{T}_2 (p,o) dp$$

where σ is sufficiently large to make the inversion integral exist.
The inversion integral is evaluated using the Cauchy residue theorem
provided all the poles can be located and their order determined. It
is assumed that there exists a set of h's which give a temperature
function which fits the data well and does not give rise to multiple
poles after the first few poles.

The problem is now reduced to choosing values for h_1, h_2 and h_3,

---

*A detailed discussion of this section has been submitted as a sep-
arate paper to the International Journal of Heat and Mass Transfer.

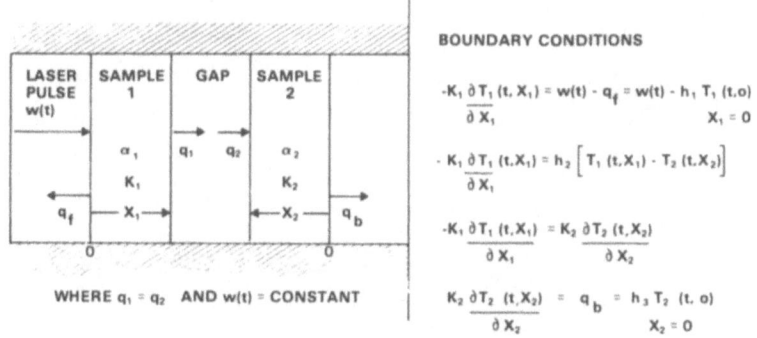

FIGURE 1. Mathematical model and associated boundary conditions where h_1 and h_3 are the unknown surface loss film coefficients and h_2 the unknown gap conductance.

deriving the corresponding expression for the back face transient temperature, and checking the analytic result against the experimental data. The h's can then be adjusted in a suitable manner until a "best fit" is obtained[12].

EXPERIMENTAL TECHNIQUE

Test Apparatus*

The MPD experimental test apparatus and sample pair-holder assembly are shown schematically in Figures 2 and 3. The sample pair is illuminated on the front face of the UO_2 sample with a laser pulse and the resultant temperature history is monitored using 0.00254cm diameter, chromel-constantan (type E) thermocouple wire spot welded to the back surface of the adjoining Zircaloy-4 sample. The ambient sample temperature is determined by a 0.0127cm diameter, chromel-alumel (type K) thermocouple spot welded at the perimeter of the back surface of the Zircaloy-4 sample. An oxygen-free gas atmosphere is maintained by pre-purifying the gas through a series of zirconium getters and placing zirconium metal sheets in the proximity of the samples and sample holder.

Heat losses during the passage of the transient heat pulse are minimized by supporting the UO_2 and Zircaloy-4 samples on the front and back faces, respectively, by three equally spaced Al_2O_3 pins as shown in Figure 3. Sample-spacer-sample heat shunting is minimized by machining a notch groove in the periphery of the Zircaloy-4 sample (as shown in Figure 3) and also by placing the gap spacers equidistant around the outer support ring in positions to maximize the heat shunting path length from the spacers to the type E monitoring thermocouple.

---

*A detailed discussion of this section has been submitted as a separate paper to the Review of Scientific Instruments.

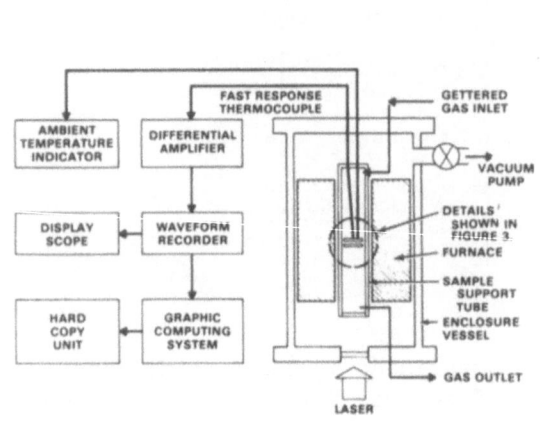

FIGURE 2. Schematic of Modified   FIGURE 3.   MPD sample holder
Pulse Design (MPD) test apparatus.   assembly.

Test Specimens

Depleted UO_2 specimens used in this study were prepared by pressing highly sinterable ceramic grade powder into discs and sintering
at 1650°C for 4 hours in a hydrogen atmosphere. The resultant 0.2cm
thick by 1.45cm diameter discs were hand ground to a nominal 0.064cm
thickness, then machine ground to produce a non-directional finish on
a flat cast iron surface using 400 grit SiC powder. Talysurf-4 pro-
filometer measurements of surface roughness yielded centerline average
(CLA) values of 0.8 ± 0.2 μm. The UO_2 stoichiometry was determined
gravimetrically to be 2.002 ± 0.002, the theoretical density to be 93%.

The Zircaloy specimens were cut from a 5cm diameter ingot of
Zircaloy-4 reactor grade material and lathe machined into 0.102cm
thick by 1.45cm diameter discs. The Zircaloy specimens were hand
ground on flat glass plates with 400 grit SiC powder in a manner producing "random direction" grinding scratches. The resulting CLA value
was determined to be 0.4 ± 0.2 μm. To remove effects of machining and
polishing induced stresses, the UO_2 and Zircaloy-4 samples were heat
treated in a gettered He atmosphere at 1273°K for 1 hour prior to surface macrowaviness measurements. To insure that only the α-phase was
present in the Zircaloy-4 samples, an additional heat treatment at
973°K for several hours in a gettered He atmosphere was used.

The values of thermal diffusivity, conductivity, and expansion
for Zircaloy-4 (required in the analytic determination of gap conductance program) were obtained in this laboratory from measurements
on representative samples. The values for UO_2 (corrected for
porosity) were taken from a round-robin study by Bates.[13]

Gap Determination*

The macrowaviness of UO_2 and Zircaloy-4 specimen mating surfaces were determined from a point-wise topographical surface map
determined using an optical height gauge. Height elevations were
taken at 25 points on the sample surfaces relative to an optically
flat base plane upon which the specimens were mounted. A computer
program re-referenced the height elevations relative to a plane defined by three gap spacer contact points common to each specimen and
calculated the separation distance between coordinates on the UO_2
and Zircaloy surface. The "gap width" was taken as the arithmetic
average of the calculated separation distances at all coordinates
within a 0.3 cm radius of the fast response thermocouple location
(5 points total).

The standard deviation in the gap width associated with repet-
itive specimen mounting and recharacterization was found to be ± 8 x
10^{-6}cm. The change in gap width values before and after each experimental run for the 6.0 x 10^{-4}cm and 2.18 x 10^{-3}cm gap widths were
determined to be +1.8 x 10^{-5}cm and -1.64 x 10^{-4}cm, respectively.

Data Acquisition and Reduction

The magnitude of the resulting emf signal produced by the type
E thermocouple during a temperature transient is approximately 50μV
(1-2°K). This signal was amplified and stored in digitized form in
a minicomputer and a program for reduction of the time-temperature
history curve executed. The curve was signal averaged over 60 cycle
time intervals to enhance the signal to noise ratio, segmented into
three overlapping ranges, and each range fitted with a 4 to 6th order
polynominal. The polynominal curves were then normalized and a table
printed from which six points were selected for use in another computer program for the analytic determination of the gap conductance.

RESULTS AND DISCUSSION

The thermal gap conductance, H_g, between UO_2 and Zircaloy-4 was
measured as a function of temperature (293-873°K), gas composition
(He, He:Ar[50:50], Ar) and gap width 6.0 x 10^{-4} and 2.18 x 10^{-3}cm).
The behavior of H_g as a function of temperature is shown in Figures
4 and 5 for various gases and gap widths measured in this study.
Each datum is the average of three or more measurements, with the
error bar representing a $\pm 10\%$ uncertainty ($\pm 3\sigma$). The primary factor
contributing to this uncertainty is the uncertainties in the thermal
property values used for UO_2 and Zircaloy-4 in the analytical program to determine H_g. A detailed assessment of experiment

---

*A detailed discussion of this section has been submitted as a
separate paper to the Review of Scientific Instruments.

uncertainties will be discussed elsewhere. Since the MPD technique
involves a transient measurement conducted at a fixed temperature;
and the CLA values of the respective adjoining sample surface are
approximately the same, the total conductance across the gap is
defined to be:

$$H_g = K_{gas}/(d+g_1+g_2) \cong K_{gas}/(d+2g); \quad g_1 = g_2$$

where d is the "gap width" between surfaces; K_{gas} the thermal
conductivity of the fill gas and g_1 and g_2 the temperature jump
distances of the UO_2 and Zircaloy, respectively. The quantity g
can be calculated from various models such as used in GAPCON
THERMAL II[14] and proposed by Lloyd et al,[15] Kennard,[16] and
more recently Loyalka.[17] The fundamental difference in the above
models is manner in which the accommodation coefficient, A, is
handled. The reader is referred to the references for details on
the calculations of the quantities of K_{gas}, g, and A.

     A qualitative comparison of H_g to predicted results is ideally
discussed in terms of two regions separated by the Kundsen domain
boundary as determined by the value of the Kundsen number. The
Kundsen number, KN, is defined by the ratio of the mean free path
of the gas molecules to the characteristic dimension (d) of the gap.
Conditions are considered to be within the Kundsen domain when KN
exceeds about 0.01-0.02. Within the Kundsen domain, normal heat
flow formulas are known to become inaccurate. The Kundsen numbers
have been calculated under conditions appropriate to each H_g datum
and are shown in Figures 4 and 5.

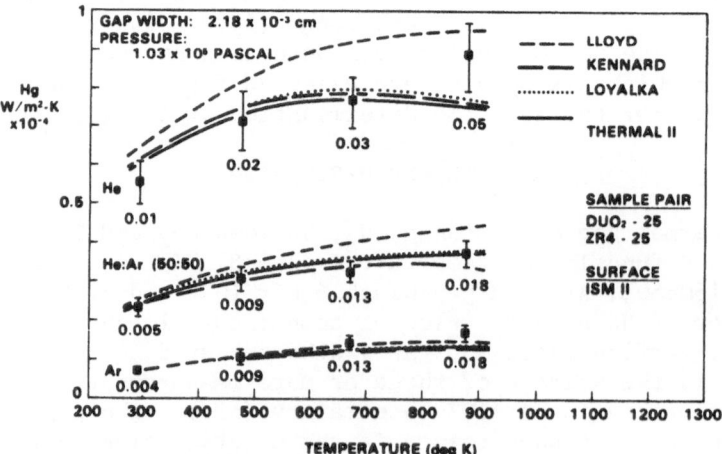

FIGURE 4. Gap conductance vs. temperature using fill gases of He(100),
He:Ar(50:50) and Ar(100) at a gap width of 2.18×10^{-3} cm. The curves
represent the calculated gap conductance predicted using different
models for the temperature jump distance.

- Region I: KN < 0.02

  In this region, normal heat flow formulas can be applied to predict the behavior of the thermal gap conductance with temperature,
  gap separation, and fill gas composition. As seen in Figure 4,
  when the KN ratio is less than ∿0.02 H_g exhibits a dependence on
  temperature and fill gas composition similar to that predicted
  using, for instance, the classical Kennard model for g.

- Region II: KN > 0.02

  As shown in Figure 5, under conditions of a narrow gap (i.e.
  6.0 x 10^{-4}cm), the calculated KN ratio ranges from about 0.02 to
  as high as 0.17 depending on fill gas composition and temperature.
  The fact that normal heat flow formulas become increasingly
  inaccurate within the Kundsen domain is clearly illustrated. The
  models predict an increase, then a continued decrease, in the
  thermal gap conductance with increasing temperature; whereas the
  H_g data is seen to continually increase with temperature from
  293°K to 873°K for all fill gas compositions. This trend is also
  observed at a gap width of 2.18 x 10^{-4}cm for He fill gas at higher
  temperatures as seen in Figure 4.

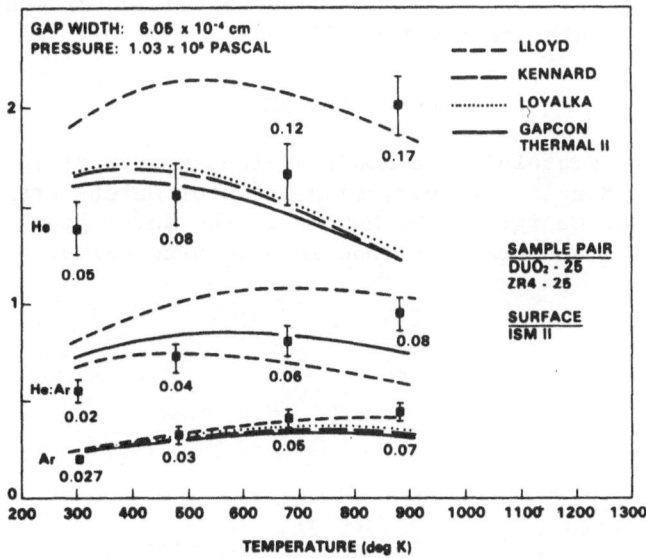

FIGURE 5. Gap conductance vs. temperature using fill gases of He(100),
He:Ar(50:50), and Ar(100) at a gap width of 6.0 × 10^{-4} cm. The curves
represent the calculated gap conductance predicted using different
models for the temperature jump distance.

The development of a model to accurately predict thermal gap conductance values when conditions are within the Kundsen domain should consider the following:

a) Energy transfer from a solid to a gas molecule (or gas molecule to the solid) will depend on the accommodation coefficient.

b) Once the energy exchange from the solid to the gas molecule has occurred, the classical concept for energy transfer across the gas medium through numerous molecular encounters within the gas media is no longer appropriate. A change in energy transfer mechanisms will occur when the KN ratio exceeds ∿0.02. Under these conditions a significant fraction of the energy can be transferred directly by individual gas molecules from one side of the body to the other side without encountering another gas molecule. If these molecules are assumed to possess a non-maxwellian velocity distribution, then a significant fraction of the total energy transferred will be carried by these higher velocity molecules. Thus, a model for energy transfer by gas molecules across a narrow gap, which accounts for a reduction in the number of molecular encounters in the gas needs to be considered.

Future investigations will be directed to ascertaining the effects of surface morphology (i.e., roughness); other gas mixtures (i.e., He:Xe) at narrow gaps, and, extension of measurements to interfacial contact:contact conditions.

ACKNOWLEDGEMENT

The authors gratefully acknowledge the support of the United States Nuclear Regulatory Commission, reactor Safety Branch. The assistance of T. George and D. McCahn in the development of the analytical model and computer code is also recognized.

REFERENCES

1. Light Water Reactor Fuel Behavior Program Project Description--Fuel and Cladding Material Property Requirements; P. MacDonald (ed.), Aerojet Nuclear Company (March 31, 1974).

2. A. C. Rapier, T. M. Jones, and J. E. McIntosh, "The Thermal Conductance of Uranium Dioxide/Stainless Steel Interfaces," International Jrn. of Heat and Mass Transfer 6, p. 397 (L963).

3. R. A. Dean, Thermal Contact Conductance Between UO_2 and Zircaloy-2, CVNA-127 (May 1962).

4. A. M. Ross and R. L. Stoute, Heat Transfer Coefficient Between UO_2 and Zircaloy-2, CRFD-1075: AECL-1552 (1962).

5. W. K. Anderson and G. L. Lechliter, LWB/LSBR Development Program--Some Input Properties for Computer Description of Fuel Properties, KAPL-3300 (June 1967).

6. C. R. Hann, C. E. Beyer, and L. J. Parchen, GAPCON-THERMAL-1: A Computer Program for Calculating the Gap Conductance in Oxide Fuel Pins, BNWL-1778 (1973).

7. C. E. Beyer, C. R. Hann, D. D. Lanning, F. E. Panisko, and L. J. Parchen, GAPCON-THERMAL-2: A Computer Program for Calculating the Thermal Behavior of an Oxide Fuel Rod, BNWL-1898 (1975).

8. J. A. Dearien, et al., FRAP-S2: A Computer Code for the Steady-State Analysis of Oxide Fuel Rods, TREE-NUREG-1107.

9. J. Wordsworth, "IAMBUS-1: A Digital Computer Code for the Design, In-pile Performance Prediction and Postirradiation Analysis of Arbitrary Fuel Rods," Nuclear Science and Engineering, 31, 309 (1974).

10. W. J. Parker, R. J. Jenkins, C. P. Butler, and G. L. Abbott, J. Appl. Phys., 32, 1679 (1961).

11. J. A. Cape and G. W. Lehman, J. Appl. Phys., 34, 1909 (1963).

12. G. Peckman, The Computer Journal, 13(4) (November 1970).

13. J. L. Bates, High Temperature Thermal Conductivity of 'Round-Robin' Uranium Dioxide, BNWL-1431, (July 1970).

14. D. D. Lanning and C. R. Hann, Review of Methods Applicable to the Calculation of Gap Conductance in Zircaloy-Clad UO_2 Fuel Rods, BNWL-1894, (April 1975).

15. W. R. Lloyd, D. P. Wilkins, and P. R. Hill, "Heat Transfer in Multi-Component Monatomic Gases in the Low, Intermediate and High Pressure Regime," (Nuclear Thermionics Conference, 1973).

16. E. H. Kennard, Kinetic Theory of Gases, McGraw-Hill, N.Y., p. 311 (1938).

17. S. K. Loyalka, "Temperature Jump in a Gas Mixture," Physics Fluids 17, p. 897 (1974).

GRAPHITE-TO-METAL BONDING TECHNIQUES *

Lloyd O. Lindquist and Richard Mah

Los Alamos Scientific Laboratory
University of California, Los Alamos, NM   87545

I. SUMMARY

Physical and thermal bonding techniques to mate pyrolytic graphite to copper were tested for suitability in extracting high heat flux loads through the bonds. The most successful bonds were titanium-copper-silver and titanium-copper-nickel braze alloys. Their heat flux capabilities were ~6 MW/m², and they appeared to withstand indefinite heat cycling at maximum heat flux.

II. INTRODUCTION

At the Clinton P. Anderson Meson Physics Facility (LAMPF) at the present time, all Line A pion production targets are radiatively cooled. As beam currents rise from present levels (~300 μA) in Line A, the surface temperature levels of the graphite targets will exceed 2300 K. At temperatures above 2300 K, graphite begins to evaporate at unacceptable rates (shown in Fig. 1). The A-5 bar proton target in its present configuration will approach this temperature at 300 μA, with an 800-MeV proton beam. Rotating target wheels of the present design installed at A-1 and A-2 will not exceed 1500 K for 800-MeV proton currents of 1 mA. If the target wheels are not spinning, then they too will experience higher temperatures, >2300 K (see Fig. 2).

The thrust of this investigation has been to develop a metal-graphite interface to transfer heat from the graphite to a coolant, pumped adjacent to the metal. We are initially working with pyrographite, since the maximum pion production per centimeter of material minimizes the dose time for experiments. Pyrographite is the most dense form of graphite, and therefore has maximum pion production per centimeter of material. Some compelling reasons for selecting this material are the low-Z, the high thermal conductivity, the high strength at elevated temperatures, and the high melting point (3500°C).

---

*Work performed under the auspices of U. S. Department of Energy.

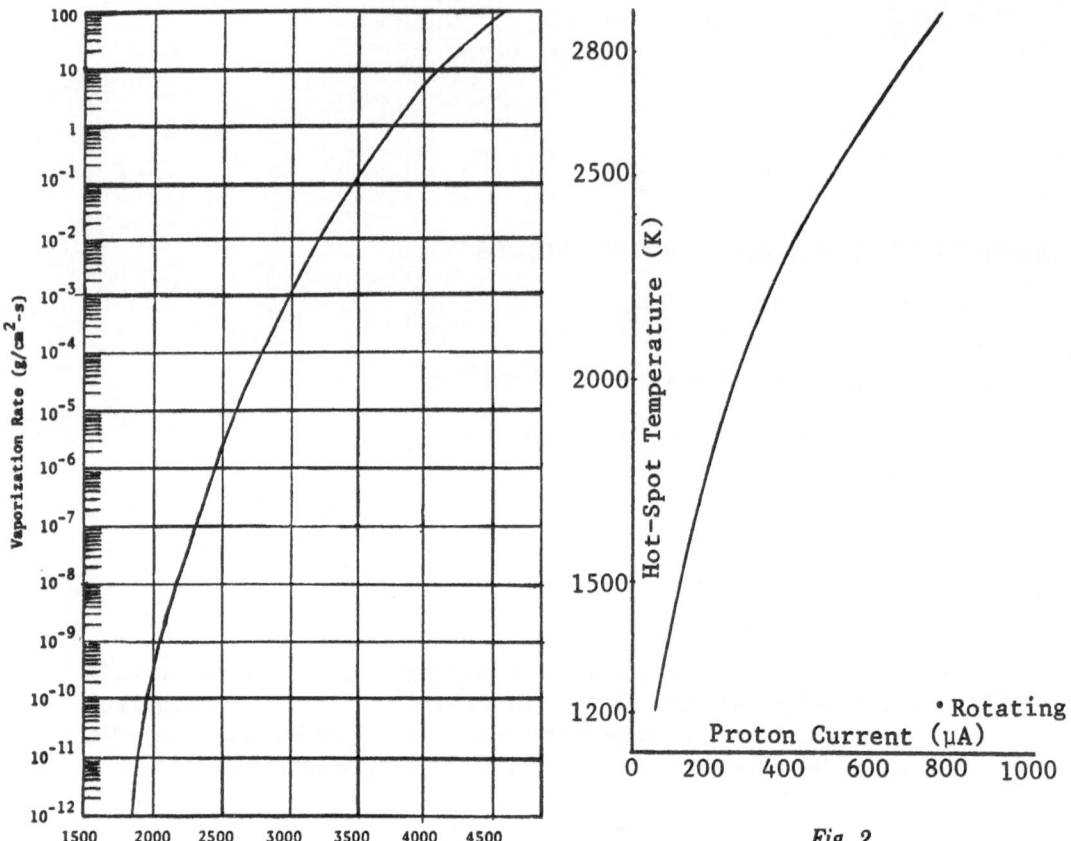

Fig. 1.

Graphite evaporation rate vs temperature of
graphite surface.

Fig. 2.

A-2 stationary target (ATJ graphite), temperature vs proton current (ρ = 1.7 g/cm³).

A. Bonding Techniques

We have investigated in parallel several types of thermal bonding approaches. To date, we
have made bonds between a metal 6.35-mm-diam (nominal) tube and a graphite cylinder with a
hole drilled for the tube (shown in Fig. 3).

The first type of bond tested was polyfurfuryl alcohol (Varcum) graphitized in the space between a copper tube and pyrographite cylinder. The graphitization process bonds the
pyrographite to the Varcum. The bonding of the Varcum to the pyrographite occurs while the
assembly is held at 200°C for 60 h to achieve graphitization. The copper tube does not react
chemically with the Varcum. Varcum merely fills the voids between the pyrographite and the
copper tube, and may even mechanically prestress the copper tube during graphitization. This
could explain the thermal flux capability of the bond (described later in this report). Two hybrid
Varcum compounds were prepared in an attempt to improve on the thermal conductivity of an
ordinary Varcum bond. One compound was composed of a heterogeneous mixture of fine
aluminum chips and graphitized polyfurfuryl alcohol (Varcum) in the fabricated form. The
assemblies that were fabricated using the hybrid compounds are described in Fig. 3.

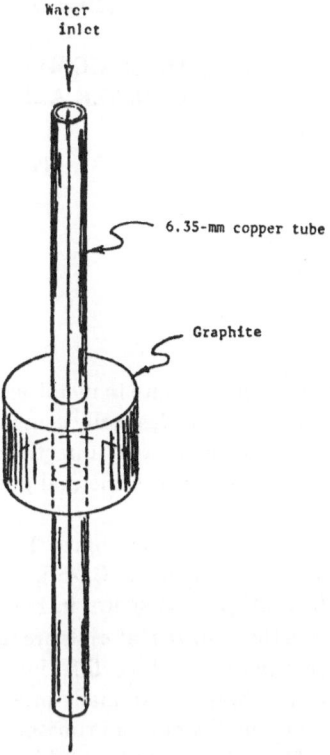

Fig. 3.
Sample for testing graphite-to-copper bond-
ing techniques in an induction heating
assembly.

The second hybrid compound used fine copper chips instead of the aluminum chips, in order to improve the thermal conductivity of the heterogeneous mixture as well as to increase the maximum operating temperature of the bond. Copper melts at 1083°C; aluminum melts at 660°C. If the bond exceeds the melting point temperature of the aluminum, it will melt out of the graphitic matrix.

The Varcum preparation technique, compositions, and fabrication procedures are described in Appendix A.

A second method of bonding graphite to metal was electroplating copper to the interior of the graphite hole, then brazing the tube to the copper-plated graphite in a conventional vacuum-brazing furnace with BT (copper-silver) brazing alloy. The electroplating operation was performed with standard copper solutions at normal plating rates for metals, to a thickness of 127 μm. No special techniques were used in the brazing operation. The braze occurred between the plated copper present on the graphite and the copper tube. This technique was attempted since the mechanical roughness of the graphite surface contributes to a mechanical bond; also, the van der Waals forces (between the plated copper atoms and the carbon atoms) added to the strength of the bond as well as to the thermal conductivity. The brazing history is outlined in Appendix B.

TABLE I

CHEMICAL COMPOSITION
OF BRAZE ALLOYS

	Ti-Cu-Sil	Ti-Cu-Ni
Titanium	4.5	70.0
Copper	26.7	15.0
Silver	68.8	----
Nickel	----	15.0

The third approach to bonding graphite to metal was to purchase* and test commercial bonding compounds. These carbon-based, thermally conductive compounds were tested and fabrication proceeded according to manufacturers' directions (see Fig. 3). One compound was Grade GC Graphite Cement and the other was a thermal putty compound. The preparation history is found in Appendix B.

A fourth bonding technique attempted to braze the graphite and metal tube directly together (Fig. 3). Two braze alloys, Ti-Cu-Ni and Ti-Cu-Sil,** that have successfully bonded copper and pyrolytic graphite (in the configuration shown in Fig. 3) were used; chemical compositions are shown in Table I. Although these braze materials are referred to as alloys, they are not solid solutions. They are actually sandwiches of three different metal foils rolled together to form what we have called braze alloys. The brazes were made in a conventional vacuum braze furnace, and studies were made to determine the optimal time-at-temperature braze. These studies showed that a temperature of 900°C for 5 min produced the best bond when brazing copper to pyrolytic graphite with Ti-Cu-Sil. When using the Ti-Cu-Ni braze, a temperature of 980°C for 2 min produced the best bond. Bond evaluations were made by macroscopically examining a longitudinal section of the brazed part for braze alloy flow and integrity. Subsequent work using the same variables with the same braze alloys showed that copper could be brazed successfully to ATJ graphite.

In the braze alloy, the titanium reacts with some of the graphite and forms an interface (TiC). The remainder of the braze alloy forms a solid solution with the TiC and the copper tube. Since titanium is a reactive metal, the same property that makes the braze alloys outstanding makes them difficult to work with. If the braze alloys — especially Ti-Cu-Ni — were allowed to react with the copper tube for too long a time, the titanium would produce holes or thinned sections in the wall of the copper tube. Nevertheless, a strong mechanical and thermal bond can be produced if carefully controlled.

Efforts are under way to braze various elemental and alloy tubes to graphite with the same braze alloys (Ti-Cu-Sil and Ti-Cu-Ni). This will widen the selection of materials that can be bonded to both ATJ and pyrolytic graphite. Other elements and alloys will further widen the selection of heat transfer fluids and, therefore, applications. Studies are being made to determine the optimal time-at-temperature conditions for this furnace with tantalum, stainless steel, and molybdenum. Copper and ATJ graphite have also been successfully brazed with Ti-Cu-Ni and Ti-Cu-Sil. The general brazing temperature has been 900°C, with typical vacuum braze times from 1-10 min (see Appendix B for details).

---

*Dylon Industries, Inc., Cleveland, OH 44162.
**Purchased from Western Gold & Platinum Company, Belmont, CA 94002.

B. Photographic Examination of Bond

Figures 4 and 5 are photomicrographs of pyrolytic graphite brazed to copper with Ti-Cu-Ni. The magnifications are 100X and 250X, respectively. The lightest section in each photograph is the copper tube; the light gray is the braze zone; and the dark gray is the pyrolytic graphite. There was no evidence of the bond breaking or being discontinuous. Figure 6 shows a broken portion in the brazed joint; the fracture is located in the pyrolytic graphite region. This clearly indicates that the braze joint is stronger than either mating material (copper and pyrolytic graphite). These photographs were taken after the braze joint had experienced more than 200 heat cycles of 6.50 MW/m² heat flux. Clearly, this bond has promise for further development and application.

C. Heat Transfer Testing Techniques

The initial heat deposition method selected was high-frequency induction heating of graphite specimens bonded to various metal tubes (see Fig. 3). The heights of the graphite cylinders varied from 12.7 to 76.2 mm. The outside diameters of the graphite cylinder and the tube were 22.2 and 6.35 mm, respectively.

The specimens were placed in the induction heating apparatus (see Fig. 3). The volume inside the quartz tube was evacuated to pressures of 1×10^{-6} to 5×10^{-4} mm of mercury.

Water flowed through the tube at rates from 1.9-11.36 ℓ/min; inlet and outlet water temperatures of this system were measured to $\pm^1/_8$ K.

The total amount of power deposited in the graphite sample is the sum of the power deposited in the water [volume/minute × density × heat capacity × (T outlet − T inlet)] plus the power dissipated radiatively through the vacuum to the surroundings [power = emissivity × σ × area × ($T^4_{surface} - T^4_{surroundings}$)], where σ = Stephan-Boltzmann constant. There is no opportunity for heat to transfer from the graphite to the surroundings in any other way in our experiment. Heating power to the graphite was controlled by adjusting the amount of power available to the radio frequency oscillator power circuit.

Power transmitted to the graphite load was measured by monitoring the water flow rate through the tube, the ΔT = (T inlet − T outlet), and the surface temperature of the graphite (measured by an optical pyrometer for temperatures >700°C).

D. Results of Heat Transfer Studies

Each bonding method was tested to determine maximum heat flux capability through the bond. The results are shown in Table II.

ACKNOWLEDGMENTS

We wish to express our appreciation to the following Los Alamos Scientific Laboratory people for their technical expertise in these tests: Donald E. Hull, high-frequency induction heating; James R. Bradberry, metallography; John Russell, vacuum furnace brazing; and Donald H. Schell, Varcum bonding techniques.

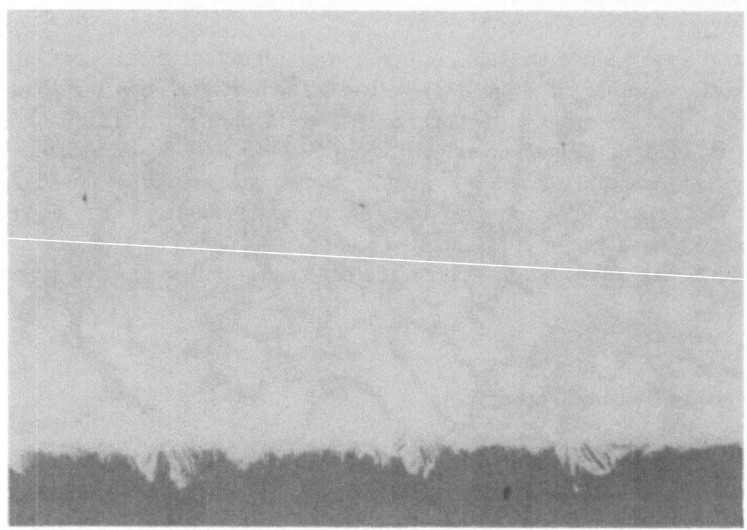

Fig. 4.
Photomicrograph of pyrolytic graphite brazed to copper with Ti-Cu-Ni; magnification = 100X.

Fig. 5.
Photomicrograph of pyrolytic graphite brazed to copper with Ti-Cu-Ni; magnification = 250X.

Fig. 6.
Pyrolytic graphite bonded to copper with Ti-Cu-Ni braze alloy.

TABLE II

RESULTS OF HEAT TRANSFER STUDIES

Material	(W/cm²)
[a]Pyrolytic Graphite To Copper/Polyfurfuryl Alcohol With Copper Hybrid	185
[a]Pyrolytic Graphite to Copper/Polyfurfuryl Alcohol With Copper Hybrid	625
[a]Pyrolytic Graphite to Copper/Copper Electroplated Hole, Then BT Braze to Tube	550
[a]Pyrolytic Graphite to Copper/Grade GC Graphite Cement	260
[a]Pyrolytic Graphite to Copper/Heat Conducting Compound	130
[b]Pyrolytic Graphite to Copper/Ti-Cu-Sil	650
[b]Pyrolytic Graphite to Copper/Ti-Cu-Ni	600

[a]Bonds partially failed.
[b]Survived more than 500 heat flux cycles without failure.

APPENDIX A

VARCUM PREPARATION PROCEDURES

A filled, partially polymerized furfuryl alcohol glue was made with 3.5 g of −400 mesh aluminum powder, stirred thoroughly into 3.5 g of catalyzed Varcum 8251.

Prior to applying the filled glue, the inner diameter of the polygraphite part was coated with unfilled Varcum by brush. This operation was possible because of the short length of the component.

The copper tubing was prepared by sanding the area where the pyrographite would be glued into place and then the roughened area wiped with a coat of unfilled Varcum. The pyrographite part was slipped into place, and rubbed back and forth along the roughened area until only a thin film of Varcum remained on each mating surface. All excess Varcum was wiped off and the Varcum was cured in an oven.

The aluminum glue was then applied to each treated surface. The pyrographite part was slipped back into place, rubbed back and forth over the roughened area until no glue was forced from the joint, and the finished product wiped clean of excess glue; it was then oven-cured for 15 h at 250°C.

The copper-filled glue was prepared in a similar manner. The copper powder used was Cu-36D, with an average particle size of ~1.5 μm. The powder was reduced in a mixture of AR + H$_2$ overnight at 280°C. The reduced powder was screened using −400 mesh.

The 46 Wt% copper/54 Wt% catalyzed Varcum 8251 was cured for 60 h at 250°C in the same manner as the aluminum-filled Varcum.

APPENDIX B

HEAT TREATMENT HISTORIES

Ti-Cu-Ni − 76.2-μm foil/6.50-mm hole/6.35-mm tube. Brazed for 20 min at 1283 K.

BT Braze − 127-μm copper plating on pyrographite/102-μm foil/6.71-hole/6.22-mm tube. Brazed 1103 K for 15 min.

Heat Compound − 6.35-mm, scribed approximately one turn per 2.5 cm, the 6.50-mm hole filled with compound. Air-dried for 24 h; 4 h in furnace at 366 K.

Graphite Cement − 6.35-mm tube, scribed approximately one turn per 2.5 cm, the 6.5-mm hole filled with compound. Baked for 4 h in a furnace at 373 K.

BT Braze − 102-μm copper plating on graphite 50.8-μm foil/6.71-mm hole/6.5-mm tube, brazed for 10 min at 1103 K.

THERMAL CONDUCTIVITY AND ELECTRICAL RESISTIVITY OF SIMULATED FUEL ELEMENTS*

J. P. Moore, R. S. Graves, and W. P. Eatherly

Oak Ridge National Laboratory

Oak Ridge, TN 37830

It is well-known that important changes may occur in the physical properties of bonded fuel rods during their lifetime within a reactor. These changes are caused by the high temperatures, environment, and lastly (and probably most importantly) by the high neutron fluxes. Neutrons have a strong effect in reducing the thermal conductivity, λ, by introducing defects such as vacancies and dislocations into the basal planes of graphitic material. The reduction of λ raises the operating temperature which lowers the lifetime of the graphite. Although the quantitative effects of neutron irradiation on the λ of polycrystalline graphites have been extensively studied, they have not been studied for carbonaceous material nor for carbonaceous material containing various quantities of spherical fuel particles which are envisioned for use in a high-temperature gas-cooled reactor (HTGR). The primary purpose of this experiment, therefore, was to measure the thermal conductivity of simulated fuel elements consisting of inert particles suspended in a carbonaceous matrix as a function of temperature, T, volume percent particle loading, and neutron fluence. The electrical resistivity, ρ, Seebeck coefficient, S, and coefficient of thermal expansion, CTE, were also measured as functions of the same parameters, but will not be discussed in this paper.

The results show that λ of extruded fuel elements with 36 vol % simulated fuel particles is higher than λ of slug injected elements by a factor of 2 to 4 depending on temperature of data comparison

---

*Research sponsored by the Department of Energy under contract with the Union Carbide Corporation.

and neutron fluence. This difference is ascribed to the greater
matrix density (1.7 Mg/m^3 versus 0.6 Mg/m^3) within the extruded
elements. Although λ of the extruded element with 0 vol % particles
decreases with neutron fluence, the λ of other loadings either
remain constant or increase with initial neutron irradiation.
Behavior of ρ and λ indicate that this effect may be due to ordering
in at least one component of the complex composites.

THERMAL DIFFUSIVITY OF LAYERED COMPOSITES

T. Y. R. Lee
Center for Information and Numerical Data Analysis and
  Synthesis (CINDAS)
Purdue University, West Lafayette, IN 47907

A. B. Donaldson
Sandia Laboratories
Albuquerque, New Mexico 87115

R. E. Taylor
Properties Research Laboratory
School of Mechanical Engineering
Purdue University, West Lafayette, IN 47907

ABSTRACT

The heat diffusion equations for three- and two-layer composites without interfacial thermal contact resistance are solved with appropriate boundary conditions, and with a heat pulse function of the form of the Dirac delta function for composites with capacitive layers. The analytical solutions are compared with those for composites with resistive layers. Of special interests are the cases where the properties of the layers are grossly different and also where a thin layer of high diffusivity material is involved.

The values of the modified Fourier number $(\alpha_2 t_{1/2}/\ell_2{}^2)$ are graphically represented as a function of the relative volumetric heat capacity $(H_{1/2})$, and the square root of the relative heat diffusion time $(\eta_{1/2})$ for two-layer composites and as a function of the relative volumetric heat capacity $(H_{3/2})$, the square root of the relative heat diffusion time $(\eta_{3/2})$, and the thickness ratio (ℓ_1/ℓ_3) for the three-layer case.

From the mathematical analysis and experimental observations, a criterion for the thin layer of high conductor being capacitive (uniform temperature) can be represented by the parameter, $\eta_{1/2}$ which is the square root of the heat diffusion time ratio between the thin film (η_1) and the substrate (η_2). The upper limiting values of $\eta_{1/2}$ are 0.1 for two-layer composites and 0.05 for three-layer composites where layers 1 and 3 are of the same thickness and material. These limits are independent of the relative volumetric heat capacity ($H_{1/2}$).

INTRODUCTION

Studies of the modes of heat transport through simple and multilayer thin films of different materials, the interface and mean free path effects in ultra-thin films, and the phonon mean free path are some of the areas of fundamental interest which have received very little attention until recently. On the technological side, heat transfer plays an important role in determining the kinetics of nucleation and growth of a thin film, and thermal aging and deterioration of thin films is important in the entire area of microelectronics employing thin films in one form or another. The thermophysical properties have been measured in a few films of Ag [1,2], Bi[3], Cu[4], and Al[5] for thermal conductivity and Au[6] for thermal diffusivity by a number of techniques. The flash technique for measuring thermal diffusivity of homogeneous materials discussed by Parker, et al. [7] has been adapted to layered composites by various investigators [8-16]. Donaldson and Schimmel [17] had proposed the theoretical model for in-situ thermal diffusivity measurement of thin films by a radial heat flow method. However, the utility of this method for accurate thermal diffusivity measuring in their films remains to be demonstrated by experiment.

It is evident from the literature that the transport properties of the film are strongly dependent on thickness below a certain thickness. This thickness dependence arises, according to the Fuchs-Sondheimer theory [18,19], because of the scattering of the electrons at the grain boundaries and the geometrical surfaces. Beyond a certain thickness, the transport properties becomes independent of thickness and approaches the bulk value. It should be noted that, as expected, the thickness dependence is more marked at low temperatures.

It is worth mentioning the distinction between a resistive layer and a capacitive layer (or thin film layer) of layered composites in the sense of heat transfer. In the one-dimensional heat transfer through a multilayer slab composite, the temperature gradient in each layer is given by Fourier's Law. The heat transfer rate may be considered as a flow, and the combination of thermal conductivity, thickness of material, and areas as a resistance to this flow. The multilayer slabs side by side act as thermal resistances in series. Thus, each slab is referred to as <u>resistive</u> <u>layer</u>. When a case arises, for example, where a thin metal sheet is in contact with a relatively poor conductor such as soil, food stuff, or insulating material, the assumption made is that the metal sheet is so thin that the temperature is constant across its thickness. This thin metal sheet is thus referred to as a <u>capacitive layer</u> (or thin film layer).

The purpose of this paper is twofold: first, to formulate the mathematical models for determining the thermal diffusivity of thin film layered composites using the flash diffusivity technique; second, to examine closely these models and those previously developed models in order to investigate the role of the individual layer of the layered composites especially for cases where a thin layer of high diffusivity material is involved.

MATHEMATICAL ANALYSIS

The schematic diagram of the geometry of a three-layer sample is shown in Figure 1.

To solve the heat diffusion equation with the appropriate boundary conditions, the following assumptions are made:

(1) one dimensional heat flow,
(2) no heat losses from the sample surfaces,
(3) no thermal contact resistance between layers,
(4) heat pulse is uniformly absorbed on the front surface,
(5) each layer is homogeneous,
(6) constant thermal properties of each layer over the range of temperature excursions.

I. With Resistive Layers

The mathematical model was derived previously by Lee [8] and Lee and Taylor [13]. For the case when the first and third layers are of the same material, the normalized temperature rise becomes a function of $H_{3/2}$, $\eta_{3/2}$, ℓ_1/ℓ_3 and modified Fourier number $(\alpha_2 t_{1/2}/\ell_2{}^2)$. Figure 2 shows one of the relationships among these four nondimensional parameters [20]. The relative volumetric heat capacity $H_{i/j}$ and the square root of relative heat diffusion time are defined as

$$H_{i/j} = H_i/H_j$$

and $$\eta_{i/j} = \eta_i/\eta_j.$$

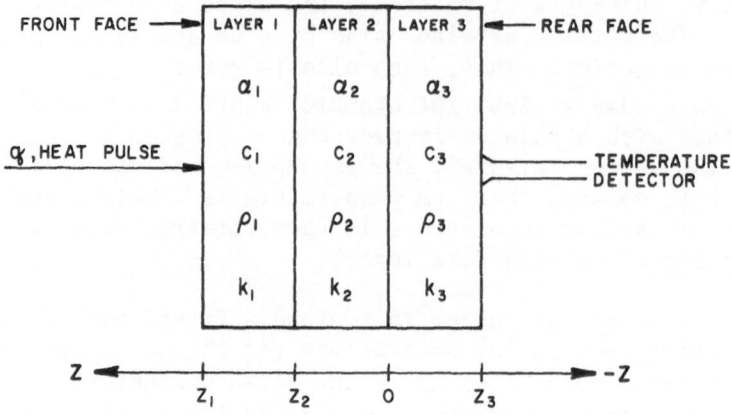

Figure 1.   Diagram of Three-Layer Sample.

H_j is the volumetric heat capacity and η_j represents the square root of diffusion time of the jth layer:

$$H_j = a\rho_j C_j \ell_j,$$
$$\eta_j = (\ell_j^2/\alpha_j)^{\frac{1}{2}},$$

where a is the cross-section area of the sample ρ_j is the density, and C_j is the specific heat capacity at constant pressure of the jth layer.

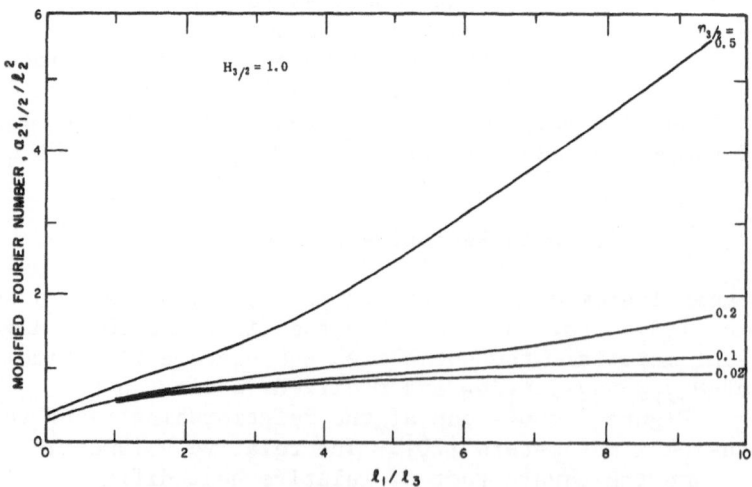

Figure 2.   Modified Fourier Number $(\alpha_2 t_{\frac{1}{2}}/\ell_2^2)$ vs Relative Thickness (ℓ_1/ℓ_3) at Constant Value of $H_{3/2}(=1.0)$ for a Family of $\eta_{3/2}$ in the Case of Three-Layer Composites.

II. With Capacitive Layers

When the first and third layers are thin and relatively good conductors compared with the middle layer, it is reasonable to assume that these two layers have infinite conductivity, i.e., uniform temperature throughout the layers. This is the case of three-layer composites with two capacitive layers (layers 1 and 3). The additional assumption to the previous six assumptions is that there is uniform temperature throughout the capacitive layer.

In Ref 20 the heat diffusion equations for three-layer composites, two-layer composites with capacitive layer are solved with appropriate boundary conditions and with a heat pulse function of the form of the Dirac delta function.

Table 1 presents the modified Fourier number as a function of $H_{1/2}$ and $H_{3/2}$ for the three-layer sample with two capacitive layers. An interesting observation is that this solution table is symmetric, which implies that reversing the sample does not change the problem as long as assumptions used in the derivation are valid. Figure 3 shows the modified Fourier number of three-layer composites as a function of $H_{1/2}$ in the special case where ℓ_1 and ℓ_3 are of the same thickness and material for both resistive and capacitive layers. In Figure 3, L_1-L_2-L_3 stand for all three layers being resistive, while c_1-L_2-c_3 stands for layers 1 and 3 being capacitive. From this figure, it is observed that the resistive layers will become the capacitive ones when the ratio of the square root of heat diffusion time becomes as small as 0.05 for $\eta_{1/2}$ and $\eta_{3/2}$. Figure 4 shows the plot for modified Fourier number as a function of $H_{1/2}$ for two-layer composites for both resistive and capacitive layers. A similar observation is obtained from this figure; the resistive layer becomes a capacitive layer when $\eta_{1/2}$ has a value of 0.1 or less.

Table 1. The Modified Fourier Number ($\alpha_2 t_{1/2}/\ell_2^2$) as a Function of $H_{1/2}$ and $H_{3/2}$ in the Case of Three-Layer Composites with Capacitive Layers

$H_{1/2}$ \ $H_{3/2}$	0.050	0.070	0.100	0.500	1.000	2.000	3.000	4.000	5.000
0.050	0.166	0.171	0.178	0.242	0.285	0.329	0.350	0.363	0.372
0.070	0.171	0.176	0.183	0.250	0.294	0.339	0.362	0.375	0.385
0.100	0.178	0.183	0.190	0.260	0.307	0.354	0.379	0.393	0.403
0.500	0.242	0.250	0.260	0.361	0.438	0.524	0.572	0.603	0.624
1.000	0.285	0.294	0.307	0.438	0.548	0.684	0.765	0.819	0.857
2.000	0.329	0.339	0.354	0.524	0.684	0.904	1.051	1.156	1.233
3.000	0.350	0.362	0.379	0.572	0.765	1.051	1.254	1.407	1.526
4.000	0.363	0.375	0.393	0.603	0.819	1.156	1.407	1.603	1.761
5.000	0.372	0.385	0.403	0.624	0.857	1.233	1.526	1.761	1.951

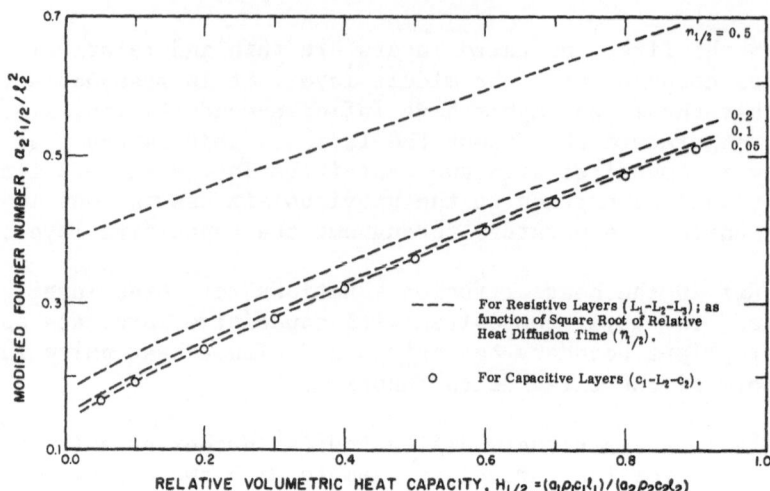

Figure 3. Modified Fourier Number ($\alpha_2 t_{1/2}/\ell_2^2$) as a Function of
Relative Volumetric Heat Capacity ($H_{1/2}$) in the Case
of Three-Layer Composites (ℓ_1 and ℓ_3 are of same
thickness and material).

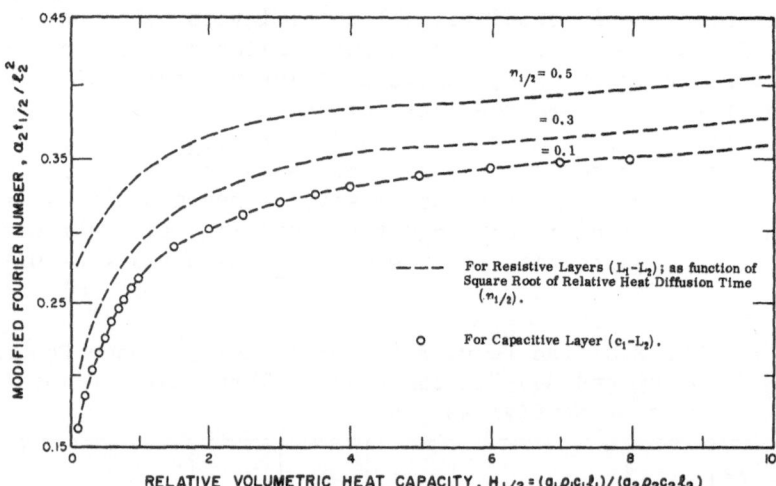

Figure 4. Modified Fourier Number ($\alpha_2 t_{1/2}/\ell_2^2$) as a Function of
Relative Volumetric Heat Capacity ($H_{1/2}$) in the Case of
Two-Layer Composites.

SAMPLE DESCRIPTIONS AND EXPERIMENTAL RESULTS

Six samples were used to examine the two-layer case. Three
different thicknesses of standard 316 stainless steel were bonded
with Scotchcast brand resin No. 251 for Samples 2-1, 2-2, and 2-3,
and a thin aluminum plate was bonded on Scotchcast brand resin

No. 251 for Sample 2-4. Samples 2-5 and 2-6 were Detaclad, which
is the registered trademark for DuPont's explosionbonded materials.

The Scotchcast brand resin No. 251 was fabricated by mixing
one part of Type A with one part of Type B and about five weight
percent Black Color Paste 502 (Thermoset Plastic, Inc.). The mixed
resin was warmed to 60°C and thoroughly blended until the color of
the mixture was absolutely uniform. The warm resin was then poured
into the preheated (60-100°C) mold with a metallic plate put in the
bottom. This mold was left in an oven for 16 hours at a curing
temperature of 150°C.

The three-layer Samples 3-1 and 3-2 were made by bonding the
stainless steel plates to the resin side of the stainless steel
resin-bonded composites (Samples 2-1 and 2-2, respectively).
Sample 3-3 was fabricated from Sample 2-4 by bonding an aluminum
plate to the resin side of the composite. The descriptions of
those layered samples are given in Table 2.

For the metal-resin two layer samples (Samples 2-1 to 2-4),
the thin metal layer was always used as the front layer facing the
laser in order to prevent the resin material from being degraded by
the laser energy. The experimentally measured half-times and the
measured thicknesses of each layer were used together with the
thermophysical properties of the individual layers to calculate
the thermal diffusivity of the front layer (or the back layer)
employing the computer program TWO1 (TWO2 for back layer) for the
mathematical model with a resistive layer. In the light of the
gross differences in the transport properties between metal-resin
layers of some of the composite samples (as high as 472 for the
ratio of thermal diffusivities of aluminum to resin), special
attention was given to the selection of the optimal experimental
half-time and to the iteration procedure to find the unknown
thermal diffusivity of the thin metallic layer. The experimental

Table 2. Layered Composite Sample Descriptions

Sample No.	Front Layer		Center Layer		Back Layer	
	Material	Thickness (cm)	Material	Thickness (cm)	Material	Thickness (cm)
2-1	SS316	0.1402	Resin	0.1280	-	-
2-2	SS316	0.0622	Resin	0.2192	-	-
2-3	SS316	0.0361	Resin	0.1631	-	-
2-4	Aluminum	0.0361	Resin	0.2532	-	-
2-5	Brass	0.3048	S.S.*	0.2802	-	-
2-6	S.S.*	0.1196	Brass	0.2149	-	-
3-1	SS316	0.1402	Resin	0.1280	SS316	0.0348
3-2	SS316	0.0635	Resin	0.1703	SS316	0.0622
3-3	Aluminum	0.0361	Resin	0.2243	Aluminum	0.0361

*Stainless steel from detaclad.

half-times for the thin layer composites (Samples 2-1 to 2-4) were
determined by a careful adjustment of the experimental value,
within measurement uncertainty, to the value of the analytical
solution of the half-times calculated from the known thermal
diffusivity values of the thin plates. The square root of the
relative heat diffusion time $\eta_{1/2}$ is the dimensionless parameter
used in the trial and error process to find the normalized
temperature V of the rear face which equals the half of the maximum
temperature rise, i.e., V = 0.5 occurring at the experimental half-
time, $t_{1/2}$. The experimental results are presented in Table 3.
Due to the awareness of the sensitivity of the half-time to the
diffusivity value of the thin metallic layer, the diffusivity
values for the thin metallic layers given in Table 3 should not be
taken as an indication that accurate diffusivity values can be
obtained for thin metallic films using present techniques. Although
there is a large uncertainty in the experimental values for the
diffusivity of the thin metallic layers, the thermal diffusivity
of the poor conducting layer (resin) was close to the value obtained
for the homogeneous sample.

Among the thin layer Samples 2-1 to 2-4, the value of $\eta_{1/2}$ for
Sample 2-1 was above the limit ($\eta_{1/2}$ = 0.1 in Figure 4) for stainless
steel 316 acting as a capacitive layer, while those for Samples 2-2
to 2-4 were below the limit. Therefore, those front thin layers
acted as capacitive instead of resistive layers. However, using
a resistive mathematical model yielded satisfactory results for
capacitive layer composites as well as for resistive ones
(Table 3).

The three-layer case was investigated using Samples 3-1 through
3-4. The purpose of this investigation was to demonstrate that the
thermal diffusivity of a poor conductor used as the middle layer
with thin high conductor layers in front and back can also be
measured satisfactorily. The computer program THRMID for resistive
layers was employed for data iteration. The results are given in
Table 4. The maximum deviation from the results obtained for a
homogeneous sample was 4.8% and the average deviation was 2.1%.
The front and back layers of these three samples were made from the
same materials (stainless steel 316 was used for Samples 3-1 and
3-2, and aluminum was used for Sample 3-3). Values of relative
thicknesses, ℓ_1/ℓ_3 of these three samples were 4.03, 1.02, and 1.00
respectively. It is noted that the value of $\eta_{3/2}$ for Sample 3-2
was 0.088, which is above the limit value of $\eta_{3/2}$ (0.05 in Figure 3)
for the thin layer acting as a capacitive layer while $\eta_{3/2}$ for
Sample 3-3 was 0.007. Thus the thin layers of this sample were
capacitive ones. Therefore, the mathematical model for resistive
layers is also valid for the capacitive case.

Table 3. Thermal Diffusivity of the Front and Back Layer and Dimensionless Parameters of Two-Layer Composites

Sample No.	Half-Time (sec)	Experimental α(cm²/sec) Front Layer	Back Layer	Material	Deviation‡ %	$\eta_{1/2}$	$H_{1/2}$	F02*
2-1	2.519	0.0360	(0.00205)‡	SS316	2.3	0.264	1.84	0.32
2-1	2.519	(0.0352)	0.00205	Resin	0.0	0.264	1.84	0.32
2-2	5.217	0.0373	(0.00205)	SS316	5.9	0.068	0.477	0.22
2-2	5.217	(0.0352)	0.00205	Resin	0.0	0.068	0.477	0.22
2-3	2.7235	0.0347	(0.00205)	SS316	-1.4	0.053	0.372	0.21
2-3	2.7235	(0.0352)	0.00205	Resin	0.0	0.053	0.372	0.21
2-4	5.4943	0.960	(0.00205)	Aluminum	-0.8	0.007	0.161	0.18
2-4	5.4943	(0.968)	0.00205	Resin	0.0	0.007	0.161	0.18
2-5	0.6569	0.292	(0.0367)	Brass	-1.7	0.382	0.971	0.31
2-5	0.6569	(0.297)	0.0365	S.S.†	-0.5	0.382	0.971	0.31
2-6	0.1568	0.0362	(0.297)	S.S.†	-1.4	1.583	0.624	0.97
2-6	0.1568	(0.0367)	0.285	Brass	-4.0	1.583	0.624	0.97

#Deviation % = $\dfrac{\alpha_{experimental} - \alpha_{homogeneous}}{\alpha_{homogeneous}}$ x 100.

*Modified Fourier number of back layer $(\alpha_2 t_{\frac{1}{2}} / \ell_2^2)$.

†Stainless steel from Detaclad.

‡The values in parentheses are the values used as input for the computer iteration.

Table 4.   Thermal Diffusivity of the Center Layer and
Dimensionless Parameters of Three-Layer Composites

Sample No.	Half-Time (sec)	Experimental α (cm^2/sec) of Center Layer	Material	Deviation[†] %	$\eta_{3/2}$	$H_{3/2}$	FO2*
3-1	3.980	0.00215	Resin	4.8	0.066	0.457	0.52
3-2	5.786	0.00207	Resin	1.0	0.088	0.613	0.41
3-3	5.579	0.00206	Resin	0.5	0.007	0.182	0.23

[†]Deviation % = $\dfrac{\alpha_{experimental} - \alpha_{homogeneous}}{\alpha_{homogeneous}}$ x 100.

*Modified Fourier number of center layer ($\alpha_2 t_{\frac{1}{2}}/\ell_2^{\,2}$).

Using the mathematical models for the capacitive layer in
two- and three-layer cases, respectively, the thermal diffusivity
values of the poorly conducting layer for Samples 2-1 through 2-5,
and the middle layer for Samples 3-2 and 3-3 were calculated and
compared with the previous values calculated using the resistive
models. These results are presented in Table 5. It is observed
from Table 5 that in using the capacitive models when the ratio
of the square root of the diffusion time was below the limit
value of $\eta_{1/2}$ (0.10) in the two-layer composite case, such as for
Samples 2-2 to 2-4, the deviation of the calculated thermal
diffusivity values from the homogeneous resin sample were within
1%. However, when the ratio of the square root of the diffusion
time was above the limit value ($\eta_{1/2}$ = 0.10), such as for Sample
2-1 ($\eta_{1/2}$ = 0.264) and Sample 2-5 ($\eta_{1/2}$ = 0.382), the calculated
thermal diffusivity values of the second layer were significantly
less than the value for the homogeneous sample (5.9% and 15.0%
for Samples 2-1 and 2-5, respectively). A similar situation
occurs for the three-layer composites Samples 3-2 and 3-3. These
experimental results confirm the theoretical prediction shown in
Figures 3 and 4.

DISCUSSION OF RESULTS AND CONCLUSIONS

The mathematical model for layered composites with resistive
layers derived by Lee [8] and Lee and Taylor [13] was closely
examined to see if it could yield satisfactory analytical solutions

Table 5.   Thermal Diffusivity at 300 K of Layer 2 of Composites
Calculated from Two Different Mathematical Models

Sample No.	Material	Resistive Model		Capacitive Model		$\eta_{1/2}$
		α	Deviation %	α	Deviation %	
2-1	Resin	0.00205	0.0	0.00193	-5.9	0.264
2-2	Resin	0.00205	0.0	0.00203	-1.0	0.068
2-3	Resin	0.00205	0.0	0.00204	-0.5	0.053
2-4	Resin	0.00205	0.0	0.00205	0.0	0.007
2-5	S.S.[†]	0.0365	-0.5	0.0312	-15.0	0.382
3-2	Resin	0.00207	1.0	0.00204	-0.5	0.090
3-3	Resin	0.00206	0.5	0.00206	0.5	0.007

[†]Stainless steel from Detaclad.

for the case in which a thin metallic layer is bonded on a substrate
of a poor heat conductor. It was found that the experimental half-
time had to be carefully adjusted (within the experimental accuracy)
in order to yield a reasonable value for the diffusivity of the thin
layer as the diffusivity value is extremely sensitive to the experimental half-time. For example, a 3% variation of the experimental
half-time caused the value of the calculated thermal diffusivity
of the thin layer to vary by two orders of magnitude. Thus, it is
not practical to measure the thermal diffusivity of a thin highly
conducting layer deposited on a substrate by using the present
technique.

The mathematical solutions for composites with capacitive
layer(s) were derived in this study. The model for three-layer
composites was readily reduced to the two-layer case and to the
single layer (homogeneous) case. From a comparison of mathematical
models for composites with capacitive and resistive layer(s), the
criteria of a thin metallic layer (or thick film) deposited on a
substrate acting as a capacitive layer in the sense of heat transfer
were obtained for the two-layer and three-layer composites. These
criteria, represented by the ratio of the square root of the heat
diffusion time between the thin layer and substrate ($\eta_{1/2}$) are independent ofthe relative volumetric heat capacity ($H_{1/2}$). These upper
limits of $\eta_{1/2}$ are 0.10 for the two-layer composites where the thin
layer is front layer and 0.05 for three-layer composites where layers
1 and 3 are thin layers and are of the same material and thickness.
For other cases of three-layer composites where layers 1 and 3 are

different in material and thickness, the limits of $\eta_{1/2}$ or $\eta_{3/2}$ for layer 1 or layer 3 acting as capacitive layer can easily be determined by comparing the two mathematical models for the resistive and capacitive layers. These criteria were verified by experimental observations. Furthermore, the mathematical models for resistive layers were shown to be valid also for capacitive cases.

The application of these mathematical models to determine the thermal diffusivity of a thin highly conducting plate of a layered composite has revealed the difficulty that the half-time would have to be determined to an unreasonable accuracy in order to give the reasonable results, whereas the thermal diffusivity of the relatively poor heat conducting material can be obtained satisfactorily from the half-time measured with the accuracy obtainable using a Digital Deta Acquisition System with crystal clock.

ACKNOWLEDGEMENTS

The authors would like to express our thanks to the staff at the Properties Research Laboratory, especially Messrs. W. E. Vaughn and H. Groot who greatly aided in the construction and maintenance of the experimental facility. This work was performed as part of an NSF grant administered by the Engineering Division.

REFERENCES

1. Abrosimov, V.M., Sov. Phys. Solid State, 11(2), 1969.

2. Boyce, T.C. and Cheng, Y.W., "A New Technique for the Measurement of Thermal Conductivity of Thin Films," in Thermal Conductivity 14 (Klemens, P.G. and Chu, T.K., Editors), 499-506, 1976.

3. Traiman, B.S. and Chudnovskii, Fiz. Tekh. Polyprov., USSR, 5, 262-6, 1971.

4. Nath. P. and Chopra, K.L., "Experimental Determination of the Thermal Conductivity of Thin Films," Thin Solid Films, 18, 29-37, 1973.

5. Boiko, B.T., Pugachev, A.T., and Bratsychin, V.M., "Method for the Determination of the Thermophysical Properties of Evaporated Thin Films," Thin Solid Films, 17, 157-61, 1973.

6. Yeh, Y., Wooten, F., and Huen, T., "Laser Reflectance Relaxation Spectroscopy: Determination of Thermal Diffusivity in Thin Films of Gold," Solid State Commun., 13, 1845-9, 1973.

7. Parker, W.J., Jenkins, R.J., Butler, C.P., and Abbott, G.L., "Flash Method of Determining Thermal Diffusivity, Heat Capacity, and Thermal Conductivity," J. Appl. Phys., 32, 1679-84, 1961.

8. Lee, H.J., "Thermal Diffusivity in Layered and Dispersed Composites," Ph.D. Thesis, Purdue University, 1975.

9. Larson, K.B. and Koyama, K., "Measurement by the Flash Method of Thermal Diffusivity, Heat Capacity, and Thermal Conductivity in Two-Layer Composite Samples," J. Appl. Phys., 39, 4408, 1968.

10. Gilchrist, K.E., "Measurement of the Thermal Conductivity of Ultra Thin Single or Double Layer Samples," in Proc. of the 1st European Conf. on Thermophysical Properties at High Temperatures (Baden-Baden), 368-92, 1968.

11. Schriempf, J.T., "A Laser Flash Technique for Determining Thermal Diffusivity of Liquid Metals at Elevated Temperatures: Applications to Mercury and Aluminum," High Temp.-High Pressures, 4, 411-6, 1972.

12. Bulmer, R.F. and Taylor, R., "Measurement by the Flash Method of Thermal Diffusivity in Two-Layer Composite Samples," International Conference on Thermal Technique of Analysis, The University of Manchester, Institute of Science and Technology, England, 1974.

13. Lee, H.J. and Taylor, R.E., "Determination of Thermal Properties of Layer Compsoites by Flash Method," National Technical Information Rept. No. PB-239, 114/256, Dec. 1974 (unpublished).

14. Chistyakov, V.I., "Pulse Method of Determining the Thermal Conductivity of Coatings," Teplofiz. Vys. Temp., 11(4), 832, 1973; English translation: High Temp., 11(4), 744-8, 1973.

15. Murfin, D., "Developments in the Flash Method for the Measurement of Thermal Diffusivity," Rev. Int. Hautes Temper., et Refract., 7, 284-9, 1975.

16. Ang, C.S., Tan, H.S., and Chan, S.L., "Three-Layer Thermal Diffusivity Problem Applied to Measurements on Mercury," J. Appl. Phys., 44, 687-91, 1973.

17. Donaldson, A.B. and Schimmel, W.P., Jr., "Theory for in-situ Thermal Diffusivity Measurement of Thin Films by a Radial Heat Flow Method," in Thermal Conductivity 14 (Klemens, P.G. and Chu, T.K., Editors), 545-6, 1976.

18. Fuchs, K., Proc. Cambridge Phil. Soc., 34, 100, 1938.

19. Sondheimer, F.H., Adv. Phys., 1, 1, 1952.

20. Lee, T.Y.R., "Thermal Diffusivity of Dispersed and Layered
     Composites," Ph.D. Thesis, Purdue University, 1977.

THE PHYSICAL PROPERTIES OF INCONEL ALLOY 718 FROM 300 TO 1000 K*

D. L. McElroy, R. K. Williams, J. P. Moore,
R. S. Graves, and F. J. Weaver

Oak Ridge National Laboratory
Oak Ridge, TN 37830

Inconel alloy 718 is a high-strength alloy which can be heat treated to be suitable for service at temperatures from 20 to 1000 K in aerospace and nuclear applications.[1] Four physical properties were measured from 300 to 1000 K on Inconel alloy 718 using the apparatuses listed in Table 1. Five right circular cylindrical samples for these apparatuses were machined from 1.2 cm thick plate of Heat No. 2180-4-9478 produced by the Stellite Division, Cabot Corporation and given a commercial heat treatment: 1230 K solution anneal plus duplex aging [985 K (8 hr) and 895 K (8 hr).] These samples had an average bulk density of 8.24 mg/m^3 and an average electrical resistivity (ρ) at 298 K of 1.182 $\mu\Omega$-m ($\pm$0.3%).

The temperature dependence of the physical properties was measured during several thermal cycles and smoothed values of these properties for Inconel alloy 718 are listed in Table 2. The ρ:T behavior is complex, and similar to that noted for other nickel-base alloys,[2-4] since ρ increases to a maximum near 900 K, then decreases to a minimum near 1000 K, and finally increases linearly to 1500 K. Between 600 and 900 K, which is below the range for age hardening reactions, ρ changes by as much as $\pm$1.5% with time at temperature. The absolute thermopower is positive, increases linearly with temperature, is 0.2 μV/K below that found by Hust and Sparks[3] at 298 K, and is sensitive to exposures to 1000 K. The thermal conductivity increases with temperature to 1000 K and check points near 475 K after obtaining data to 1000 K are 2 to 3% above the original λ near 475 K. A similar hysteresis effect was noted in Inconel 702 λ:T data.[4] The present λ:T data are bracketed to $\pm$5% by other

*Research sponsored by the Department of Energy under contract with the Union Carbide Corporation.

data[1,3] on Inconel 718, and are about 8% below λ-values for Inconel
702. The thermal expansion data (α_m and α_o) are 5 to 6% below
values reported by Manson[1] and reflect structural instabilities in
Inconel alloy 718. The α_o:T data were obtained from steady-state
length-temperature measurements and show a stable and reproducible
behavior from 300 to 625 K and from 800 to 1000 K. But from 625 to
800 K the α_o values were unstable with time and after various
thermal cycles showed 10% differences. In the range 800 to 1000 K
the α_o:T slope is nearly 4 times that of the 300 to 625 K range.
The electrical resistivity results and the Sommerfeld Lorenz constant were used to separate the total thermal conductivity into
electronic and lattice components. The lattice thermal resistivity
increases to 600 K and then exhibits a broad maximum. Collectively,
these physical property data suggest that Inconel alloy 718 is
subject to structural changes in the range 600 to 800 K that may be
associated with short range ordering.

Table 1. Characteristics of Apparatuses Used on Inconel
Alloy 718 (52.6 Ni—18.2 Cr—18.8 Fe—5.1 Nb
+ Ta—3 Mo—1.0 Ti—0.5 Al—0.05 C)

Apparatus	Temperature Ranged Spanned (K)	Measurement Accuracy (±%)
1. Knife-Edge Electrical Resistivity (ρ)	293—303	0.2
2. High Temperature Electrical Resistivity (ρ)	293—1200	0.4
3. Comparative Cut Bar Thermal Conductivity (λ)	298—373	5
4. High Temperature Guarded Longitudinal Heat Flow Thermal Conductivity (λ) Electrical Resistivity (ρ) Thermopower (S)	298—973	3 0.4 0.5 μV/K
5. Quartz Push Rod Dilatometer Differential Thermal Expansion Coefficient (α_o) Mean Thermal Expansion Coefficient (α_m)	293—1000	3 2

Table 2. Smoothed Values of the Physical Properties
of Inconel Alloy 718

Temperature (°K)	ρ $\mu\Omega$–m	λ W/m·K	$|S|$ μV/K	$\alpha_o \times 10^6$ (K^{-1})	$\alpha_m \times 10^6$ (K^{-1})
300	1.183	10.6	1.41	12.68	12.61
400	1.215	12.2	1.93	13.40	13.00
500	1.243	13.8	2.41	14.02	13.37
600	1.268	15.3	2.87	14.62	13.64
700	1.291*	17.0	3.29	15.08*	13.85*
800	1.310*	18.6	3.69	16.95*	14.08*
900	1.327*	20.2	4.05	18.60	14.75
1000	1.317	21.9	4.39	21.03	15.45

*Unstable.

References

1. S. S. Manson, Aerospace Structural Metals Handbook, Volume 4, pp. 1–72, Code 4103 Inconel Alloy 718, Revised March 1974, Belfour Stulen, Inc.

2. H. E. McCoy, Jr. and D. L. McElroy, "Electrical Resistivity Anomaly in Nickel-Base Alloys," Trans. ASM 61, 730–741 (1968).

3. J. G. Hust and L. L. Sparks, Thermal Conductivity, Electrical Resistivity, and Thermopower of Aerospace Alloys from 4 to 300 K: III. Annealed Inconel 718, NBS Report 9775 (November 24, 1970).

4. J. P. Moore and W. P. Murray, "The Thermal Conductivity of Inconel 702," pp. 23–27, Proceedings of the 4th Conference on Thermal Conductivity, October 13–16, 1964. (Limited distribution) U.S. Naval Radiological Defense Laboratory, San Francisco, CA.

SESSION D   TECHNIQUES, DATA ANALYSIS

Session Chairman: R.E. Taylor
                   Purdue University
                   West Lafayette, Indiana

USE OF A GENERAL BESSEL SOLUTION IN THE COMPARATIVE MEASUREMENT OF UO_2 THERMAL CONDUCTIVITY

M. A. Aragones

Research Scientist
Instituto Nal. de E. Nuclear
Mexico, D.F.

ABSTRACT

The thermal conductivity of UO_2 is usually measured in the laboratory using two different methods: longitudinal or radial. For the comparative longitudinal method, in this paper, we present a general solution to the problem, taking into account the small radial flux always existing in the thermal conductivity measurement. A general solution for the isotropic UO_2 cylindrical pellets temperature distribution was obtained

$$\Theta(\rho,z) = \sum_0^\infty J_0(\alpha_n\rho)(C_n \exp(-\alpha_n z) + D_n \exp(\alpha_n z))$$

and a similar formula for the upper cylindrical rod used as secondary standard was obtained. With an adequate thermocouples distribution, which is described in the paper, it is possible to evaluate the heat fluxes in any sections of the selected secondary standard and over the specimen whose thermal conductivity we are interested in: UO_2 in our case. If we designate for α_n, A_n, B_n, and β_n, C_n, D_n the parameters which characterize the temperature distribution in the upper rod (secondary standard) and the UO_2 cylindrical pellet, for two sections at the very neighbourhood, we obtain $\phi_1 = \phi_2$ and

$$\lambda_{UO_2} = \lambda_s (\Sigma_1^2 + \Sigma_2^2)^{1/2} / (\Sigma_1'^2 + \Sigma_2'^2)^{1/2}$$

where:

$$\Sigma_1 = \frac{\delta}{\delta\rho} \sum_0^\infty J_0(\alpha_n\rho)(A_n \exp(-\alpha_n z) + B_n \exp(\alpha_n z))$$

$$\Sigma_2 = \frac{\delta}{\delta z} \sum_0^\infty J_0(\alpha_n\rho)(A_n \exp(-\alpha_n z) + B_n \exp(\alpha_n z))$$

and similarly for $\Sigma_1^!$ and $\Sigma_2^!$

$\lambda_x = \lambda_{UO_2}$ = unknown conductivity of UO_2

λ_s = known thermal conductivity of secondary standard

We conclude that the radial flux associated with the longitudinal flux in the comparative method can be very important and in any case, it can be accounted for by using the Bessel solution.

OBJECTIVE

We realize the convenience of improving the analytical model based on the longitudinal UO_2 thermal method, taking into account in the model the radial heat flux which in practice is always present. This work can be justified as follows: The thermal conductivity of UO_2 is one of the most important design parameters in a UO_2 nuclear fuel element, and a great portion of the theory on mechanisms of thermal conductivity published recently is waiting to be confirmed or rejected through good experimentation. We also believe that the longitudinal conductivity method in the future can provide reliable comparative tools for experimental work.

THEORY

In the comparative longitudinal method most researchers prefer using guards to try to avoid the radial flux always present in the method. For their calculation they use simplified solutions (1,2,3,4).

We shall start from the general equation of heat transport in a solid:

$$D \, C_p \, \frac{\delta\theta}{\delta t} = \nabla(\lambda \, \nabla \, \theta) \qquad\qquad \text{Eq. I}$$

$$\underline{\phi} = - \, \lambda\nabla\theta \qquad\qquad \text{Eq. II}$$

where: $\underline{\phi}$ = heat flux density
$\overline{\theta}$ = temperature
λ = thermal conductivity of the solid
D = mass density
C_p = specific heat

Under steady state conditions

$$\nabla^2\theta = 0 \qquad\qquad \text{Eq. III}$$

Laplace equation gives us for cylindrical geometry and isotropic UO_2 cylindrical pellets

$$\frac{1}{\rho}\left(\frac{\partial \Theta}{\partial \rho} + \rho\frac{\partial^2 \Theta}{\partial \rho^2}\right) + \frac{\partial^2 \Theta}{\partial z^2} = 0 \qquad \text{Eq. IV}$$

After some derivation[5] we obtain a general solution for the isotropic UO_2 cylindrical pellets temperature distribution

$$\Theta(\rho, z) = \sum_{o}^{\infty} J_0(\alpha_n \rho)(C_n \exp(-\alpha_n z) + D_n \exp(\alpha_n z)) \quad \text{Eq.V}$$

and a similar formula for the upper cylindrical rod used as secondary standard. Figure 2 gives the thermocouple distribution in the UO_2 pellet or in the 304 stainless steel secondary standard. The procedure of embedding the thermocouples and general techniques used are described elsewhere (5). If we designate for α, A_n, B_n and β_n,C_n,D_n the parameters which caracterize the temperature distribution in the upper rod (secondary standard) and the cylindrical UO_2 pellet, for two cylindrical cross sections in the same neighborhood, we obtain (see Figure 1)

$$\underline{\phi}_1 = \underline{\phi}_2 \qquad \text{Eq. VI}$$

but

$$\underline{\phi}_1 = \lambda_s \left\{ \frac{\partial \Theta}{\partial \rho}\hat{\rho} + \frac{\partial \Theta}{\partial z}\hat{z} \right\} \qquad \text{Eq. VII}$$

at the secondary standard rod, and

$$\underline{\phi}_2 = \lambda_{UO_2} \left\{ \frac{\partial \Theta}{\partial \rho}\hat{\rho} \quad \frac{\partial \Theta}{\partial z}\hat{z} \right\}$$

for the UO_2 pellet. If we designate Σ_1 and Σ_2 variations of temperature according to ρ direction and z direction in the upper secondary standard rod

$$\Sigma_1 = \frac{\delta}{\delta z} \sum_0^{\infty} J_0(\alpha_n\rho)(A_n \exp(-\alpha_n z) + B_n \exp(\alpha_n z))$$

$$\Sigma_2 = \frac{\delta}{\delta \rho} \sum_0^{\infty} J_0(\alpha_n\rho)(A_n \exp(-\alpha_n z) + B_n \exp(\alpha_n z))$$

Eq. VIII

and similarly for Σ_1' and Σ_2' for the UO_2 pellets temperature we get

$$\lambda_{UO_2} = \lambda_s \frac{|\Sigma_1^2 + \Sigma_2^2|^{1/2}}{|\Sigma_1'^2 + \Sigma_2'^2|^{1/2}} \qquad \text{Eq. IX}$$

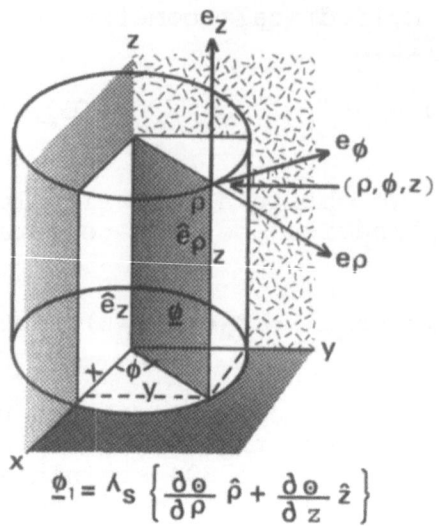

$$\underline{\phi}_1 = \Lambda_s \left\{ \frac{\partial \Theta}{\partial \rho} \hat{\rho} + \frac{\partial \Theta}{\partial z} \hat{z} \right\}$$

Figure 1

If we have a reliable function of temperature for the thermal conductivity of the secondary standard, we can get a very precise function of the UO_2 thermal conductivity also or any other one unknown.

THE BOUNDARY VALUE PROBLEM

Our analytical model gives us the possibility to solve the UO_2 thermal conductivity function after selecting temperatures at several different points which serve as boundary conditions to determine the constants of the function.

In the simplest case of keeping only the first term in the above series expansion, the temperature at three different points serve as boundary conditions to determine the three constants β, C and D. Two intermediate points where the temperatures were measured also during the experiments were used as testing points for the resulting expression. Then, the equation V and the values β, C, and D can be calculated as:

$$\Theta(0,0) = C + D$$

$$\Theta(r,0) = J_0(\beta r)(C+D)$$

$$\Theta(r,z) = J_0(\beta r)(C \exp(-\beta z) + D \exp(\beta z))$$

Even in this last case our analytical model reproduced the temperatures used as testing ones within 1.5% of agreement (5).

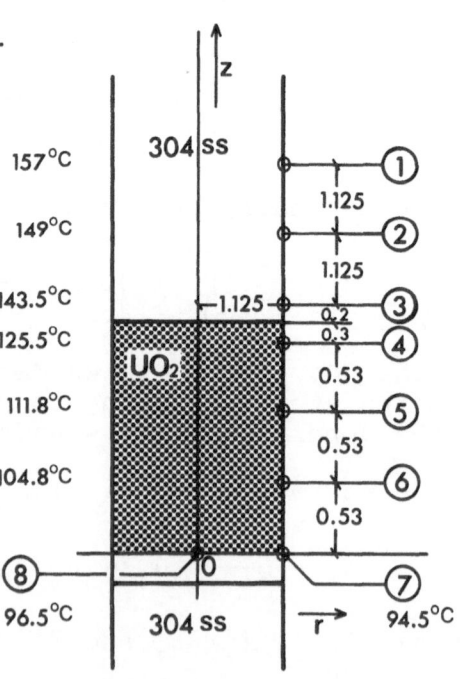

Thermocouples - O

Figure 2

EXPERIMENTAL

Much of our earlier experimentation was presented elsewhere (5), in some detail. We shall give here some typical tables which illustrate temperature distributions,
longitudinal and radial fluxes, and the thermal conductivity of UO2 obtained. Table I gives typical temperature distributions during the experiments. Table II gives
experimental radial and longitudinal fluxes and shows
typical UO_2 thermal conductivity measurements.

CONCLUSION

This method provides the possibility of precise estimation of the radial flux associated with the longitudinal
flux, and of direct correlation with the true heat flux
used to evaluate the thermal conductivity of UO2, or
other unknowns.

Table I

O/U ratio	Exp. No.	Distribution of temperatures in the UO2 sintered pellets and stainless steel							
		Temperatures of the termocouples							
		in the stainless steel rod				in the UO2 sintered pellets			
		1	2	3	4	5	6	7	8
2.045 ±0.004	II-1	214.8	198.6	194.0	167.8	142.4	122.0	113.2	155.8
	2	216.6	201.4	196.8	169.8	144.4	123.6	115.6	118.0
	3	218.4	203.6	198.6	171.6	144.6	124.6	117.2	119.0
	4	249.6	231.0	225.6	191.4	157.6	134.8	127.2	129.7
	5	258.4	238.8	233.2	197.8	162.8	139.0	131.8	134.7
	II)Av.	231.7	214.6	209.6	175.7	150.4	128.8	121.0	124.5

We have found the radial flux associated with the longitudinal method without guards, is rather appreciable (5).

If we have a realiable function of temperature for the
thermal conductivity of the secondary standard, we would
get a very precise function of temperature of UO_2, $\lambda(\theta)$,
or any other unknown also.

Table II

THERMAL CONDUCTIVITY OF THE UO2 SINTERED PELLETS

Exp. No. Av.	A	B	α	$\frac{\partial\theta}{\partial z}$ °C/cm	$\frac{\partial\theta}{\partial\rho}$ °C/cm	$\frac{\phi}{W/°C}$ cm^{-1}	λ_m λ_c W°/cm **	$-T_o-$
I.	23.82	59.41	0.276	24.93	5.04	0.899	0.036 0.040	116°C
II.	38.55	85.95	0.297	40.24	8.70	1.482	0.036 0.040	175°C
III.	35.11	18.35	0.287	55.43	10.61	1.794	0.032 0.035	229°C
IV.	18.58	78.71	0.245	24.95	4.20	0.950	0.035 0.039	127°C

**λ_c corrected for porosity W/°C cm

This method gives the possibility of estimating absolute values of thermal conductivity which has apparently cast in doubt by some authors (3).

ACKNOWLEDGEMENTS

I wish to thank to J. R. Mc Ewan for his helpful criticism and J. A. L. Robertson. I also, wish thank Dr. A. Morales his help making possible to attend the 15th Conference.

REFERENCES

1. A.M. Ross. ".The Dependence on The Thermal Conductivity of Uranium Dioxide on Density, Microstructure, Stoichiometry and Thermal-Neutron Irradiation", Atomic Energy of Canada Limited, AECL 1096, Chalk River, Ontario. Canada, September 1960.

2. J.L. Daniel et al., "Thermal Conductivity of UO_2". Hanford Atomic Products Operation Richland, Washington HW 69945, September 1962.

3. J.R. McEwan and R.L. Stoute. "Annealing Irradiation-Induced Thermal Conductivity Changes in ThO_2-1.3 wt% UO_2". The American Ceramic Society, Vol. 52 No. 3, March 1969 (160-165).

4. Dynatech. "Comparative Thermal Conductivity Measuring System Model TCFCM Series. Dynatech R/D Company 99 Erie St., Cambridge, Mass. 02139.

5. M. A. Aragones."An Analytical Model for the Measurement of UO_2 Thermal Conductivity". 14th International Conference on Thermal Conductivity, June 1-4, 1975, Storrs, Conn. USA.

GRAPHITE AS A STANDARD REFERENCE MATERIAL*

J. G. Hust

Institute for Basic Standards
National Bureau of Standards
Boulder, Colorado   80302

ABSTRACT

The Cryogenics Division (Boulder, Colorado) in conjunction
with the Office of Standard Reference Materials (Gaithersburg, MD)
of the National Bureau of Standards has been investigating graphite
as a potential, extended temperature range, Standard Reference
Material (SRM). A large number of isotropic, fine-grained graphite
rods in various diameters have been obtained for homogeneity and
stability investigations. Electrical resistivity and density
measurements have been performed on numerous rods at temperatures
from 4 to 300 K. Thermal conductivity measurements have been
performed on thirteen specimens at about 20°C. These measurements
show that transport property variations both between and within
these rods is relatively large (approximately 10%). However, a
correlation between these variables is shown to exist which will
allow the calculation of thermal conductivity from simple and
inexpensive electrical resistivity and density measurements to
within about ± 2%.

INTRODUCTION

Considerable interest has been shown in establishing a fine-
grained, isotropic graphite for use as a Standard Reference Material. Preliminary work on the Air Force Materials Laboratory-

---

*This work was sponsored by the National Bureau of Standards, Office
of Standard Reference Materials (NBS-OSRM), Washington, D.C. Not
subject to copyright.

Advisory Group for Aerospace Research and Development, NATO (AFML–AGARD) program showed that graphite is a promising material. It is especially interesting because of its relatively low cost, ease of fabrication, and its wide temperature range.

The AFML–AGARD program (1965 to 1975) resulted in extensive thermophysical property measurements on several materials including graphite. These measurements have been reported in detail by Fitzer [1] and Minges [2]. The graphite portion of that program is summarized below:

AFML–AGARD PROGRAM (1965–1975)

Material

Fine-grained-isotropic graphite   Ave. density = 1.757 g/cm^3
4" dia. x 12" long cylinders   Max. variations = ± 1.3%
2" x 4" x 6" blocks   Heat treated at 2500°C

Measurements

Property	Temp. Range	Variations	No. of Investigators
Thermal Diffusivity	400–2600K	± 7%	9
Thermal Conductivity	300–2500K	± 10%	5
Thermal Expansion	300–2800K	± 3%	10
Heat Capacity	1900–2800K	--	1
Electrical Resistivity	400–2600K	± 4% @ 1300K	2

It was concluded as a result of the program that this graphite is very promising but that further work should be performed. The remaining stock of this material was donated to NBS–OSRM for further study and/or use for SRM's. The remaining quantity of material, however, was quite limited. Unfortunately, also, the specimens that were distributed to participants of the program were neither individually characterized prior to measurement, nor were they collected after measurement for post-characerization. Because of this, NBS–OSRM decided to purchase a supply of the same fine-grained graphite (AXM-5Q1) for further study. To establish a reasonable supply, we ordered 400 6.4 mm diameter rods, 150 12.7 mm diameter rods, and 70 25.4 mm diameter rods. All rods are 30 cm in length. This material was specified to have a density in the range 1.72-1.75 g/cm^3 and selected to assure a high degree of homogeneity. The final graphitization is performed at 2500°C. These rods were machined from 2" x 4" x 12" plates.

This characterization investigation has been divided into three phases. Phase I includes measurement of electrical resistivity and density to determine the variability and correlation of these variables. Phase II is the measurement of ambient temperature thermal conductivity and correlation with other variables. Phase III includes selection of the most homogeneous specimens and characterization at lower and higher temperatures.

PHASE I

Specimens for electrical resistivity and density measurements were prepared by machining 6.4 mm diameter by 50 mm long rods from various portions of the graphite stock, including the AFML material. These specimens were selected to study both the inter-rod (long range) and intra-rod (short range) inhomogeneities. A total of 39 specimens were prepared: 21 from the 6.4 mm diameter rods, 12 from three 25.4 mm diameter rods and six from a single slab of AFML material.

Electrical resistivity measurements were performed at 4, 76, 273, and 293 K with an uncertainty of about ± 0.5%. The 293 K measurements were performed on 17 specimens at Southern Research Institute through the cooperation of C. Pears (SoRI) and M. Minges (AFML). The measurements at SoRI were performed both before and after a one hour anneal at 3180°C.

Densities of 39 specimens were measured at 20°C with an estimated uncertainty of near 0.1% at NBS and on 17 specimens at SoRI.

Typical electrical resistivities as a function of temperature are illustrated in figure 1. Electrical resistivities plotted as a function of density are shown in figure 2.

The conclusions of Phase I of this work are:

1. The inhomogeneity in electrical resistivity (and in thermal conductivity by association) of this lot of graphite rods is excessively high for use as SRM's unless each piece is certified.
2. A strong intra-rod correlation exists between electrical resistivity and density. This correlation is practically non-existent over this range of densities for inter-rod specimens.
3. Annealing the specimens at 3180°C produces an average decrease of 16% in electrical resistivity and 0.2% in density, but statements (1) and (2) above are still applicable to essentially the same degree for the annealed specimens.
4. The validity of the above statements is independent of temperature from 4 to 300 K.

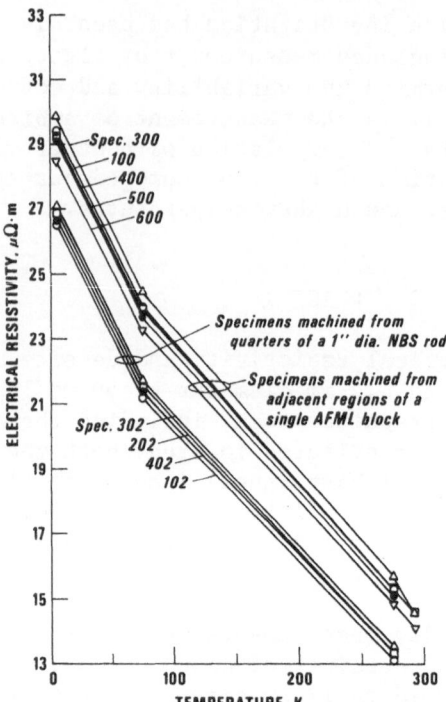

Figure 1. Typical electrical resistivities as a function of temperature for AXM graphite (illustrating both short range and
long range inhomogeneities).

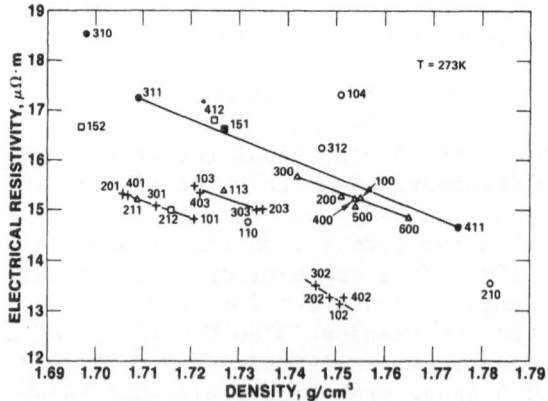

Figure 2. Electrical resistivity versus density at 273 K for AXM
graphite (the solid lines illustrate the short range correlations).

PHASE II

Specimens for thermal conductivity measurements 6.4 mm in length were cut from the previously measured 50 mm long specimens. A comparative apparatus with an accuracy of ± 1% was used to measure the thermal conductivity of these small specimens near ambient temperature. The thirteen measured specimens were chosen to span the entire range of electrical resistivities and densities measured in Phase I. The correlation between thermal conductivity and electrical resistivity is best illustrated through the variance of the Lorenz ratio, $\rho\lambda/T$. These specimens exhibited a Lorenz ratio variation of ± 6% at 296 K. Thus electrical resistivity can not be used by itself to characterize the thermal conductivity to within the desired maximum inaccuracy of ± 2%. To further analyze the interdependence of the measured variables we decided to plot Lorenz ratio versus density, figure 3. This figure reveals that by measuring electrical resistivity and density of a given specimen, thermal conductivity can be calculated at ambient temperature, to within ± 2%. This remarkable result allows one to perform inexpensive piecewise certification of graphite specimens from this lot. Note that one of the points in figure 3 was independently determined at Oak Ridge National Laboratory through the cooperation of P. Moore (ORNL) and M. Minges (AFML). The scatter of the data in figure 3 can be almost totally accounted for by experimental uncertainties (± 1% in thermal conductivity and ± 0.8% in density of the small specimens).

PHASE III

The final phase of this work includes extension of the above encouraging result to a wide range of temperatures and final certification of this material for various thermophysical properties. In preparation for this extensive task, we decided to first select those rods which exhibit the least intra-rod variability (preferably less than ± 2% in electrical resistivity). To accomplish this task for the near term we measured the average density and electrical resistivity as a function of position (2.5 cm intervals) of 150 rods (50 from each diameter available; 6.4 mm, 12.7 mm, and 25.4 mm). Typical electrical resistivities versus position are illustrated in figure 4. As a result of these measurements approximately 12 of each size rod were found which satisfy the ± 2% requirement. Shorter portions of other rods also showed an adequate homogeneity. The 1500 electrical resistivities were measured in a thermally lagged enclosure whose temperature drifted slowly with room conditions. The temperatures at which each rod was measured is in parentheses in figure 4. These values can be corrected to a common temperature of 20°C by using the slope illustrated in figure 1, 0.034 $\mu\Omega$·m/K.

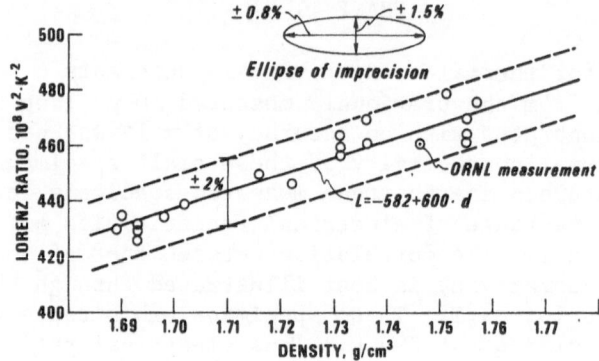

Figure 3. Lorenz ratio versus density correlation of AXM graphite at 296 K.

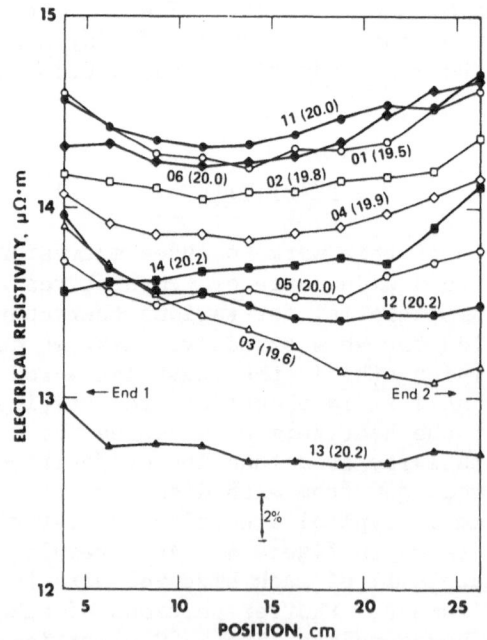

Figure 4. Typical electrical resistivity versus position variations along the 25.4 mm diameter AXM graphite rods.

Measurements of electrical resistivity and thermal conductivity below ambient conditions are scheduled at NBS-Boulder during 1978. Specimens will also be distributed to other participants for further characterization at higher temperatures. These data will be collected and analyzed at NBS-Boulder.

ACKNOWLEDGEMENTS

I would like to acknowledge the assistance, suggestions, and encouragement of several associates involved in this program. In particular, I wish to thank Greg Ruff, Susan Fiske, Kevin Kayse, and Bruce Howrey of this laboratory for the numerous measurements and data analysis they have performed. My appreciation is extended to Peyton Moore and his associates at ORNL for advice and assistance on the construction of the ambient-temperature comparative thermal conductivity apparatus. The continued support, encouragement, and suggestions of M. Minges (AFML) and R. E. Michaelis and R. K. Kirby (NBS, OSRM) is also acknowledged. The interest and cooperation of Wayne Fagan (POCO Graphite, Inc.) has helped to assure the success of this program. The current status and future success of this program are highly dependent on the cooperation of these associates and I express my deepest appreciation to each of them.

REFERENCES

1.   E. Fitzer, Thermophysical Properties of Solid Materials, Advisory Report 12 (1967); Advisory Report 38 (1972); Report 606 (1972), AGARD, NATO, France.

2.   M. L. Minges, Evaluation of Selected Refractories as High-Temperature Thermophysical Property Calibration Materials, AFML Technical Report TR-73-278 (1974); Int. J. Heat and Mass Transfer 17, 1365-1382 (1974).

CYCLIC HEAT FLOW METER APPARATUS TO MEASURE THERMAL CONTACT

RESISTANCE, CONDUCTIVITY AND DIFFUSIVITY*

C.J. Shirtliffe, D.G. Stephenson and W.C. Brown

Division of Building Research, National Research

Council of Canada, Ottawa, Ontario. K1A 0R6

ABSTRACT - A cyclic heat meter apparatus has been developed for
measuring thermal diffusivities or thermal conductivity and diffusivity of flat slab specimens. The specimens may be moist or dry.
The assumption that a moist soil may be considered a normal heat
conducting solid has been validated for a moist clay. The reproducibility of the apparatus is about 3 per cent. The measurement
of contact resistance is possible but results are not presented.

INTRODUCTION

When a temperature gradient exists through a porous material
which contains some moisture, heat is transferred by two mechanisms.
There is the normal thermal conduction and a parallel heat transfer
due to the migration of the moisture toward the region of lowest
temperature.

Studies of the annual variations of ground temperature suggest
that the ground can be represented as a normal heat conducting
solid whose apparent thermal diffusivity and conductivity are
functions of the soil type, density, and mean moisture content.
This raises the problem of how to determine the appropriate thermal
conductivity and diffusivity of moist soils and other porous
materials. Kirkham and Jackson (1) and Higashi (2) have reported
methods that use periodic (cyclic) boundary conditions to determine

*Based on presentation given at the Second International CIB/RILEM
 Symposium "Moisture Problems in Buildings," Rotterdam, Sept. 1974.

the diffusivity of moist samples. In 1959 the Division of Building
Research developed a laboratory apparatus similar in principle to
these but measuring thermal conductivity as well as diffusivity.
This paper describes the current form of the DBR apparatus and
method and discusses the accuracy of the results for a neoprene
rubber slab and a moist clay soil.

THE CYCLIC HEAT METER

To maintain a constant mean temperature and mean moisture
content through a sample during test it is necessary to use a
temperature gradient that is periodically reversed. The Cyclic
Heat Meter used at DBR consists of two 30- by 30-cm aluminum heat
exchanger plates which can be kept at a controlled temperature by
circulating controlled temperature liquid through them. The
liquid is supplied by two circulating laboratory baths each of
which can be controlled at any temperature between -70C and 90C.

The baths and plates are connected to a valve matrix as shown
in Fig. 1. A two circuit patch board allows one set of valves to
be energized while the other is de-energized. Powering of the
groups can be reversed by activating a remote switch. A computer
acts as a cycle timer, closing the switch for 50 per cent of the
period. The baths are set at different temperatures. The
resulting temperatures on the surface of the specimens are equal
amplitude near-square waves, once initial transients disappear. The

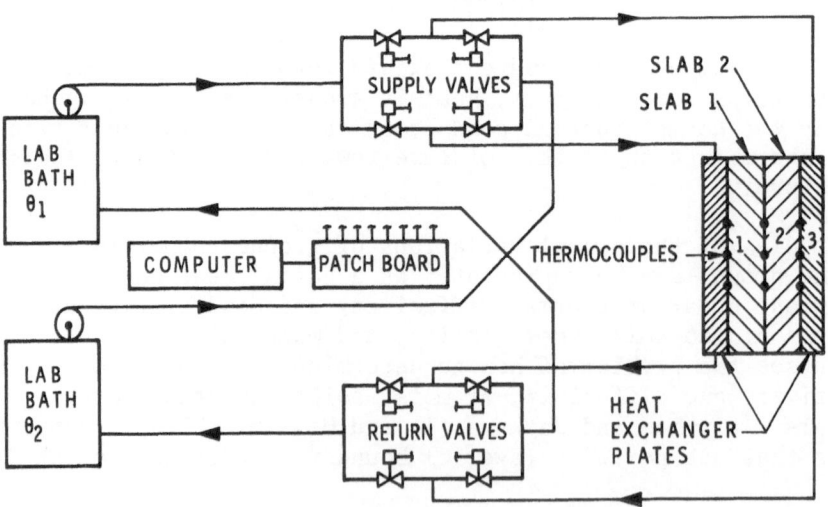

Fig. 1   Cyclic heat meter apparatus

waves can be either in-phase or 180° out-of-phase as selected by
the programmer.

Two slabs of materials are placed between the heat exchanger
plates (Fig. 1). Temperatures are measured at planes 1, 2 and 3
with 36-gauge butt-welded thermocouples. Three couples are used at
each plane; one at the centre of the area and one 5 cm to each side
of centre. When the couples indicate the same temperature it shows
that the heat flow in this central region is one dimensional. The
requirement for one-dimensional heat flow limits the total thicknesses of specimen and standard slab to about 8 cm. When the
properties of one slab of material are known and those of the other
unknown, the configuration is identical to a heat flow meter
apparatus. Because in this case the plate temperatures can be
cycled, the apparatus is termed the cyclic heat meter.

DATA ACQUISITION AND CONTROL SYSTEM

A computer-based, data acquisition and control system controls
the apparatus, acts as the cycle timer, collects, preprocesses,
displays and records the data. A programmable gain, 15-bit analogue
to digital converter measures the filtered output from the nine
thermocouples in 9 milliseconds at intervals of 0.1 second. The
data are filtered digitally with a single-stage binary filter, then
recorded at a selected interval on magnetic tape. An off-line
computer is used for final processing. A simpler system, used
successfully during part of the study, was based on a multi-point
strip chart recorder with punched paper tape output.

PERIODIC HEAT FLOW IN A COMPOSITE SLAB

For a homogeneous slab with sinusoidal temperature variation
at both surfaces, the surface temperatures and heat flows are
related by linear equations which may be expressed conveniently in
the matrix notation (3). For the situation shown in Fig. 1 this
is

$$
\begin{vmatrix} \theta_1 \\ \\ q_1 \end{vmatrix} = \begin{vmatrix} A_1, & B_1 \\ \\ C_1, & A_1 \end{vmatrix} \cdot \begin{vmatrix} \theta_2 \\ \\ q_2 \end{vmatrix} \quad (1)
$$

$$
\text{and} \quad \begin{vmatrix} \theta_2 \\ \\ q_2 \end{vmatrix} = \begin{vmatrix} A_2, & B_2 \\ \\ C_2, & A_2 \end{vmatrix} \cdot \begin{vmatrix} \theta_3 \\ \\ q_3 \end{vmatrix} \quad (2)
$$

where: θ_j, q_j = harmonic components of temperature and heat flow
with period t_o at position j. The matrix elements for slab k are:

$A_k = \text{Cosh} (1+i) \phi_k$; $B_k = (R_k/(1+i)\phi_k) \cdot \text{Sinh} (1+i)\phi_k$;

$C_k = ((1+i)\phi_k/R_k) \cdot \text{Sinh} (1+i)\phi_k$; $R_k = d_k/\lambda_k$, thermal resistance;

d_k = thickness; λ_k = thermal conductivity; $\phi_k = (\pi d_k^2 / D_k t_o)^{\frac{1}{2}}$,

nondimensional frequency; D_k= thermal diffusivity = $\lambda_k/\rho_k c_k$;

c_k = heat capacity; ρ_k = density.

If the periodic driving temperature is not a simple sinusoidal variation it can be expressed as the sum of a series of components, each a sine wave, with periods that are simple fractions of the fundamental period. When λ and D are not temperature dependent, the effect of each component can be calculated as if it alone was present and the effects added to give the total effect. When the materials are slightly temperature dependent, as they usually are, the fundamental component of temperature inside the slab is almost unaffected (4,5). The higher harmonic components are changed in amplitude and phase and a mean temperature shift is generated in the material. The amplitude and phase of the fundamental component is masked by these effects. This is true even when pure sine waves are applied to the surfaces. An accurate Fourier Analysis or Fast-Fourier Transformation is necessary to recover the fundamental component. This is not practical without automatic data recording equipment and computer analysis.

DETERMINATION OF PROPERTIES FROM HARMONIC COMPONENTS

Equations (1) and (2) can be rearranged to give

$$(A_1 - \theta_1/\theta_2)/B_1 + q_2/\theta_2 = 0 \tag{3}$$

$$q_2/\theta_2 = (A_2 - \theta_3/\theta_2)/B_2 . \tag{4}$$

Combining these gives

$$(A_1 - \theta_1/\theta_2)/B_1 + (A_3 - \theta_3/\theta_2)/B_2 = 0 \tag{5}$$

Equation (5) is a complex equation. It can be split into either modulus and argument or real and imaginary parts. The two equations can then be solved for any two unknowns. A_1, A_2, B_1 and B_2 contain the parameters R_1, R_2, ϕ_1, ϕ_2, d_1 and d_2. If four of these plus θ_1, θ_2 and θ_3 are known, the remaining two can be determined.

Should equal contact resistances, R_f, occur on each interface, and if the thermocouples measure a temperature at a mid-point in

the resistance and the two slabs are identical, the equation
equivalent to equation (5) is

$$(\theta_1 + \theta_3) / \theta_2 = 2 \cdot A + R_f \cdot C/2 \tag{6}$$

The most pertinent combinations of knowns and unknowns used in
solving equations (5) and (6) are as follows:

Given	Solve for
1. ϕ_1, R_1	ϕ_2, R_2
2. R_1, R_2	ϕ_1, ϕ_2
3. d_1, d_2, $\lambda_1 = \lambda_2$	ϕ_1, ϕ_2
4. materials identical	ϕ, d_1 / d_2
5. materials identical, equal contact resistances	ϕ, R_f
6. materials identical, d_1, d_2	ϕ (2 solutions)

Several procedures have been used for solving these
simultaneous, non-linear equations. In case No. 6 the equations
have been solved explicitly for ϕ. In cases Nos. 1, 2, and 5 the
equations were divided into modulus and argument to yield one
equation in ϕ and the other for R or R_f in terms of $\phi(6)$. The
first equation has been solved for ϕ graphically, or to a more
satisfactory precision, numerically by the Regula Falsi (7),
Newton-Raphson (7), and Mueller (7) methods. The value of ϕ was
then used to determine R or R_f. All these numerical methods have
been found to experience problems occasionally when singularities,
caused by experimental errors, occurred near the solution. In
certain cases even small experimental errors would preclude a
solution. All cases have been solved using Powell's Method (7).
This method worked whether or not the equations were divided into
modulus and argument or real and imaginary parts.

Powell's Method can either solve the simultaneous equations
for the unknowns or, where no exact solution exists, find the set
of values of the unknown that came closest to satisfying the
equations. In the latter case, the sum of the least-square errors
in the solutions is minimized. The process is termed "finite
dimensional optimization." It requires an initial guess at the
solution. This is obtained from one of the methods already
mentioned or from a knowledge of the material's properties. The

method converges very rapidly on the best least-squares solution. The solution obtained by this method has in many cases differed significantly from those obtained by the other methods. Powell's method has been used for case No. 6 to find the single value of ϕ that best satisfies two equations, again in the least squares sense.

EXPERIMENTAL PROCEDURE

The experimental procedure is simple: the mean temperature and amplitude are set by adjusting the set points for the temperatures of the two baths and the frequency is set by entering the desired value into the computer. The phasing is selected manually. The start and termination of recording on tape is controlled by the computer. The tape is later processed on the off-line computer. The conversion of the readings to temperature and the Fourier Transformation are performed first. Then thermal properties are determined using one or more of the numerical methods described. At least five independent sets of answers are calculated for each test. The mean and standard deviation are then calculated. When desired, the results for other harmonics, principally the third and fifth, are calculated. In general only the first harmonic is used because of sensitivity of the higher harmonics to sources of errors. The total amount of computer time used varies depending on the number of cycles processed, but is normally less than 15 minutes.

EXPERIMENTAL STUDIES AND RESULTS

An experimental program was carried out to check the validity of the assumptions used in the analysis of the results. The experiments can be divided into two series: tests with rubber slabs to prove that the method works for dry materials, and tests with moist Leda clay to prove that the method works for moist materials.

The neoprene rubber slabs were approximately 2.5 cm thick. Two pairs with differing density were selected. The thermal resistance of each set was determined with a 30-cm guarded hot plate apparatus to an accuracy of about ½ per cent. The Leda clay slabs were prepared in the laboratory with a thickness of one inch and moisture content of 21 per cent. The maximum coupling of heat and mass transfer was predicted to occur near this moisture content.

In the first series of tests, the slabs of rubber were tested in pairs to determine ϕ, then one of each pair was used to test the method of obtaining R and ϕ. In the second series of tests one slab of rubber was used with the slab of soil and the R and ϕ of the soil were determined. In each series the amplitude and frequency of the

temperature oscillations were varied. The per cent deviation from
the mean of D for both the rubber and soil and of λ for the soil
obtained in the tests are plotted in Fig. 2. Most of these data
were obtained before the apparatus was improved. The reproducibility of the simpler form of the apparatus was not as good as the
current version. In spite of this, there is little variation in
either parameter. The thermal conductivity of the rubber slabs
determined from the cyclic heat meter results agreed with the
guarded hot plate result to within ½ per cent.

The accuracy of the diffusivity measurements could not be
determined due to a lack of a suitable standard. A comparison with
the measurements from a calorimeter indicated an agreement to within
3 per cent. This was less than the sum of the possible errors in
the apparatuses.

A separate sensitivity analysis too long to report here
indicated the reproducibility of the current apparatus to be better
than 1 per cent with the type of specimens used in these tests.

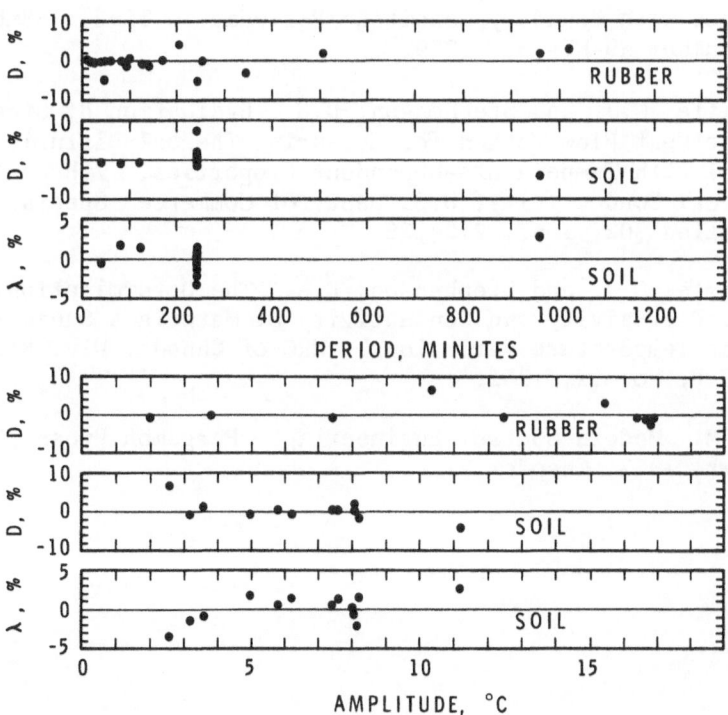

Fig. 2. Variation of D and λ of rubber and soil with largest
temperature on surfaces. (Data from current and former apparatuses.)

ACKNOWLEDGEMENTS

The authors wish to thank Mr. J.G. Theriault and Mr. J.P.R. Raymond for their assistance in constructing the apparatus, conducting the tests, and processing the results and Mr. J.G. Arseneault for preparing many of the computer programs. This paper is a contribution from the DBR/NRC, and is published with the approval of the Director of the Division.

REFERENCES

1. Kirkham, R.D., and Jackson, D. Method of Measurement of the Real Thermal Diffusivity of Moist Soil. Soil Science of America Proceedings, 22, No. 6, Nov-Dec. 1958, 479-82.

2. Higashi, A. On the Thermal Conductivity of Soil, with Special Reference to that of Frozen Soil. Trans., Amer., Geophys. Union, 34, No. 5, Oct. 1953, 737-48.

3. Carslaw, H.S. and Jaeger, J.C. Conduction of Heat in Solids. 2nd ed. Clarendon Press, Oxford, 1959.

4. Roy, D.N., and Thompson, J.S. One-Dimensional Quasilinear Heat Flow with Boundary Condition Periodic in Time. ASME Paper Number 59-HT-26. 1959.

5. Shirtliffe, C.J. and Stephenson, D.G. Evaluation of Steady-Periodic Heat Flow Method for Measuring Thermal Diffusivity of Material with Temperature-Dependent Properties. 7th Conference on Thermal Conductivity, U.S. Dept. of Commerce, Special Publication 302, 1967, 219-228.

6. Shirtliffe, C.J. and Stephenson, D.G. The Determination of Thermal Diffusivity and Conductivity of Materials Subjected to Periodic Temperature Variations. NRC of Canada, Div. Bldg. Res., C.P. No. 12, 1962.

7. Noton, M. Modern Control Engineering. Pergamon Press, New York, 1972, Chapter 2.

ON CORRECTING THERMOPHYSICAL PROPERTY DATA FOR THERMAL EXPANSION EFFECTS

R. E. Taylor

Purdue University

ABSTRACT

Most researchers do not correct thermophysical property data published in the literature for thermal expansion effects. Even values reported by the National Bureau of Standards for standard reference materials (SRM's) are not corrected for dimensional changes. Users of the data and, in fact, many researchers assume that corrections have been made or that they are negligible compared to experimental errors. However, the corrections may be several times larger than they are assumed to be because customary thinking concerning dimensional correction is based on the units of the thermophysical properties. The present paper shows that there is not necessarily a one-to-one relation between the expansion correction and the power to which length enters into the property units. In the case of thermal conductivity "λ" for example, the units are often expressed as W cm^{-1} K^{-1} and one might think that $\lambda_{corrected} = (1 + \Delta L/L_0)^{-1} \lambda_{uncorrected}$. Actually the correction term may be $(1 + \Delta L/L_0)^{-n}$ where n=0, 1, 2 or 3 depending upon the experimental technique used - or the correction term may even be based on the expansion of a material other than the sample. Similarly the hemispherical total emittance (a dimensionless quantity) may require a $(1 + \Delta L/L_0)^{-2}$ correction term. Thus the units of the property involved do not necessarily provide a good guide to the correction term.

Basically one must correct each experimentally measured value for dimensional changes rather than merely using property units and correcting the final value. Also in the case of anisotropic materials, one must use the appropriate values of expansion for the directions in question. In this case λ is W cm cm^{-2} K^{-1} and not W cm^{-1} K^{-1}.

INTRODUCTION

Very little thermal property data reported in the literature is corrected for thermal expansion. Even recommended values for standard reference materials do not include this correction [1]. In part this is due to the general feeling that the uncertainties in experimental values for thermophysical properties are sufficiently large that expansion corrections are not warranted. It is hoped that the currect trend of computerization of test facilities will soon result in an increase in accuracy in addition to the increase in ease of data generation. If this hope is realized, then expansion corrections would become meaningful. This is particularly true since the expansion corrections for some experiments may be several times larger than generally recognized. Customary thinking concerning dimensional corrections is based upon the units of the physical property, i.e., the power to which length occurs in the units. For example, thermal conductivity (λ) expressed in derived SI units is W m^{-1} K^{-1}. Thus one would tend to think that $\lambda_{corrected}=(1+\Delta L/L_0)^{-1} \lambda_{uncorrected}$. If one carried the argument further, the base units of λ are kg mS^{-3} K^{-1}. Thus the correction would be $\lambda_{corrected}=(1+\Delta L/L_0) \lambda_{uncorrected}$, or a factor of $(1+\Delta L/L_0)^2$ different from that obtained using derived units. Actually the correction term for λ will be $(1+\Delta L/L_0)^{-n}$ where n = 0, 1, 2 or 3 depending upon the experimental technique used. The correction term may even be based on the expansion of a material different from that of the sample. Similarly the correction term for other thermophysical properties are not necessarily based on a one-to-one relation between the expansion correction term and property units. Even dimensionless ratios such as total emittance may require dimensional corrections. Basically one must correct each experimentally measured value for dimensional changes, rather than basing the correction on property units. Also in the case of anisotropic materials, one must use the appropriate values of expansion for the directions in question. In this case λ is W cm cm^{-2} K^{-1} and not W cm^{-1} K^{-1}.

THERMAL EXPANSION CORRECTIONS FOR RADIAL HEAT FLOW THERMAL CONDUCTIVITY MEASUREMENTS

First let us consider thermal conductivity measurements using radial outward heat flow. The sample configuration is shown in Figure 1A. The sample is in the form of a thickwalled right circular cylinder with several temperature measurement holes located at radii r_1 and r_2. An electrical core heater is inserted

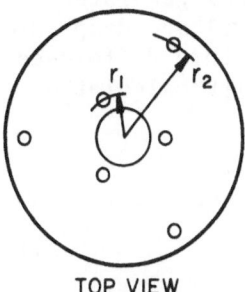

TOP VIEW

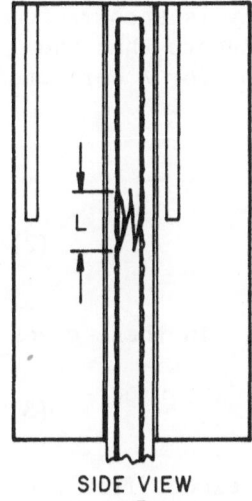

SIDE VIEW
I B

CASE I.
INTERNAL WATER – COOLED HEAT SINK

$$\lambda_c = \frac{Q \ln \dfrac{r_2(1+\Delta r_2/r_2)}{r_1(1+\Delta r_1/r_1)}}{2\pi L \Delta T}$$

$$\lambda_c \cong \lambda_u$$

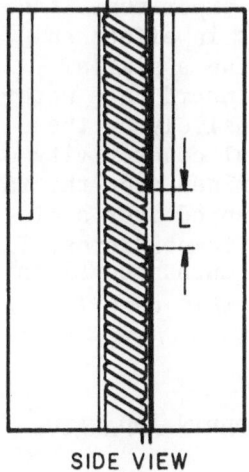

SIDE VIEW
I A

CASE 2.
INTERNAL HEATER

$$\lambda_c \cong \frac{Q \ln \dfrac{r_2(1+\Delta r_2/r_2)}{r_1(1+\Delta r_1/r_1)}}{2\pi L (1+\Delta L_h/L_h) \Delta T}$$

$$\lambda_c \cong \lambda_u (1+\Delta L_h/L_h)^{-1}$$

Figure 1

in the ID of the cylinder. Additional heaters may surround the
sample to raise the ambient temperature. The power generated along
a length L of the heater is determined from the voltage drop across
L and the current flowing. The resulting temperature difference
($\Delta T = T_1 - T_2$) is measured and the thermal conductivity calculated
from the equation

$$\lambda_u = \frac{Q \ln \frac{r_{2u}}{r_{1u}}}{2 \Pi L_u \Delta T} \tag{1}$$

where the subscript u designates that the quantity is uncorrected
for thermal expansion. The subscript c will designate that the
quantity is corrected for expansion. The equation for λ corrected
for expansion is

$$\lambda_c = \frac{Q \ln \frac{r_{2c}}{r_{1c}}}{2 \Pi L_c \Delta T} = \frac{Q \ln \frac{r_{2u}(1 + \Delta r_2/r_2)}{r_{1u}(1 + \Delta r_1/r_1)}}{2 \Pi L_u (1 + \Delta L/L_0) \Delta T} \tag{2}$$

Often ΔT is small and $(1 + \Delta r_2/r_2)/(1 + \Delta r_1)r_1 \cong 1$. In these cases

$$\lambda_c = \lambda_u (1 + \Delta L/L_0)^{-1}. \tag{3}$$

We note that Eq (3) is of the form that we would expect from the
derived property units. The suprising thing is that the correction
is based upon the expansion of the heater and not the sample. Thus
the correction for the thermal expansion in this experiment is not
based upon the sample characteristics but becomes in effect an
instrument calibration factor. Next let us examine a similar
experimental setup except that the heat flow is inward to a water-
cooled heat sink. This setup has been used extensively by the
author and colleagues [1,2] to measure the thermal conductivity of
a variety of materials. The heat input is determined from the rate
of water flow and the water temperature rise measured over a distance L. When the sample is heated to very high temperatures, T_2
may be 2000°C and T_1 may be 1950°C but L remains unchanged since
the thermopile is immersed in flowing water. In this case $\lambda_c = \lambda_u$,
i.e., there is no expansion correction.

THERMAL EXPANSION CORRECTIONS FOR THE MULTI-PROPERTY APPARATUS

The multi-property apparatus at PRL is capable of measuring
at least ten physical properties on the same sample up to high

temperatures (~ 3000K). The sample is heated by direct passage of DC current through it while the sample is suspended between water-cooled electrodes. One of the electrodes is connected to a strain gage mechanism which permits the sample to expand and contract under nearly stress-free conditions. The apparatus has been discussed in a number of technical reports and publications of which ref [3,4] are typical. The device has been used to generate state-of-the-art data at high temperatures on thermal conductivity, specific heat (C_p), thermal expansion, electrical resistivity (ρ), hemispherical total emittance (ε_H), normal spectral emittance (0.3-11 um), Wiedemann-Franz-Lorenz ratio (L), and thermal diffusivity (α). The governing equation is

$$\lambda_u \frac{d^2T}{dZ^2} + \frac{d\lambda_u}{dT} \left(\frac{dT}{dZ}\right)^2 + \frac{I^2\rho_u}{A_u^2} - \frac{\Pi \, D_u \sigma(\varepsilon_H)_u \, (T^4 - T_0^4)}{A_u}$$

$$+ \mu \frac{I}{A_u} \frac{dT}{dZ} = C_p \, d_u \frac{dT}{dt} \tag{4}$$

where I is the current, μ is the Thomson coefficient, σ is the Stefan-Boltzmann constant, d_u is the sample density, A_u is the sample cross-sectional area and D_u is the sample diameter uncorrected for expansion. For the central region of an infinitely long ($\frac{d^2T}{dZ^2} = \frac{dT}{dZ} = 0$) sample at steady state, Eq 4 becomes

$$\frac{I^2\rho_u}{A_u^2} = \frac{\Pi D_u \sigma(\varepsilon_H)_u \, (T^4 - T_u^4)}{A_u} \tag{5}$$

i.e., all the power generated per unit volume is radiated to the surroundings. From these equations, we may derive the appropriate expansion corrections.

For electrical resistivity:

$$\rho_u = \frac{A_u}{L_u} \frac{E}{I} \tag{6}$$

and

$$\rho_c = \frac{A_c}{L_c} \frac{E}{I} = \frac{A_u(1+\Delta \, L/L_0)^2}{L_u(1+\Delta \, L/L_0)} \frac{E}{I} = \frac{A_u}{L_u} \frac{E}{I} (1+\Delta \, L/L_0)$$

$$= \rho_u(1+\Delta \, L/L_0). \tag{7}$$

Thus the usual expansion correction applies.

For hemispherical total emittance:

$$(\varepsilon_H)_u = \frac{E \cdot I}{L_u \Pi D_u \sigma (T^4 - T_0^{\,4})} \tag{8}$$

and

$$(\varepsilon_H)_c = \frac{E \cdot I}{L_u(1 + \Delta\,L/L_0)\,\Pi D_u(1 + \Delta\,L/L_0)\,\sigma(T^4 - T_0^{\,4})}$$

$$= \frac{E \cdot I}{L_u\,\Pi D_u \sigma(T^4 - T_0^{\,4})\,(1 + \Delta\,L/L_0)^2}$$

$$= \frac{(\varepsilon_H)_u}{(1 + \Delta\,L/L_0)^2} = (\varepsilon_H)_u(1 + \Delta\,L/L_0)^{-2} \tag{9}$$

Thus the hemispherical total emittance should be corrected by a $(1 + \Delta\,L/L_0)^{-2}$ term, even though the emittance is a dimensionless ratio.

For thermal conductivity:

$$\lambda_u \frac{d^2 T}{dZ^2} + \frac{d\lambda_u}{dT}\left(\frac{dT}{dZ}\right)^2 + \frac{I^2 \rho_u}{A_u^{\,2}} - \frac{\Pi D_u \sigma(\varepsilon_H)_u(T^4 - T_0^{\,4})}{A_u}$$

$$+ \mu \frac{I}{A_u} \frac{dT}{dZ} = 0 \tag{10}$$

and

$$\lambda_c \frac{d^2 T}{dZ^2} + \frac{d\lambda_c}{dT}\left(\frac{dT}{dZ}\right)^2 + \frac{I^2 \rho_u(1 + \Delta\,L/L_0)}{A_u^{\,2}(1 + \Delta\,L/L_0)^4}$$

$$- \frac{\Pi D_u(1 + \Delta\,L/L_0)\,\sigma(\varepsilon_H)_u(1 + \Delta\,L/L_0)^{-2}(T^4 - T_0^{\,4})}{A_u(1 + \Delta\,L/L_0)^2}$$

$$+ \frac{\mu\,I}{A_u(1 + \Delta\,L/L_0)^2} \frac{dT}{dZ} = 0 \tag{11}$$

thus

$$\lambda_c \frac{d^2T}{dZ^2} + \frac{d\lambda_c}{dT}(\frac{dT}{dZ})^2 + \frac{I^2\rho_u}{A_u^2}(1+\Delta\ L/L_0)^{-3} - \frac{\Pi D_u\sigma(\epsilon_H)_u(T^4-T_0^4)}{A_u}$$

$$(1+\Delta\ L/L_0)^{-3} + \mu\frac{I}{A_u}\frac{dT}{dZ}(1+\Delta\ L/L_0)^{-2} = 0 \qquad (12)$$

and

$$\lambda_c = \lambda_u(1+\Delta\ L/L_0)^{-3} \qquad (13)$$

The first and second temperature derivations, dT/dZ and d^2T/dZ^2 have no expansion corrections associated with them because the temperature distribution is measured with an automatic optical pyrometer mounted on a movable pedestal external to the vacuum system. The correction to area in the last term on the left hand side affects the computed value of μ(the Thomson coefficient) but does not affect the value of λ.

For specific heat:

$$\frac{I_e^2\rho u}{A_u^2} = \frac{P_u\sigma(\epsilon_H)_u(T_e^4-T_0^4)}{A_u} \qquad \text{(at equilibrium condition)}$$

$$= \frac{I_e^2}{A_u^2}\frac{A_u}{L_u}\frac{E_e}{I_e} = \frac{(EI)_e}{A_uL_u} \qquad (14)$$

and

$$\frac{I^2\rho_u}{A_u^2} - \frac{P_u\sigma(\epsilon_H)_u(T_e^4-T_0^4)}{A_u} = (C_p)_u d_u \frac{dT}{dt} \qquad \text{(transient conditions)} \qquad (15)$$

thus

$$\frac{EI}{A_uL_u} - \frac{(EI)_e}{A_uL_u} = (C_p)_u d_u \frac{dT}{dt} = (C_p)_u \frac{W}{A_uL_u}\frac{dT}{dt} \qquad (16)$$

and

$$(C_p)_u = \frac{EI-(EI)_e}{W\frac{dT}{dt}} = (C_p)_c \qquad (17)$$

Thus no correction term is needed for specific heat.

For thermal diffusivity:

$$\alpha_u = \frac{\lambda_u}{C_p d_u} \tag{18}$$

and

$$\alpha_c = \frac{\lambda_u (1+\Delta L/L_0)^{-3}}{C_p d_u (1+\Delta L/L_0)^{-3}} = \frac{\lambda_u}{C_p d_u} = \alpha_u \tag{19}$$

Thus no correction factor is involved for thermal diffusivity even though the units are $m^2 s^{-1}$.

For the Wiedemann-Franz-Lorenz function L,

$$L_u = \lambda_u \rho_u / T_u \tag{20}$$

and

$$L_c = \lambda_u (1+\Delta L/L_0)^{-3} \rho_u (1+\Delta L/L_0)/T. \tag{21}$$

Thus

$$L_c = L_u (1+\Delta L/L_0)^{-2} \tag{22}$$

and a $(1+\Delta L/L_0)^{-2}$ factor is required even though the units of L do not include a length term.

These results are summarized in Table 1.

Table 1.   Factors Used to Correct for Thermal Expansion Effects
Using the Multiproperty Apparatus

Property	Factor	Property Units
ρ, Electrical Resistivity	$1+\dfrac{\Delta L}{L_0}$	ohm-m
λ, Thermal Conductivity	$(1+\Delta L/L_0)^{-3}$	Watt $m^{-1} K^{-1}$
C_p, Specific Heat	1	Watt sec $kgm^{-1} K^{-1}$
L, Wiedemann-Franz-Lorenz Ratio	$(1+\Delta L/L_0)^{-2}$	Watt ohm K^{-2}
α, Thermal Diffusivity	1	$m^2 \cdot sec^{-1}$
ε_H, Total Hemispherical Emittance	$(1+\Delta L/L_0)^{-2}$	-----

Note that there is no apparent relation between the power to which length enters into the property units and the correction factor for four of the six quantities (λ, L, ε_H and α). This emphasizes the fact that the correction must be based on measured quantities and not on property units.

DISCUSSION

The magnitude of the thermal expansion correction is illustrated in ref [5], which gives uncorrected and corrected results for the thermal conductivity of SRM tungsten measured with the multiproperty apparatus. The correction term increases to 5.1% at 3050K. This correction is about twice the stated accuracy of the results.

Although there is no expansion correction for thermal diffusivity values derived from the multiproperty apparatus, often the correction term is $(1+\Delta L/L_0)^2$. This is the case, for example, with the flash technique, which is currently used to generate over 75% of the diffusivity results. For a material expanding 1.5%, the correction term for the flash method would amount to about 3%. The latter figure is often stated, but not necessarily realized, as the accuracy of this technique - so expansion corrections should definitely be included when reporting results for materials expanding more than 1%.

There is currently a great deal of interest in composite materials. Such materials often exhibit a great deal of anistropy. In some cases the expansion may vary by order of magnitude along different directions. In such cases, thermal expansion effects may be quite large and significantly alter the reported values.

Along with the hoped-for increase in accuracy resulting from computerization of test facilities, the involvement with so-called "exotic" materials with possibly large expansions, the increase in high-temperature testing of materials, and the realization that expansion corrections may be significantly larger than a casual examination would indicate, it behoves researchers to carefully examine expansion effects.

REFERENCES

1. R. E. Taylor and J. Morreale, "Thermal Conductivity of Titanium Carbide, Zirconium Carbide, and Titanium Nitrite at High Temperatures," J. Am. Ceram. Soc. 47[2] pp. 69-73, 1964.

2. N. S. Rasor and J. D. McClelland, "Thermal Property Measurements at Very High Temperatures," Rev. Sci. Inst. 31 [6] pp. 595-604, 1960.

3. R. E. Taylor, F. E. Davis, and R. W. Powell, "Direct Heating Methods for Measuring Thermal Conductivity of Solids at High Temperatures," High Temp. - High Pressures $\underline{1}$ pp. 663-673, 1969.

4. R. E. Taylor, "Survey on Direct Heating Methods for High-Temperature Thermophysical Property Measurements of Solids," High Temp. - High Pressures $\underline{4}$ pp. 523-531, 1972.

5. R. E. Taylor, "Thermal Properties of Tungsten SRM's 730 and 799" (submitted for publication, Journal of Heat Transfer).

MEASUREMENT OF THE LORENZ FUNCTION BY THE ELECTRICAL METHOD

M.J. Laubitz, P. van der Meer and J.G. Cook

National Research Council of Canada

Ottawa, Canada K1A 0S1

Direct measurement of the Lorenz function by the electrical method, in which the sample is heated directly by an electrical current, is in principle a fast and convenient way to determine this parameter. However, recent theoretical treatments of this method are somewhat misleading; we discuss this and a number of other problems of the method, present the design of a new system which eliminates some of them and is suitable for small specimens, and compare the results obtained with it for Pb with selected values from the literature.

SESSION E   THEORY, SOLIDS AT LOW TEMPERATURE

Session Chairmen: C.R. Levens
                     National Research Council
                     Ottawa, Ontario

                     E. Taylor
                     National Research Council
                     Ottawa, Ontario

PHONON LIMITED RESISTIVITY TO $O(T^2)$[†]

R.C. Shukla, E.R. Muller, and Martin VanderSchans

Brock University

St. Catharines, Ontario, Canada, L2S 3A1

ABSTRACT

The theory of phonon-limited resistivity (ρ) of metals as developed by Baym and others has been extended to include the anharmonic effects using Green's function technique. The higher order correlation functions involving 3 and 4 operators have been obtained from the corresponding Green's functions. The physical significance of the contributions from these correlation functions to ρ is pointed out. The contribution to ρ from the third order correlation functions can be identified with the interference term and the fourth order correlation functions can be identified with the correction to ρ from the Debye Waller factor and the first term of the multiphonon series. The derived expressions are valid for all temperatures. In the high temperature limit the Debye Waller factor and multiphonon contributions are found to be of $O(T^2)$. All other terms of $O(T^2)$ come from the cubic, quartic shifts, phonon widths and the interference term. Thus the formula for ρ in the high temperature limit is found to be $\rho = AT + BT^2$. Numerical estimates for potassium and sodium of the Debye-Waller factor and multiphonon contributions to ρ indicate that there is a strong cancellation between them for potassium (correction to ρ is of the order of 1.3% at $T \simeq T_m$), but for sodium this correction is as large as 2-5% in the temperature range 160°K to 370°K.

[†] Work supported by the National Research Council of Canada.

1. INTRODUCTION

It is well known that the phonon limited resistivity (ρ) of metals when computed in the harmonic approximation varies linearly with temperature in the high temperature limit $T \geqslant \Theta_D$). Beyond the Debye temperature (Θ_D) and up to the melting temperature (T_M) the experimental results indicate roughly a temperature dependence for ρ of the form

$$\rho = AT + BT^2 \tag{1}$$

In a recent semi-empirical analysis of the experimental phonon-limited resistivity data of various metals (k, Na, Cu, Ag and Au) at constant volume, Grimvall[1] found that the second term in Eq. (1) is of the order of 10% of the total in the high temperature limit. Although Grimvall ascribed that the contribution to BT^2 term will come from the Debye-Waller factor, lattice anharmonicity and multiphonon scattering, etc. no attempt was made to derive an explicit expression for the coefficient B in Eq.(1).

More recently a first principle calculation of ρ for k and Na was performed by Shukla and Taylor[2] in the quasiharmonic approximation for the temperature range 20°K to the room temperature. At no stage of the calculation were any parameters in the theory adjusted to fit any experimental data. The agreement between the calculated and experimental data was better than 3% for K at all $T \geqslant 40°K$ and better than 4% for Na at $T \geqslant 60°K$. These calculations indicate that the effect of anharmonicity, multiphonon scattering, Debye Waller factor etc. is rather small at least in K and Na. However, it has been shown before that anharmonic effects beyond the quasiharmonic approximation are necessary to explain a number of other physical properties such as anharmonic shifts of the phonon frequencies (Duesbery et al[3], Glyde and Taylor[4]) and free energy (Shukla and Taylor[5]). It would therefore appear that there is a strong cancellation between the various contributions to ρ from the Debye-Waller factor, multiphonon scattering, and anharmonic effects, etc.

In order to have the quantitative estimate of these contributions from a first principle or a parametric type calculation it is necessary to derive explicit expressions for them. In this paper we present explicit expressions for the contribution to the phonon limited resistivity of metals from the Debye-Waller factor, multiphonon scattering, and anharmonic effects. Obviously, because of space limitations it is not possible to present a complete derivation of all these expressions and we therefore present a summary of the results.

2. THEORY

The phonon limited resistivity of simple metals (spherical Fermi surface) is given by the formula [6,7]

$$\rho(T) = C \int_{<2K_F} d^3q \; q|W(q)|^2 \int_{-\infty}^{+\infty} d\omega \; S(q,\omega) \beta\omega/(e^{\beta\hbar\omega}-1)$$

where in self evident notation $C = 3\Omega_0/16e^2 v_F^2 K_F^4$ and the dynamical structure factor is

$$S(q,\omega) = (1/2\pi N) \int_{-\infty}^{+\infty} dt e^{i\omega t} \left\langle \sum_{\ell\ell'} e^{-iq\cdot x_\ell(t)} e^{iq\cdot x_{\ell'}(0)} \right\rangle_T$$

The angular bracket denotes the thermal average with respect to the anharmonic Hamiltonian involving the cubic and quartic terms in the Taylor's expansion of the crystal potential energy.

Substituting for the instantaneous displacement $X_\ell(t)$ for the ℓth site, $X_\ell(t) = R_\ell(0)+u_\ell(t)$, where $R_\ell(0)$ denotes the equilibrium position and $u_\ell(t)$ is the displacement defined by

$$u_\ell(t) = (\hbar/2NM)^{\frac{1}{2}} \sum_{qj} \left[\varepsilon(qj)/(\omega(qj))^{\frac{1}{2}}\right] e^{iq\cdot R_\ell(0)} A_{qj}(t)$$

The terms in the thermal bracket of Eq.(2) can be expanded and we find

$$S(q,\omega) = S_0(q,\omega) + S_2(q,\omega) + S_3^A(q,\omega) + S_3^B(q,\omega)$$

$$+ S_4^A(q,\omega) + S_4^B(q,\omega) + S_4^C(q,\omega)$$

where the subscript indicates the number of operators to be averaged, $S_0(q,\omega)$ is the elastic part of the scattering which does not contribute to ρ, and

$$S_2(q,\omega) = \frac{\hbar}{2M} \sum_{j_1 j_2} \frac{[q\cdot\varepsilon(qj_1)][q\cdot\varepsilon^*(qj_2)]}{[\omega(qj_1)\omega(qj_2)]^{\frac{1}{2}}} \times \frac{1}{2\pi} \int_{-\infty}^{\infty} e^{i\omega t} \langle A_{qj_1}(t)A_{qj_2}(0)\rangle dt$$

$$S_3^A(q,\omega) = - \sum_{1,2,3} f^3(1,2,3)\delta(-q+q_1+q_2)\delta(q+q_3)J_3^A$$

$$S_3^B(q,\omega) = \sum_{1,2,3} f^3(1,2,3)\delta(-q+q_1)\delta(q+q_2+q_3) \; J_3^B$$

$$S_4^A(q,\omega) = -\frac{1}{3!} \sum_{1,2,3,4} f^4(1,2,3,4)\delta(-q+q_1+q_2+q_3)\delta(q+q_4) \ J_4^B$$

$$S_4^B(q,\omega) = \left(\frac{1}{3!}\right) \sum_{1,2,3,4} f^4(1,2,3,4)\delta(-q+q_1+q_2)\delta(q+q_3+q_4)J_4^B$$

$$S_4^C(q,\omega) = \left(\frac{1}{2!}\right)^2 \sum_{1,2,3,4} f^4(1,2,3,4)\delta(-q+q_1+q_2)\delta(q+q_3+q_4)J_4^C$$

where we have introduced the notation $q_i j_i \equiv i$ for $i=1,2,3,4$ and

$$f^3(1,2,3) = \frac{iN}{2!}\left(\frac{\hbar}{2NM}\right)^{3/2} \frac{[q \cdot \varepsilon(1)][q \cdot \varepsilon(2)][q \cdot \varepsilon(3)]}{[\omega(1)\omega(2)\omega(3)]^{\frac{1}{2}}}$$

$$f^4(1,2,3,4) = N\left(\frac{\hbar}{2NM}\right)^2 \frac{[q \cdot \varepsilon(1)][q \cdot \varepsilon(2)][q \cdot \varepsilon(3)][q \cdot \varepsilon(4)]}{[\omega(1)\omega(2)\omega(3)\omega(4)]^{\frac{1}{2}}}$$

$$J_3^A = \frac{1}{2\pi}\int_{-\infty}^{+\infty} e^{i\omega t} <A_1(t)A_2(t)A_3(0)> dt$$

$$J_3^B = \frac{1}{2\pi}\int_{-\infty}^{+\infty} e^{i\omega t} <A_1(t)A_2(0)A_3(0)> dt$$

$$J_4^A = \frac{1}{2\pi}\int_{-\infty}^{+\infty} e^{i\omega t} <A_1(t)A_2(t)A_3(t)A_4(0)> dt$$

$$J_4^B = \frac{1}{2\pi}\int_{-\infty}^{+\infty} e^{i\omega t} <A_1(t)A_2(0)A_3(0)A_4(0)> dt$$

$$J_4^C = \frac{1}{2\pi}\int_{-\infty}^{+\infty} e^{i\omega t} <A_1(t)A_2(t)A_3(0)A_4(0)> dt$$

In order to evaluate each of the above terms in $S(q,\omega)$ it is necessary to determine the Fourier transform of thermal~average of two, three, and four operators. These can be obtained from their corresponding Green's functions (see for example Shukla and Muller[8] where a detailed derivation for $<A_1(t)A_2(0)>$ and $<A_1(t)A_2(t)A_3(0)>$ arising in $S_2(q,\omega)$ and $S_3^A(q,\omega)$, respectively is presented). Similar procedures using the equation of motion method were used to obtain the Green's functions and corresponding correlation functions arising in S_3^B, S_4^A, S_4^B and S_4^C.

It is found that the contribution to ρ from $S_2(q,\omega)$ comes from several sources:

(a) the usual harmonic result,

(b) the quartic shift of phonon frequency,

(c) the cubic shift of phonon frequency,

(d) an explicit contribution from the phonon width. The contribution to ρ from S_3^A and S_3^B terms are equal and correspond to the so called interference effect. Among the group of S_4 contributions to ρ it is found that S_4^A and S_4^B contribute equally but are different from that of S_4^C. The former can be identified with the Debye-Waller factor and the latter with the multiphonon correction to ρ.

In the high temperature limit (e.g. $T > \theta_D$, the Debye temperature) it is possible to expand the thermal factors arising in all the expressions of ρ obtained from $S_2(q,\omega)$, $S_3^A(q,\omega)$, $S_3^B(q,\omega)$...etc. in powers of T. The high temperature expansion of $\rho_2(T)$ arising out of $S_2(q,\omega)$ can be written as

$$\rho_2(T) = [\rho_2]_H T + [\rho_2]_C T^2 + [\rho_2]_q T^2 + [\rho_2]_W T^2$$

where

$[\rho_2]_H, [\rho_2]_C, [\rho_2]_q$, and $[\rho_2]_W$ are the corresponding temperature independent contributions to ρ from the harmonic, cubic shift, quartic shift, and phonon width respectively.

Similarly the other high temperature expansions yield

$$\rho_3(T) = [\rho_3^A + \rho_3^B]_I T^2 = [\rho_3]_I T^2$$

$$\rho_4(T) = [\rho_4^A + \rho_4^B]_{D_W} T^2 + [\rho_4^C]_{M_P} T^2 = [\rho_4]_{D_W} T^2 + [\rho_4]_{M_P} T^2$$

where $[\rho_3]_I$ is the temperature independent contribution to ρ from the interference term; $[\rho_4]_{D_W}$ and $[\rho_4]_{M_P}$ are the respective temperature independent contributions to ρ from the Debye-Waller factor and the first term of the multiphonon series.

In terms of the anharmonic ordering parameter λ, the anharmonic terms $[\rho_2]_C$, $[\rho_2]_q$, and $[\rho_2]_W$ are of order λ^2 while $[\rho_3]_I$ is of the order λ. Since the terms of $O(\lambda^2)$ are by an order of magnitude smaller than λ term we would expect the most important anharmonic contribution to ρ of the $O(T^2)$ will come from $[\rho_2]_I$.

The other two terms of $O(T^2)$ do not contain the anharmonic coefficients and are essentially harmonic in nature. Some numerical estimates of these two terms, namely $[\rho_4]_{D_W}$ and $[\rho_4]_{M_P}$ indicate that they almost cancel each other out in potassium and the maximum correction to ρ is of the order of 1.5% at $T \simeq T_m$, where T_m is the melting temperature, whereas for sodium the correction is as large as 2-5% in the temperature range 160°K to 375°K.

ACKNOWLEDGEMENT

The authors would like to thank Mrs. J. Cowan for the careful job of typing this manuscript.

REFERENCES

1. G. Grimvall, Phys. Cond. Matter 17, 135 (1974).
2. R.C. Shukla and R. Taylor, J. Phys.F: Metal Phys.6, 531 (1976).
3. M.S. Duesbery, R. Taylor and H.R. Glyde, Phys. Rev. B8, 1372 (1973).
4. H.R. Glyde and R. Taylor, Phys. Rev. B5, 1206 (1972) and J. Phys. F1, L46 (1971).
5. R.C. Shukla and R. Taylor, Phys. Rev. B9, 4116 (1974).
6. G. Baym, Phys. Rev. 135, A1691 (1964).
7. M.P. Greene and W. Kohn, Phys. Rev. 137, A513 (1965).
8. R.C. Shukla and E.R. Muller, Phys. Stat. Sol. (b) 43, 413 (1971).

COMPETITION BETWEEN THE ELECTRON-ELECTRON AND ELECTRON-PHONON MECHANISMS IN THE ELECTRICAL AND THERMAL RESISTIVITIES OF SIMPLE METALS

W. E. Lawrence

Dartmouth College

Hanover, New Hampshire 03755

I. INTRODUCTION

The purpose of this paper is to discuss the use of the thermal (and electrical) resistivities as tools for studying the electron-electron interaction in metals. We focus on the simple and noble metals, where only tentative evidence exists at present, and we hope to indicate where further evidence might best be sought. The difficulty in observing electron-electron scattering is that it tends to be dominated by electron-phonon scattering, particularly in the simple metals. To observe the electron-electron mechanism, one is thus pushed to extremes of temperature (high or very low) where it tends to compete more favorably: Electron-electron contributions to ρ and WT increase quadratically with temperature, while the electron-phonon contributions increase linearly at high temperatures ($T \gtrsim \Theta_D$), but much faster ($\rho \sim T^5$ and WT $\sim T^3$) at extremely low temperatures $T \ll \Theta_D$. The competition in the high and low temperature regimes will be discussed in sections II and III.

For the remainder of this section we discuss electron-electron scattering exclusively. Its contribution to the electrical resistivity may be written as [1]

$$\rho_{ee} = m\Delta/ne^2\tau_{ee} \tag{1}$$

where

$$\Delta = \langle |\vec{v}_1 + \vec{v}_2 - \vec{v}_3 - \vec{v}_4|^2 \rangle \langle |2\vec{v}|^2 \rangle^{-1} \tag{2}$$

is the appropriate measure of umklapp character in the scattering (angular brackets denote the Fermi surface average). τ_{ee}^{-1} is the average scattering rate [2] of quasiparticles near the Fermi surface. Calculated within the Born approximation it has the simple expression

$$\tau_{ee}^{-1} = \pi^3 (k_B T)^2 \Gamma / 12 \hbar E_F \ ,$$ (3)

where the averaged scattering probability

$$\Gamma = \int_0^{2k_F} dq \ W(q) / 2k_F W(o)$$ (4)

is of order unity; in particular, for Thomas Fermi screening it is within 10% of the value $\Gamma \approx 0.6$, for all cases studied here. Many-body corrections have been studied by Kukkonen [3], and these increase the values of τ_{ee}^{-1} over the ones given by (3), for the case of alkali metals. These corrections will be discussed later.

The corresponding thermal resistivity is

$$W_{ee} = 18m(1 - \langle\cos\theta\rangle + \tfrac{1}{2}\Delta) / 5\pi^2 nk_B^2 T \tau_{ee} \ .$$ (5)

The only additional parameter which enters here is the Fermi surface average,

$$\langle\cos\theta\rangle = \langle \vec{v}_1 \cdot \vec{v}_2 \rangle \langle |\vec{v}|^2 \rangle^{-1} \ .$$ (6)

The Lorenz number for electron-electron scattering is given by (1) and (5):

$$L_{ee} \equiv \rho_{ee} / W_{ee} T = \tfrac{5}{6} L_o \Delta / (1 - \langle\cos\theta\rangle + \tfrac{1}{2}\Delta) \ ,$$ (7)

where $L_o = \pi^2 k_B^2 / 3e^2$ (8)

is the "classical" Lorenz number, which one would expect if the relaxation time approximation were valid (in the sense that the same relaxation time should apply to both ρ and W). Typical values of the parameters Δ and $\langle\cos\theta\rangle$ are shown on Table I, together with the resulting Lorenz numbers. These are significantly smaller than the classical value L_o, reflecting the fact that the effective relaxation rate Δ/τ_{ee} for the electrical resistivity, is suppressed over the corresponding rate $\tfrac{5}{6}(1 - \langle\cos\theta\rangle + \tfrac{1}{2}\Delta)/\tau_{ee}$ for the thermal resistivity. In fact if we were to somehow remove umklapp processes, then (assuming a spherical Fermi surface) Δ and ρ_{ee} would vanish, while W_{ee} would remain relatively unchanged. This point is essential for observing electron-electron scattering at high temperatures.

Table I.   Electron-electron quantities, taken from ref. 1.   Lorenz number is calculated from eqn. 7.

	$\tau_{ee}^{-1}T^{-2}(10^6 sec^{-1}K^{-2})$	Δ	$-\langle\cos\theta\rangle$	L_{ee}/L_o
Al	1.3	2/5	0.4	1/5
Cu	2.4	3/4	0.37	3/8
Ag	2.9	3/4	0.35	3/8
Au	2.9	3/4	0.35	3/8
K	9.2	0.06	0.3	0.04

II.   LORENZ NUMBERS AT HIGH TEMPERATURES

The quadratic (T^2) electron-electron contributions to ρ and WT begin to compete more favorably with the (approximately) linear electron-phonon contributions, as temperature is increased above Θ_D.   However the electron-phonon contributions are not readily subtracted, because certain effects (such as temperature-dependent phonon spectrum and thermal expansion of the lattice) distort their otherwise linear temperature-dependence.   The resolution of this difficulty was suggested and implemented by Laubitz, Cook, and Van der Meer; it is based on the idea that in the presence of electron-phonon scattering alone, the Lorenz number approaches its classical value at high temperatures.   Specifically,

$$W_{e\phi} - \rho_{e\phi}/L_oT \rightarrow A/T^2 , \tag{9}$$

independently of the aforementioned effects.   Electron-electron scattering now manifests itself in the Lorenz number, as a departure (from the classical value) which increases with temperature: Using the total measured resistivities, we find, in place of (9),

$$W - \rho/L_oT \rightarrow A/T^2 + BT , \tag{10}$$

where the coefficient B has the theoretical value (1, 5, and 8)

$$B = \frac{W_{ee}}{T} \cdot \frac{1-\langle\cos\theta\rangle-\Delta/3}{1-\langle\cos\theta\rangle+\Delta/2} . \tag{11}$$

(C is independent of temperature).   Previous analyses have assumed $\Delta = 0$ in lieu of theoretical estimates; we have redone the analysis (Table II) using the parameter values shown in Table I.   The alkali

Table II.   Values of $\tau_{ee}^{-1} T^{-2}$ in 10^6 sec^{-1} K^{-2}.

	Simple Theory	Theory with manybody corrections[a]	Measured "B" (eqn.11)[b] (Δ=0)	(actual Δ)	Infrared Absorptivity[c]
Cu	2.4	4.2	2.1	3.3	7.7
Ag	2.9	6.4	1.4	2.2	5.3
Au	2.9	6.4	2.4	3.7	9.5
Na	5.8	20	17		
K	9.2	62			

a. Ref. 3,   b. Ref. 4,   c. Ref. 5.

metal results are essentially unchanged, while the noble metal
results are changed by a factor of about 50%, as shown. We have
chosen to tabulate the values of $\tau_{ee}^{-1} T^{-2}$ (as inferred from measured values of B via eqns. 5 and 11) in order to facilitate
comparison with other measurements. In particular Beach, Christy,
and Parkins [5] have inferred values from the measured intraband
absorptivity by fitting to the frequency-dependent scattering
rate [6]

$$\Delta\tau_{ee}^{-1}(\omega) = \Delta\tau_{ee}^{-1}\left[1+(\hbar\omega/2\pi k_B T)^2\right] ,$$

to be used in the Drude formula. (Note that Δ appears in this
scattering rate, as it did in the d.c. case, eqn. 1).

    Manybody corrections to the simple model presented here were
studied in ref. 3. These become substantial in the alkali metals
(because of their low electron densities) and they improve the
agreement with experiment. In the case of the noble metals, the
corrections were made in ref. 3 neglecting the ionic core polarizabilities. In fact these polarizabilities are sufficiently large
in the noble metals [5] to negate the corrections, so that the
"simple model" results may be a fair estimate. Clearly we must
admit some uncertainty in the theoretical magnitudes. It is
particularly interesting to note that the two different measurements yield fairly similar relaxation rates.

III.   RESISTIVITIES AT LOW TEMPERATURE

At low temperatures the electron-phonon contributions W_{ee} and particularly ρ_{ee} have complicated temperature-dependences; pure power laws are expected only at the very lowest temperatures $T < \Theta_G \approx \Theta_D V_G/E_F$. The quantity V_G is a Fourier component of the pseudopotential which characterizes the scale of Fermi surface distortion $\Delta k_G \approx k_F V_G/E_F$ near its intersection with the zone boundary G. This condition on the temperature is much more stringent than the naive one $T \ll \Theta_D$, but it may still be satisfied at the very low temperatures where electron-electron scattering becomes competitive. Therefore we shall use the asymptotic low temperature forms to estimate the electron-phonon contributions. Moreover, since even these expressions are too involved to reproduce here [7], we give only the qualitative dependences and simply tabulate the appropriate numbers later.

The electrical resistivity is proportional to T^5, and inversely proportional to the Fourier component (or band gap) V_G. Written as an effective relaxational rate,

$$\frac{ne^2}{m} \, \rho_{e\phi} \sim \frac{k_B T^5 E_F}{\hbar \Theta_T^4 V_G} \quad . \tag{12}$$

The "transverse" Debye temperature is used, since these modes overwhelmingly dominate the longitudinal ones. The thermal resistivity is proportional to T^2; the effective relaxation rate is

$$\frac{nk_B^2 T}{m} \, W_{e\phi} \sim \frac{k_B T^3}{\hbar \Theta_L^2} \left[1 + \frac{V_G \Theta_L^2}{E_F \Theta_T^2} \right] \quad . \tag{13}$$

Here, the contributions from longitudinal and transverse modes are comparable. We may now estimate the temperature T_ρ below which the electron-electron contribution (1) dominates the electron-phonon one (12) in the electrical resistivity:

$$T_\rho \sim \Theta_T \cdot (k_B \Theta_T V_G/E_F^2)^{\frac{1}{3}} \quad . \tag{14}$$

The thermal resistivity crossover temperature may similarly be deduced from (5) and (13):

$$T_W \sim \text{smaller of } k_B \Theta_L^2/E_F \, , \quad k_B \Theta_T^2/V_G \quad . \tag{15}$$

The estimated crossover temperatures listed on Table III show that the electrical resistivity is preferred over the thermal resistivity, for observing the electron-electron effect. Suggestive (though not conclusive) evidence has been seen in potassium [8], copper and silver [9]. The noble metal data is not inconsistent

Table III. Crossover temperatures for electrical and thermal
resistivities (in degrees K)

	Al	In	Cu	Ag	Au	K
T_ρ	2.2	0.16	3.6	2.4	2.2	1.3
T_W	0.13	.007	0.47	0.57	0.42	0.43

with that from other sources shown in Table II. In the case of
aluminum, Van Harlingen and Garland [10] have measured both ρ and
WT from 1.5 to 6.5 K. The find quadratic temperature-
dependences in both, with a Lorenz number of about 0.15, close to
the predicted value for electron-electron scattering. However the
magnitudes of ρ and W   are much too large, and the corresponding
effect is absent in the high-temperature measurements [11].

REFERENCES

*This work was supported by the U.S. Department of Energy through
Grant No. EY-76-S-02-2315.

1. W.E. Lawrence and J.W. Wilkins, Phys. Rev. B7, 2317(1973), and
W.E. Lawrence, Phys. Rev. B13, 5316(1976).
2. τ_{ee}^{-1} denotes the average over energies near the Fermi level
$\int d\varepsilon(-\partial f^\circ/\partial\varepsilon)\tau_{ee}^{-1}(\varepsilon)$. This microscopic rate is nearly isotropic
in the simple metals (see ref. 1).
3. C.A. Kukkonen, Ph.D. thesis, Cornell University, Ithaca, New
York (1975).
4. M.J. Laubitz, Phys. Rev. B2, 2252(1970); J.G. Cook, M.P. Van
der Meer, and M.J. Laubitz, Can. J. Phys. 50, 1386(1972); and
M.J. Laubitz and J.G. Cook, Phys. Rev. B6, 2082(1972).
5. R.T. Beach and R.W. Christy, Phys. Rev. B16, 5277(1977); and
G.R. Parkins, Ph.D. thesis, Dartmouth College, Hanover, New
Hampshire (1978).
6. R.N. Gurzhi, Soviet Phys. JETP 35, 673(1959), and Ph.D. thesis,
Physics-Techn. Inst. Acad. Sci. Ukrain. S.S.R. (1958), establishes
the relationship between the frequency-dependent and d.c. scattering rates.
7. For exact expression see second paper of ref. 1.
8. H. Van Kempe et al., Phys. Rev. Lett. 37, 1574(1976).
9. E.R. Rumbo, J. Phys. F5, 1(1975), and J. Phys. F3, L9(1973).
10. D. Van Harlingen and J. Garland, to be published in J. Phys. F.
11. M.J. Laubitz and J.G. Cook, Phys. Rev. B7, 2867(1973).

DEVIATIONS FROM MATTHIESSEN'S RULE AND THE ELECTRONIC THERMAL CONDUCTIVITY OF ALLOYS

P. G. Klemens

Dept. of Physics and Institute of Materials Science

University of Connecticut, Storrs, Connecticut 06268

ABSTRACT

In order to separate the electronic and lattice components of the thermal conductivity of alloys, one calculates the electronic component from the measured electrical resistivity. The residual electrical and thermal resistivities are connected by the Wiedemann-Franz law. The ideal thermal resistivity can be obtained from measurements of the parent metal in the impure limit, or from measurements in the pure limit and the use of the theoretical deviations from the thermal Matthiessen's rule for "vertical" motion. The additional deviations from the thermal Matthiessen's rule are related to the electrical deviations by the Wiedemann-Franz law provided the effects of alloying on the band structure are small, and provided the electron relaxation time for electrical conduction varies over the Fermi surface primarily because of the topology of the Fermi surface rather than owing to a variation of the electron-phonon interaction probability. Phonon assisted impurity scattering processes also lead to thermal and electrical deviations related by the Wiedemann-Franz law.

INTRODUCTION

The thermal conductivity of alloys is additively composed of the electronic and lattice components. To study the thermal conductivity, and to interpolate or extrapolate it as function of temperature and composition, one must know each component separately. There is no certain way of achieving this separation. One can calculate the electronic component from measured values of the electrical resistivity, then deduce the lattice component K_g by subtracting the calculated electronic component K_e from the measured total thermal conductivity, and the only check one has on this procedure is whether K_g thus obtained varies with temperature and composition in a manner consistent with the theory of lattice thermal conduction.

The electrical resistivity ρ of alloys usually shows strong deviations from Matthiessen's rule, i.e.

$$\rho = \rho_0 + \rho_i(T) + \Delta\rho \tag{1}$$

where ρ_0 is the residual resistivity of the alloy, $\rho_i(T)$ is the ideal resistivity of the parent metal, and $\Delta\rho$ is the deviation from Matthiessen's rule, and is a function of T, of ρ_0 and of the nature of the solute. The electronic thermal resistivity, $1/K_e$, shows similar deviations from an additive resistance rule, i.e.

$$1/K_e = W = W_0 + W_i + \Delta W \tag{2}$$

Here W_0 is the thermal resistivity due to the scattering of electrons by defects and impurities, and is related to ρ_0 by

$$W_0 = \rho_0/L_0 T \tag{3}$$

where L_0 is the Sommerfeld value of the Lorenz ratio. The second term in (2), W_i, is the ideal thermal resistivity and arises from the electron-phonon interaction, while ΔW is the deviation from additivity, analogous to $\Delta\rho$ in (1).

Equation (3) is a consequence of the fact that scattering of electrons by solutes is elastic, so that the same relaxation time enters ρ_0 and W_0.

The ideal thermal resistivity departs from the Wiedemann-Franz law, that is

$$W_i > \rho_i/L_0 T \tag{4}$$

except in the limit of highest temperatures, where equality is approached in (4). One knows W_i because it can be measured in the parent metal. Also, one can measure $\Delta\rho$ as function of T for the alloy under consideration. The real question is how ΔW may be related to $\Delta\rho$.

ELECTRON SCATTERING

To discuss the effectiveness of electron scattering, one can divide the motion of an electron in momentum space into two kinds: horizontal and vertical.[1] Horizontal motion changes the direction of the electron, but not its energy; vertical motion changes the energy but not the direction. Elastic scattering is horizontal. Electron-phonon scattering is both horizontal and vertical. At low temperatures the vertical component is larger than the horizontal one; at very high temperatures the reverse holds. Electrical resistivity is due only to the horizontal motion, while both horizontal and vertical motions contribute to the thermal resistivity. Thus

$$\rho_i = \rho_H \tag{5}$$

and

$$W_i = W_H + W_V \tag{6}$$

where

$$W_H = \rho_H/L_oT \quad . \tag{7}$$

The presence of the term W_V in (6) leads to deviations from the Wiedemann-Franz law, and at low temperatures, where W_V exceeds W_H, these deviations are very pronounced.[1]

DEVIATIONS FROM MATTHIESSEN'S RULE

The deviations $\Delta\rho$ in equation (1) arise from three types of causes[2]:-

(a) changes in the electronic band structure, in the phonon spectrum and in the electron-phonon interaction produced by alloying.

(b) additional scattering processes involving both defects and phonons (phonon-assisted impurity scattering).[2]

(c) two band effects and their generalization (i.e. τ_i/τ_o varies over the Fermi surface, where τ_i and τ_o are the electron relaxation times as limited by electron-phonon and by defect scattering respectively).

(d) In the case of thermal resistivity (equation 2) there is another deviation, similar to (c), because τ_i varies rapidly with electron energy near the Fermi surface while τ_o does not.[1] The corresponding effect in the electrical resistivity is very small.[3] For the thermal resistivity at low temperatures, i.e. for vertical motion,

$$W_V = 0.5 \ \alpha(W_o/W_i)W_V \tag{8}$$

where $\alpha(W_o/W_i) = 0$ as $W_o/W_i = 0$, increases with increasing W_o/W_i and reaches a saturation value $\alpha = 1$ in the limit of large residual resistivities.[1] In most alloys used in the determination of lattice thermal conductivities one can take $\alpha = 1$, while the determination of $W_i = W_V$ at low temperatures is usually done on specimens of such purity that $\alpha \ll 1$.

We are unable to give a good account of deviations due to (a), because we do not know enough about the basic properties of the parent metal and its alloys. Therefore we cannot relate $\Delta\rho$ and ΔW due to this cause. We have reason to believe, however, that these deviations are small relative to the others, unless we consider alloying concentrations which change the topology of the Fermi surface. We shall disregard these sources of deviation.

Additional processes, such as phonon-assisted impurity scattering, tend to randomize the electron direction, so that

$$W_{(b)} = \quad \rho_{(b)}/L_o T \tag{9}$$

Deviations of type (c) are more difficult to treat. However, in many cases the variation of τ_i/τ_o of the Fermi surface arises not because the intrinsic scattering probability varies, but because the emphasis on small-angle scattering at low temperatures and the shape of the Fermi surface make τ_i depend on the location of an element of the Fermi surface relative to the zone boundaries. This implies that deviations of type (c) are important for horizontal motion but not for vertical motion, so that

$$\Delta W_{(c)} = \Delta W_H = \quad \Delta\rho_{(c)}/L_o T \tag{10}$$

Thus, neglecting $\Delta W_{(a)}$ and the effects of deviations of type (c) (two-band effects) on vertical motion, we obtain for ΔW

$$\Delta W = \Delta W_{(b)} + \ \Delta W_{(c)} + \ \Delta W_V$$
$$= \Delta\rho/L_o T + \ \Delta W_V \tag{11}$$

Here $\Delta\rho$ is simply the deviation obtained by measuring the electrical resistivity on the same alloy, while $\Delta W_V = 0.5 \ W_i$ at low temperatures where $W_i \ \alpha \ T^2$, and decreases with increasing temperatures according to the calculations of Sondheimer.[3]

Equation (11) was suggested some time ago[4] and was then justified on the basis of mechanism (b), together with the idea that changes of type (a) would affect horizontal motion more sensitively than vertical motion. In cases where the variation of the scattering probability over the Fermi surface is not too different for phonon and for defect scattering (as distinct from the topology-dependent relaxation time), this can also be extended to the two-band effect, which is known to be a major contributor to $\Delta\rho$.[5] Only experience will tell us whether lattice thermal conductivities derived from (11) will be consistent in concentration and temperature dependence, and thus whether the assumptions underlying (11) are justified.

REFERENCES

1. P. G. Klemens in "Handbuch der Physik", ed. by S. Fluegge, vol. 14, p. 198, Springer, Berlin, 1956.
2. D. H. Damon, M. P. Mathur and P. G. Klemens, Phys. Rev. 176, 876 (1968).
3. E. H. Sondheimer, Proc. Roy. Soc. (London) A203, 75 (1950).
4. P. G. Klemens, Austral. J. Phys. 12, 199 (1959).
5. J. Bass, Adv. in Physics 21, 431 (1972).

SESSION F   GASES AND LIQUIDS

Session Chairman: R. Taylor
                      University of Manchester
                      Manchester, England

THERMAL CONDUCTIVITY OF DISSOCIATING HYDROGEN-ARGON MIXTURES AT HIGH TEMPERATURES

J. Maštovský

Sen. Research Scientist, Institute of Thermomechanics

Czechosl. Acad. of Sci., Puškinovo 9, 16000 Prague 6

INTRODUCTION

Thermal conductivity of many simple gases and gas mixtures which is needed in high temperature technology can be calculated in wide temperature range with sufficient accuracy on the basis of the rigorous kinetic theory. The reliability of this theory was in many cases experimentally verified. In special cases, particularly for gases, the structure of which changes in the considered temperature range due to chemical reactions (dissociation), additional information is needed to check the applicability of theoretical considerations.

In continuation of the research into thermal conductivity of high temperature gases, the binary mixture with extremely different molecular masses and concentrations of the species was investigated in order to get new experimental data for this mixture, to compare them with the results of theoretical computations and so to check both common calculations and the shock tube method under extreme conditions.

EXPERIMENTS

The Method

For the measurement the method following the well-known idea [1], based on the study of heat transfer in the thermal boundary layer between hot gas and the end plate of a shock tube

after the shock wave reflection was used. For the evaluation of
thermal conductivity can be used either the optically measured
temperature across the thermal boundary layer or the surface
temperature increase of the end plate measured by a thin film
thermometer.

For the latter case, in [2], an explicit, quantitative and
accurate evaluation of thermal conductivity of slightly dissoci-
ated (chemically reacting) gases was achieved using a complex
expression for

$$\rho C_p = \frac{p}{RT}\left[R + \sum_i x_i \frac{\partial U_i}{\partial T} + \sum_i U_i \left(\frac{\partial x_i}{\partial T}\right)_p - \frac{RT + \sum_i x_i U_i}{\sum_i x_i M_i} \sum_i M_i \left(\frac{\partial x_i}{\partial T}\right)_p \right] \quad (1)$$

in the energy equation. Here R is the gas constant, U is the internal energy, M the molecular mass and x the mole fraction of
the component. Total thermal conductivity λ /W.m^{-1}.K^{-1}/ related
to the pressure p /Pa/ follows from the relation

$$\lambda = \frac{k_w}{\rho\, C_p}\left[d\left(\frac{\Delta E\sqrt{p}}{E\sqrt{p_5}}\right)/dT_5 \right]^2 , \quad (2)$$

where ΔE /V/ corresponds to the measured surface temperature increase of the end plate of a shock tube after the shock wave reflection, E /V/ is the voltage of the measuring bridge, p_5 /Pa/
and T_5 /K/ are the equilibrium gas pressure and temperature behind the reflected shock, ρ /mol.m^{-3}/ and C_p /J.mol^{-1}.K^{-1}/ are
the molar density and heat capacity of the equilibrium gas mix-
ture at temperature T_5. The material constant of the measuring
thin film thermometer is $k_w = k_w(\lambda_w, \rho_w, c_w, \alpha)$ /J^2.s^{-1}.m^{-4}/,
where λ_w, ρ_w, c_w /SI units/ are thermal conductivity, density
and heat capacity of the device and α /K^{-1}/ is its temperature
coefficient of resistance.

Further elaboration of this procedure and of the experimental technique along with better knowledge of the influence of
subtle fast transient effects (thermal diffusion, thermal accommodation, thermal inertia of the gauge) on the measurement of
thermal conductivity [3,4] enables us to measure this property
even in gas mixtures containing strongly dissociating (chemically reacting) component with reasonable relaxation time.

The Equipment and Diagnostics

The experiments were carried out in a conventional hypersonic shock tube [5] modified into a single diaphragm constant a-
rea facility (Fig.1). The length of the driver (1) and the nickel
plated test section (2) was 1 and approx. 6 m respectively, the
I.D. was 0.05 m. Hydrogen and hydrogen-argon mixtures were used

as driver gases (12). The scribed aluminum diaphragms burst at
0.7-6 MPa. The shock tube was evacuated to 1.07 Pa by a two stage
oil pump (11) and filled with the binary test gas mixture (10) to
approx. 0.4-6 kPa. Hydrogen was 99.9 % and argon 99.98 % pure.

The initial static pressures were measured by calibrated
Bourdon-type and Wallace-Tiernan pressure gauges (13,14). The
primary shock wave velocity was measured (accuracy 0.5 %) digitally (4,5) by means of thin-film gauges (3) located along the test
section. The pressure jump at the end plate after the reflection
of the shock wave was measured (accuracy better than 5%) by means
of a fast quartz pressure gauge (7) (Kistler 603B) and a calibrated piezo-amplifier (8) (Vibrometer TA-3/C). The surface temper.
increase of the end plate after the shock wave reflection was measured (accur. 2-5 %) by a precision thin film thermometer [6]
and a special bridge circuit (6). Pressure and temperature vs.
time histories were displayed on an oscilloscope (9).

Both measured initial and shocked gas parameters, the Mach
number of the primary shock wave and thermodynamic tables [7] were entries for the iterative computation [8] of the equilibrium
composition and thermodynamic and gas dynamics parameters and
properties of the reacting mixture behind the primary and reflected shock waves.

Results

Thermal conductivity values were measured for three initial
concentrations, i.e. 0.1 H_2 - 0.9 Ar, 0.5 H_2 - 0.5 Ar and 0.75 H_2
- 0.25 Ar in the temperature ranges approx. 1380-5000 K, 1800-
4000 K and 1100-3100 K resp. and at nearly atmospheric pressure.
The plot of measured $\Delta E\overline{Vp}/(E\overline{Vp_5})$ vs. temperature T_5 values e.g.
for the concentration 0.5 H_2 - 0.5 Ar is shown in Fig. 2 (o = ex-
perimental points; solid curve = exper. points processed by the
least squares method). Experimental thermal conductivities resulting from similar plots and evaluated from eq.(2) are given for
all three concentrations in Fig. 3. The constants k_w have been
determined so that the measured λ values fit at the beginning of

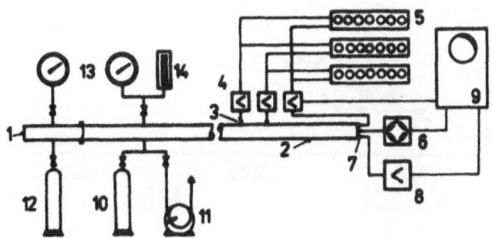

Fig. 1 Experimental arrangement (numbers explained in the text)

the relevant temperature range (prior to dissociation) our reference values of λ computed from the rigorous kinetic theory.The
values of this constant for the individual concentrations (i.e.
sets of experimental points and used thermometer gauges) are:
For 0.1 H_2-0.9 Ar k_w = 4.8.10^{13}, for 0.5 H_2-0.5 Ar k_w = 10.9.10^{13}
and for 0.75 H_2-0.25 Ar k_w = 3.88.10^{13} /$J^2.s^{-1}.m^{-4}$/.

The estimated error of experimental λ values is about $\pm$30 %
as shown in Fig. 3. The accuracy seems rather poor in comparison
with values measured by steady state methods and at common temperatures; we find it quite reasonable, however, considering the
unusual temperature range and gas behaviour and the difficulties
inherent in the experimental method and measuring technique.

THEORETICAL CALCULATIONS

The frozen and equilibrium thermal conductivities were calculated for 5 initial concentrations (0.1 H_2-0.9 Ar, 0.25 H_2-0.75
Ar, 0.5 H_2-0.5 Ar, 0.75 H_2-0.25 Ar and 0.9 H_2-0.1 Ar) in the temperature range 300-5000 K and at pressures 0.1, 0.2 and 0.3 MPa,
using chemical thermodynamics and the rigorous kinetic theory.
The parameters of the Lennard-Jones (12-6) potential function were: σ_{H_2} = 2.827 Å, ε_{H_2}/k = 59.7 K, σ_H = 2.708 Å, ε_H/k = 37.0 K,
σ_{Ar} = 3.408 Å, ε_{Ar}/k = 119.9 K. The results are shown in Fig. 4.
The comparison of these frozen λ_F values with those [9] for the
L.-J. parameters from [10] and for other potential functions (Buckingham exp-6 with parameters [11] and Morse with parameters

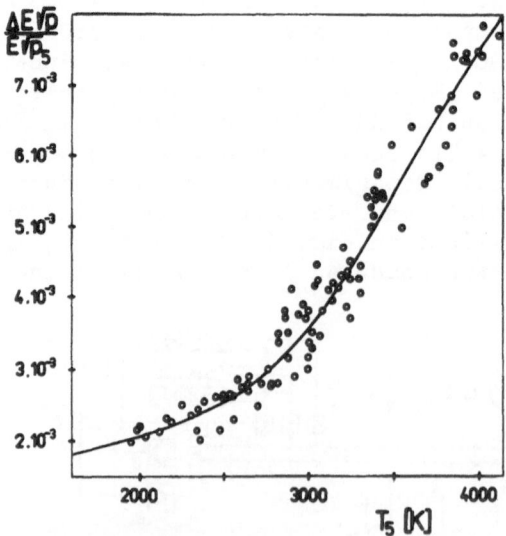

Fig. 2 Measured $\Delta E\overline{Vp}/(E\overline{Vp_5})$ values vs. temperature T_5 for the
        0.5 H_2 - 0.5 Ar mixture

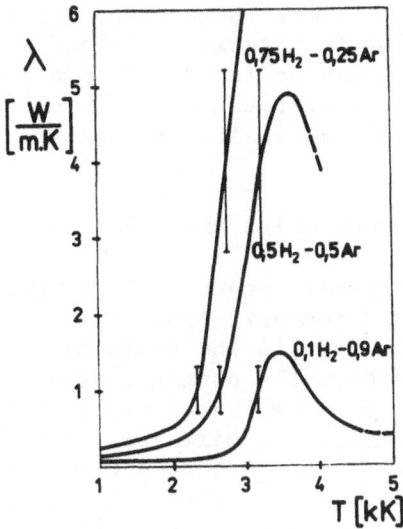

Fig. 3 Experimental thermal conductivity for three H_2 – Ar concentrations at atmospheric pressure

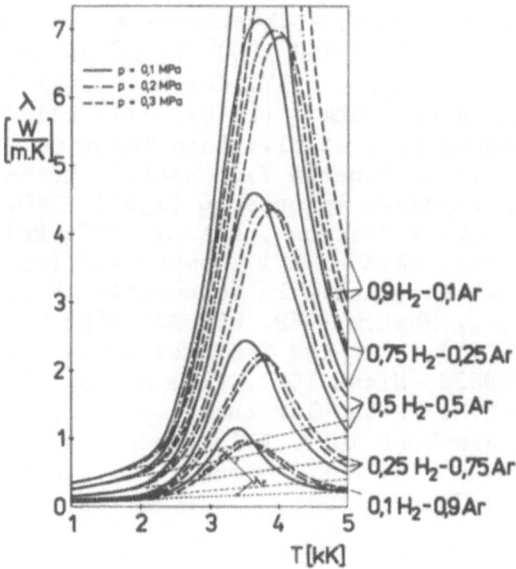

Fig. 4 Calculated thermal conductivity vs. temperature for five H_2 – Ar concentrations and various pressures

[12]) shows that our values vs. temperature are slightly steeper,
at lower temperatures (up to approx. 1500-2500 K) being a few
percent lower and just crossing the cited data in this important
reference range. For the comparison with our equilibrium calculations no data are available in the literature.

CONCLUDING REMARKS

High temperature thermal conductivity of gas mixtures with
very different masses and concentrations of the constituents was
measured in a shock tube even in the range of strong dissociation with reasonable accuracy. The obtained experimental values agree within experimental error with our rigorous kinetic theory
calculations for the equilibrium mixtures. Somewhat greater discrepancy was found only for the 0.1 H_2 - 0.9 Ar mixture between
approx. 2500-3000 K. However, for the present, the attainable experimental accuracy does not make it possible to distinguish the
suitability of the individual potential functions and parameters
for the description of the high temperature thermal conductivity
of this mixture.

REFERENCES

[1] Smiley E.F.: The measurement of the thermal conductivity of
gases at high temperature with a shocktube; experimental results
in argon at temperatures 1000 and 3000° K. Ph.D. Thesis, Catho-
lic Univ. of Amer., Washington 1957. [2] Maštovský J., Slepička
F.: Wärme- und Stoffübertragung 3 (1970), 237. [3] Shashkov A.G.,
Maštovský J., Abramenko T.N. et al.: Acta technica ČSAV 20 (1975)
564. [4] Maštovský J.: Inzhenerno-fizicheskii zhurnal (in press).
[5] Maštovský J.: Strojnícky časopis 16 (1965), 148. [6] Kmoníček V., Korejs B. et al.: Czechosl. Pat. 123307; Brit. Pat.
1,085506; Schweiz. Pat. 447652. [7] Glushko V.P.(ed.): Termodi-
namicheskie svoistva individualnikh veshchestv, Vol. 2 (Tables),
Publ. USSR Acad. Sci., Moskva 1962. [8] Kmoníček V. et al.: Acta
technica ČSAV 17 (1972), 542. [9] Shashkov A.G., Abramenko T.N.:
Int. Rept. ITMO AN BSSR, Minsk 1972. [10] Hirschfelder J.O.,
Curtiss C.F., Bird R.B.: Molecular theory of gases and liquids;
4th print., Wiley, New York 1967. [11] Mason E.A.: J. Chem. Phys.
23 (1955), 49. [12] Bahethi O.P., Gambhir R.S., Saxena S.C.: Z.
Physik 19a (1963), 1479.

ON THE PRINCIPLE OF CORRESPONDING STATES FOR TRANSPORT
PROPERTIES OF SIMPLE DENSE FLUIDS

B. Le Neindre, R. Tufeu, Y. Garrabos & B. Vodar

LIMHP - CNRS

Université de Paris Nord, Villetaneuse 93430 France

I - INTRODUCTION

It was observed from experiments that the excess thermal conductivity :

$$\tilde{\lambda} = \lambda\,(\rho,T) - \lambda\,(0,T) = \lambda - \lambda_o \tag{1}$$

and the excess viscosity :

$$\tilde{\mu} = \mu\,(\rho,T) - \mu\,(0,T) = \mu - \mu_o \tag{2}$$

of noble gases were temperature independent in a restricted temperature range. By application of the principle of corresponding states to noble gases according to the relation :

$$\tilde{\lambda}^* = \frac{\sigma^2\,m^{1/2}}{\varepsilon^{1/2}\,k}\,\tilde{\lambda} \tag{3}$$

$$\tilde{\mu}^* = \frac{\sigma^2}{(m\varepsilon)^{1/2}}\,\tilde{\mu} \tag{4}$$

$$T^* = \frac{k}{\varepsilon}\,T \tag{5}$$

$$\rho^* = \frac{\sigma^3}{m}\,\rho \tag{6}$$

where σ and ε are the parameters of the Lennard-Jones potential, T the absolute temperature and ρ the density, we consider to what extend $\tilde{\mu}^*$ and $\tilde{\lambda}^*$ can be considered as temperature independent. As we will show we get one curve in reduced coordinates for each transport property which can be compared to the Enskog theory. The excess viscosity is found to be temperature independent. The excess thermal conductivity is slightly temperature dependent in $T^{*1/6}$.

General formulae are proposed to correlate both thermal conductivity and shear viscosity coefficients of noble gases.

These formulae expressed in terms of parameters of the Lennard-Jones potential are used to predict the transport properties of real gases in the temperature range above $1.3\ T_c$, where T_c is the critical temperature. The calculated data are compared to experimental data.

II - THERMAL CONDUCTIVITY MEASUREMENTS

Most of the data that we use in the analysis of the excess reduced thermal conductivity coefficient were measured by ourselves. We have determined the thermal conductivity coefficients of He(1)(2) up to $\rho^* = 0.300$, of Ne(3) up to $\rho^* = 0.300$, of Ar(1)(2) up to $\rho^* = 0.550$, of Kr(4) up to $\rho^* = 0.700$, of Xe(5) up to $\rho^* = 0.840$, and at different temperatures ranging from room temperature up to 500°C. Other data which can be useful for the analysis are those of Sengers et al. These authors reported data on the thermal conductivity coefficients of argon(6) up to $\rho^* = 0.720$ and neon(7) up to $\rho^* = 0.560$ in the temperature range from 0°C to 75°C.

In Figure 1 is represented the reduced excess thermal conductivity in terms of reduced density. The values for helium at high density have been extrapolated assuming that the pressure effect on λ is linear as that was observed below 100 MPa, of course such an approximation is questionable. Figure 1 shows that with the reduction that we have used the excess thermal conductivity is slightly temperature dependent. If we divide the reduced excess thermal conductivity $\tilde{\lambda}^*$ by $T^{*1/6}$ we get one curve to a first approximation as is shown in Figure 2.

III - VISCOSITY MEASUREMENTS

Most of the data that we use for the analysis of the excess reduced viscosity for noble gases were measured at room temperature by Vermesse and Vidal. They have reported the viscosity of helium(8), neon(9), and argon(10) up to reduced densities of 0.793, 0.804, 0.951 respectively. The viscosity of krypton was determined by Trappeniers et al.(11) along three isotherms up to a reduced density of 0.806. The viscosity of xenon measured by Reynes and Thodos(12) up to a reduced density of 0.788 which seems less accurate was not used in the analysis.

In Figure 3 the reduced excess of viscosity is represented in

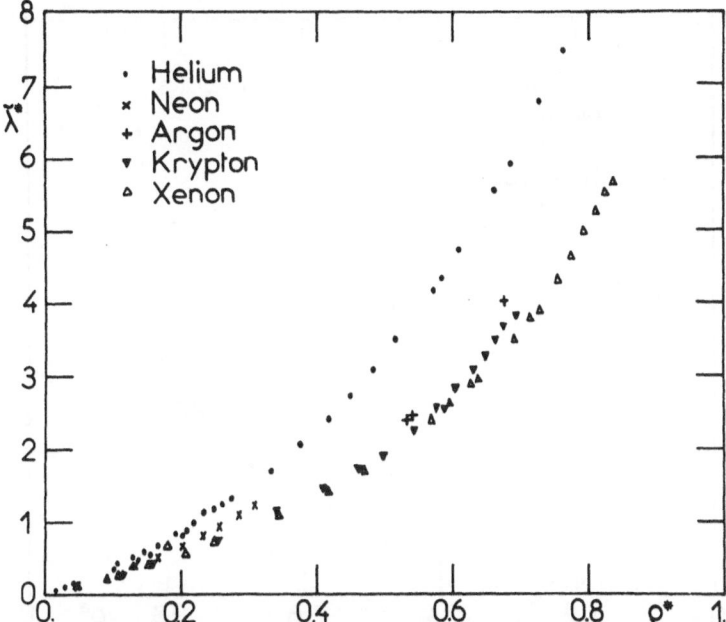

Fig. 1 Reduced excess thermal conductivity of noble gases at
25°C in terms of the reduced density.

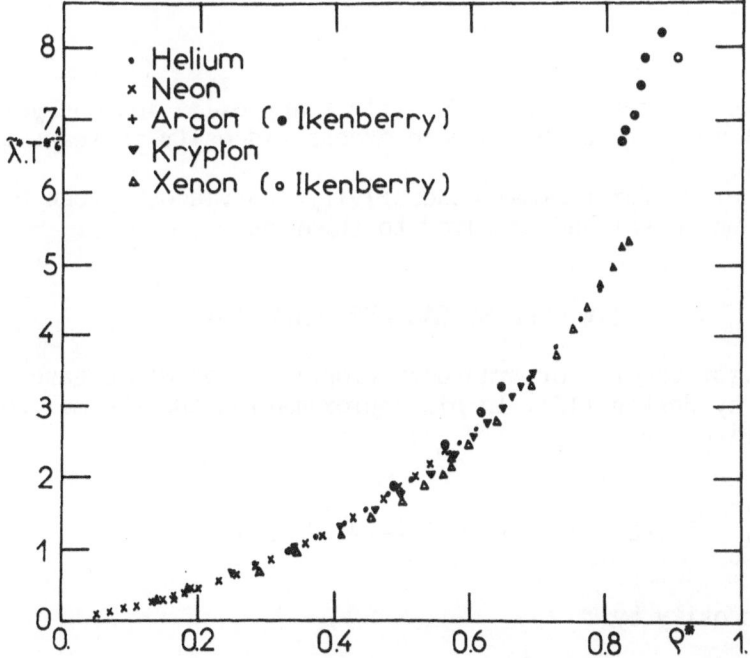

Fig. 2 Variation of $\lambda^* T^{-1/6}$ of noble gases at 25°C in terms
of the reduced density.

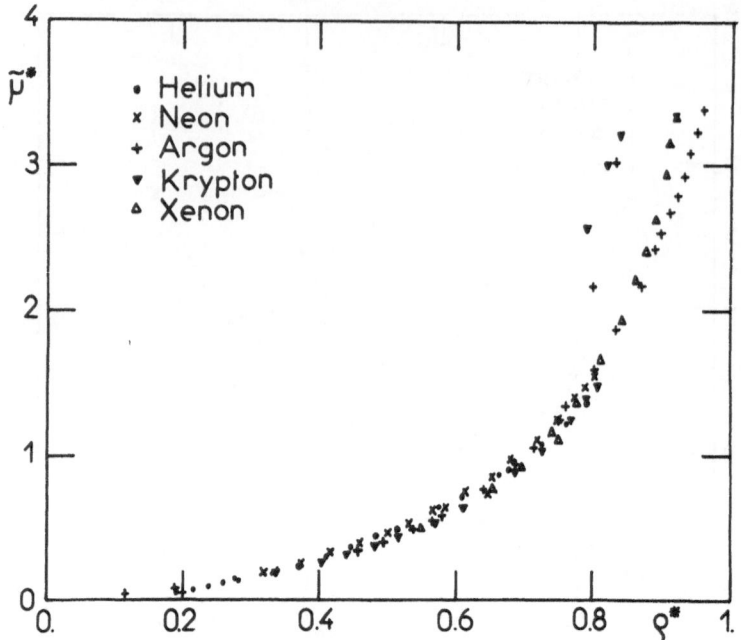

Fig. 3 Reduced excess viscosity of noble gases at 25°C in terms of the reduced density.

terms of the reduced density. To a first approximation one gets one curve. That means that the reduced excess viscosity is temperature independent.

Both curves for thermal conductivity and viscosity can be considered as universal and compared to theories.

IV - DENSE GAS APPROXIMATION

The first theory for transport properties of dense gases was formulated by Enskog.(13). In his approximation the thermal conductivity is given by :

$$\lambda^* = \frac{\lambda^E}{\lambda_0} = 1/g\ (\sigma) + 4.8\ n + 12.08\ n^2\ g\ (\sigma) \qquad (7)$$

and the viscosity by :

$$\mu^* = \frac{\mu^E}{\mu_0} = 1/g\ (\sigma) + 3.2\ n + 12.18\ n^2\ g\ (\sigma) \qquad (8)$$

where : $n = \pi N \sigma^3 \rho / 6$ for a spherical molecule of diameter σ.
g (σ) is the radial distribution at contact which is satisfactory approximated by the relation of Starling and Carnahan.

$$g \ (\sigma) = (1 - 0.5 \ n) \ (1 - n)^{-3} \tag{9}$$

λ_0 is the dilute gas approximation for the thermal conductivity and μ_0 is the dilute gas approximation for the viscosity.
One gets by limiting the expansion of equations (7) and (8) to the fourth power :

$$\frac{\lambda^E - \lambda_0}{\lambda_0} = 1.2043 \ \rho^* + 3.7915 \ \rho^{*2} + 4.3171 \ \rho^{*3} + 4.0904 \ \rho^{*4} \tag{10}$$

and

$$\frac{\mu^E - \mu_0}{\mu_0} = 0.3665 \ \rho^* + 3.8196 \ \rho^{*2} + 4.3539 \ \rho^{*3} + 4.1251 \ \rho^{*4} \tag{11}$$

V - COMPARISON WITH EXPERIMENTS

We have fitted the right terms of equations 10 and 11 to the universal curves of thermal conductivity and viscosity. The best fits lead to the following equations :

$$\lambda_c = \lambda_0 + 0.860 \ \frac{\varepsilon^{1/2}}{\sigma^2} \frac{kT^{*1/6}}{m^{1/2}} \ (1.2043 \ \rho^* + 3.7915 \ \rho^{*2} + 4.3171 \ \rho^{*3}) \tag{12}$$

and

$$\mu_c = \mu_0 + 0.222 \ (\frac{m \ \varepsilon}{\sigma^2})^{1/2} \ (0.3665 \ \rho^* + 3.8196 \ \rho^{*2} + 4.3539 \ \rho^{*3}$$
$$+ 4.1251 \ \rho^{*4}) \tag{13}$$

Equations (12) and (13) have been used to calculate the thermal conductivity and viscosity of molecular fluids. The parameters of the Lennard-Jones potential which are used in the equations (12) and (13) are taken from (14).
We have compared the calculated values of the thermal conductivity and viscosity coefficients with some experimental values from the literature and we report the maximum deviations. The agreement is generally better than 10% except for the viscosities of xenon(12) and carbon dioxide. Our model is too simple to represent the correct behavior of the thermal conductivity of polar substances as NH_3 and

H_2O. But it shows that an additional type of heat conduction seems to be present in these substances due for instance to molecular association or heat conduction along H-bonds.

VII - CONCLUSION

Near the liquid-solid transition and for polar substances, extra temperature dependent terms must be added. The present equations can also be applied to mixtures.

REFERENCES

1. Le Neindre B., PhD. Thesis, University of Paris, 1969.
2. Tufeu R., PhD. Thesis, University of Paris, 1971.
3. Tufeu R., Le Neindre B. and Bury P., CR Acad. Sc., Paris, 271 (1970), 589.
4. Tufeu R., Le Neindre B. and Bury P. CR Acad. Sc. Paris 273 B (1971), 61.
5. Tufeu R., Le Neindre B. and Bury P., CR Acad. Sc., Paris 273 (1971), 113.
6. Michels A., Sengers J.V. and Van der Kleindert L.J.M., Physica 29 (1963), 149.
7. Sengers J.V., Bokl W.T. and Stigter C.J., Physica 30 (1964), 1018
8. Vermesse J. and Vidal D., CR Acad. Sc. Paris 282 B (1976), 5.
9. Vermesse J. and Vidal D., CR Acad. Sc. Paris 280 B (1975), 749.
10. Vermesse J. and Vidal D., CR Acad. Sc. Paris 277 B (1975), 191.
11. Trappeniers N.J., Botzen A., Van Oosten J. and Van den Berg H.R., Physica 31 (1965), 945.
12. Reynes E.G. and Thodos G., Physica 30 (1964), 1529.
13. Enskog D., Svensk Akad Handl 4 (1922) 63.
14. Hirschfelder J.O., Curtiss C.F., Bird R.B., Molecular Theory of Gases and Liquids, Wiley, New-York 2 ed.(1964)
15. Ikenberry L.D. and Rice S.A., J. Chem. Phys. 39 (1963), 1561.

THERMAL CONDUCTIVITY OF NEON, ARGON AND THEIR MIXTURES AT HIGH TEMPERATURES (400-1500 K)[*]

A.G. Shashkov, E.I. Marchenkov and V.I. Aleinikova

Luikov Heat and Mass Transfer Institute

15 Podlesnaya, Minsk, BSSR, USSR

and

R. Afshar, R.K. Joshi and S.C. Saxena

Department of Energy Engineering, University of Illinois

Chicago, Illinois 60680, USA

ABSTRACT

Experimental thermal conductivity values of neon, argon and their mixtures are reported in the temperature range of 400-1500 K. These values are compared with the theoretical predictions of kinetic theory based on Lennard-Jones (12-6), exp-six and (11-6-8) intermolecular potentials. Two versions of the Sutherland-Wassiljewa expression for the mixture conductivity are considered and one scheme is found preferable for purposes of correlation and extrapolation of data.

INTRODUCTION

Thermal conductivity of 99.882% pure neon and 99.987% pure argon and their mixtures has been measured in the temperature range of about 400-1500 K employing a molybdenum hot-wire cell. The highest temperature at which the thermal conductivity for this binary

* This work is supported in part by the United States National Science Foundation under grant No. ENG-02992 under a USA-USSR cooperative research program.

system has been measured earlier is 739.2 K [1]. The details of the experimental arrangement and data processing scheme are given by Marchenkov and Shashkov [2,3] and therefore only a brief reference to such details is given here. The experimental conductivity values have been compared with the Chapman-Enskog kinetic theory predictions based on Lennard-Jones (12-6), modified Buckingham (exp-six) and (11-6-8) intermolecular potentials. The combination rules for the latter potential giving the interaction between unlike molecules in terms of the interactions between like molecules are derived. The mixture conductivity values are also employed to assess the adequacy of the semi-empirical expressions of the Sutherland-Wassiljewa type.

EXPERIMENTAL

The schematic of the lower end of the hot-wire cell, potential lead type, is given in Figure 1. It consists of a vacuum smelted molybdenum type MChVP tube having an axial wire of MR-50 molybdenum alloy and held taut by tungsten springs. The three potential leads are of tungsten (VR-20), 0.05 mm in diameter with aluminum oxide insulation. The necessary constants of the cell are given in Table 1. The internal diameter of the cell tube is determined by filling it with a known amount of water. The wire and external diameter of the tube are measured with a microscope within an accuracy of ±0.001 mm. The length of the working sections of the wire is obtained with a comparator to an accuracy of ±0.001 mm. The degree of alignment and eccentricity of the wire are checked before and after the experiments by taking x-ray photographs in two mutually perpendicular directions.

The cell tube is heated by passing a regulated alternating current upto a maximum of 200 amperes. Coaxial cylindrical molybdenum screen and protected end heaters are independently energized from a regulated A.C. power supply and enable to get a uniform temperature in the central region of the wire. The wire is heated by a direct current from a regulated power supply and is measured on a potentiometer within an accuracy of ±0.01 mA. Tube temperature is measured by the hot-wire itself and the temperature distribution over the test section by a pyrometer and three platinum-rhodium thermocouples installed at the center and at the ends. The temperature of the hot-wire is kept about 6-14 K above the tube temperature.

Centering bushing and collar are made of ceramic materials. The design permits the expansion of the molybdenum tube due to heating at the top end only as the lower end is fixed.

The thermal conductivity, k, is calculated from the following relation:

$$k(\bar{T}) = \frac{\ln(d_2/d_1)}{2\pi \ell_{eff}} \cdot \frac{Q_c}{\Delta T_{gas}} \tag{1}$$

where

$$\bar{T} = T_{wall} + (\Delta T_{gas}/2) . \tag{2}$$

d_1 and d_2 are the outer and inner diameters of the wire and the tube respectively, ℓ_{eff} is the effective length of the wire, Q_c is conductive heat transfer, ΔT_{gas} is the true temperature difference across the gas layer, and T_{wall} is the temperature of the molybdenum tube. The temperature drop across the tube thickness is negligibly small. The energy fed to the wire, Q, is related to Q_c and to the energy radiated by the wire, Q_r, by

$$Q_c = Q - Q_r . \tag{3}$$

Further,

$$Q = I^2 R_{eff} - I_o^2 R^o_{eff} , \tag{4}$$

and

$$Q_r = \sigma_s \varepsilon \pi d_1 \ell_{eff} \left[T^4_{wire} - T^4_{wall} \right] . \tag{5}$$

Here σ_s is the Stefan-Boltzmann constant, I and I_o are the values of the heating and nonheating currents respectively when the corresponding resistances of the effective wire length are R_{eff} and R^o_{eff}. ε is the emissivity of the wire whose temperature is T_{wire}. The temperature difference between the wire and the wall, ΔT, is given by:

$$\Delta T = \frac{R_{eff} - R^o_{eff}}{R^{293}_{eff} \frac{d}{dT}\left(\dfrac{R^T_{eff}}{R^{293}_{eff}}\right)} . \tag{6}$$

R^T_{eff} and R^{293}_{eff} are the resistance of the effective length of the wire at temperatures T and 293 K, respectively. Finally,

$$\Delta T = (B/P) + \Delta T_{gas} . \tag{7}$$

P is the gas pressure and B is a quantity which is dependent on the wire material, test gas, cell geometry and thermal flux.

In these experiments, the contribution of thermal energy trans-

ported by radiation to the total heat flux increased with temperature and attained a maximum value (27%) in a mixture of 80% argon
at a temperature of 1474 K. As the proportion of the lighter component, neon, increased in the mixture, the radiated power decreased
and was only about 17% for pure neon. The magnitude of the tempera-
ture-jump correction is estimated on the basis of equation (7) and
procedure outlined by Marchenkov and Shashkov [2]. At 373 K, the
correction is always less than one percent for neon and argon-neon
mixtures. However, the correction increases with temperature and
is 13% for neon at 1503 K. As the concentration of the heavier component increased, the correction decreased and is always less than
3% for a mixture of 20% neon at 1474 K. The corrections due to temperature drop in the molybdenum tube wall, thermal diffusion, convection and eccentricity in the wire mounting are found to be negligibly small.

The experimentally determined values of thermal conductivities
of neon, argon and their mixtures as a function of temperature are
given in Table 2. The estimated uncertainty in these data is about
3% and varies slightly with temperature and composition of the mixture as shown in Figures 2 to 6. These data are correlated on the
basis of least square analysis with the following cubic polynomial
in temperature for each mixture to facilitate interpolation:

$$k(T) = C_1 + C_2 T + C_3 T^2 + C_4 T^3 . \tag{8}$$

The constants C_i are listed in Table 3.

DISCUSSION

The thermal conductivity of neon-argon mixtures has been measured
by Srivastava and Madan [4], at 273 K, Srivastava and Saxena [5] at
311 K, Thornton and Baker [6] at 291 K, von Ubisch [7] at 302 and 793
K, and Mathur, Tondon and Saxena [8] in the range 313-363 K employing
some variant of the hot-wire cell. It will be noted that the data
at high temperatures are non-existent and the values of von Ubisch
[7] are generally greater than our measured values, the disagreement
increasing with the neon concentration in the mixture. However, the
difference is not large and for the four concentrations the mean disagreement is only about 1.5% which lies within the limits of experimental uncertainties. A more detailed comparison of the present data
for pure gases is possible with the measurements of other workers.
The thermal conductivity of these gases have been measured over an
extended temperature range using the column method by Saxena and
Saxena [9], Springer and Wingeier [10], and Jody and Saxena [11] for
neon, and by Saxena and Saxena [12], Faubert and Springer [13],
Springer and Wingeier [10] and Chen and Saxena [14] for argon. The
neon values of Saxena and Saxena [9] and Jody and Saxena [11] agree

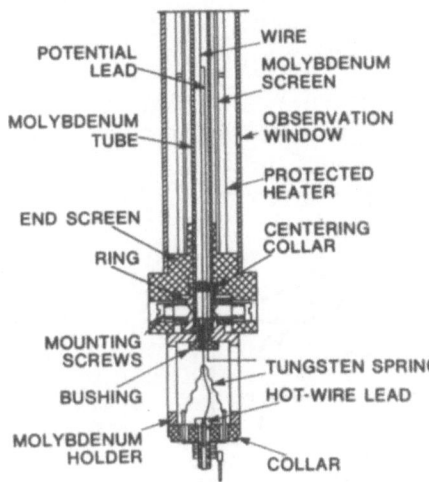

Figure 1. The schematic of the lower end of the molybdenum potential lead type hot-wire cell.

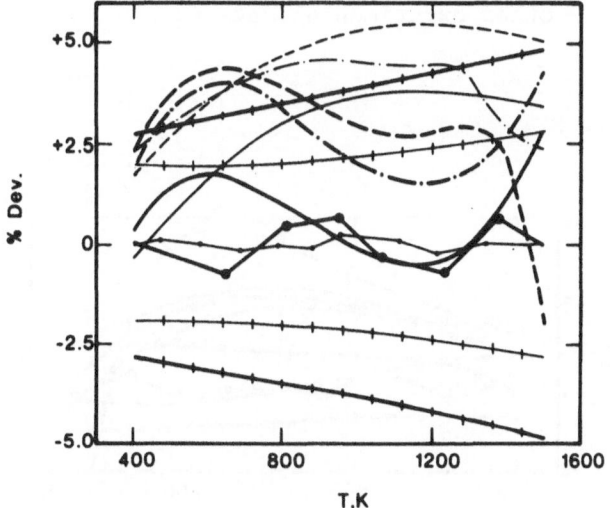

Figure 2. Deviation curves for the thermal conductivity of neon (thin) and argon (thick) from smoothed experimental data points.*

* Figures 2-6. ● experimental data points, ___ (11-6-8) potential, - - - L-J(12-6) potential, .—.—.—. exp-six potential, ▰▰▰ SWPI, ▲▲▲ SWPII, and ╫╫╫ experimental error limits.

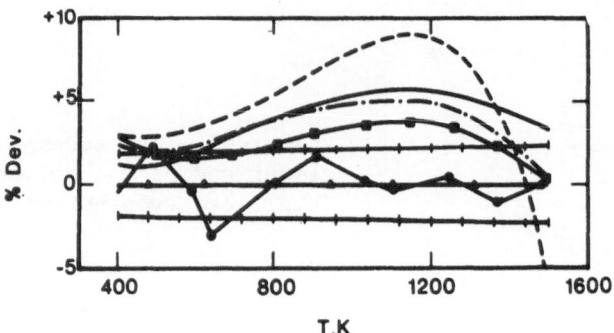

Figure 3. Deviation curves for
the thermal conductivity of neon-
argon mixture (20% Ar) from smoothed experimental data points.

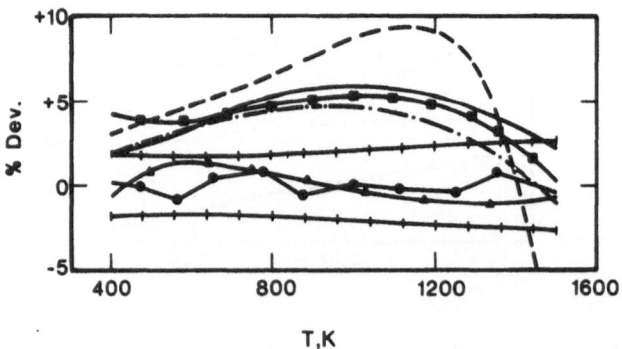

Figure 4. Deviation curves for
the thermal conductivity of neon-
argon mixture (40% Ar) from smoothed experimental data points.

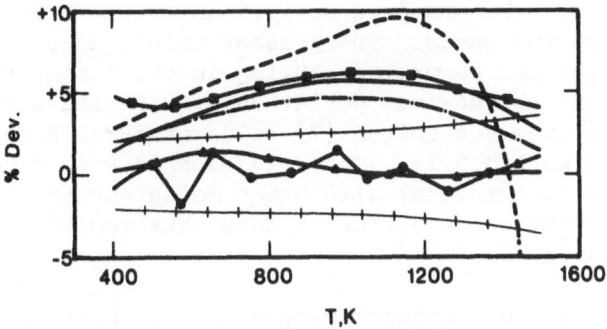

Figure 5. Deviation curves for the thermal conductivity of neon-argon mixture (60% Ar) from smoothed experimental data points.

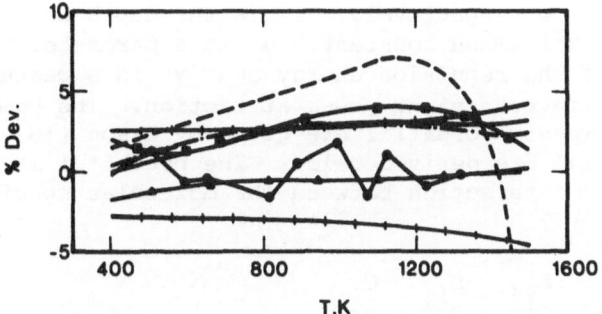

Figure 6. Deviation curves for the thermal conductivity of neon-argon mixture (80% Ar) from smoothed experimental data points.

with the present data in the temperature range of 400-1500 K within
the maximum deviation of 1.9% and 1.4% respectively. The corresponding standard deviations are 1.2% and 0.9%. The experimental data of
Springer and Wingeier [10] agree with the present data in the overlapping temperature range, 900-1500 K, within the maximum and standard deviations of 1.2% and 0.9% respectively. The argon data of
Saxena and Saxena [12] and Chen and Saxena [14] in the temperature
range 400-1500 K agree with the present results within the maximum
deviation of 2.3%. The standard deviations are 1.86 and 1.36% respectively for the two cases. The present values agree with the
values of Springer and coworkers [10,13] in the temperature range of
1000-1500 K within the maximum deviation of 2.8% and a standard deviation of 1.6%. The von Ubisch [7] value at 793.2 K agrees with
the present data within 3.1%. Thus, in each case we find that the
present data are in agreement with other measurements within the
estimated uncertainties of different data sets including more recent
work of Stefanov [15].

The present thermal conductivity data have been compared with
the predictions of Chapman-Enskog kinetic theory [16,17] in conjunction with Lennard-Jones (12-6), modified Buckingham exp-six, and
(11-6-8) intermolecular potentials. The parameters for these potentials and for these two gases are given in references [17-19] and
are reproduced in Table 4. The parameters for the unlike interaction are determined for the (12-6) potential by adopting the arith-
metic rule for σ and geometric mean rule for ε/k. σ and r_m are
the values of the molecular separations where the interaction energy
is zero and minimum respectively. ε is the depth of the potential
well and k is Boltzmann constant. α is a parameter which measures
the steepness of the repulsion energy and γ is a measure of the
strength of the inverse eight power attraction. The combination
rules for the exp-six potential are given by Mason [18] and for the
(11-6-8) potential are derived below. The potential may be written
to represent the interaction between the molecular species i and j
as [20]

$$\phi_{ij}(r_{ij}) = \frac{A_{ij}}{r_{ij}^{11}} - \frac{B_{ij}}{r_{ij}^{6}} - \frac{C_{ij}}{r_{ij}^{8}} . \tag{9}$$

Here ϕ_{ij} is the interaction energy at an intermolecular separation
distance r_{ij}, and A_{ij}, B_{ij} and C_{ij} are quantities representing
the strengths of the repulsive and attractive components of the potential energy. This potential has been found to be a good correlating
potential for the transport properties of these two gases by Nain et
al. [19]. The potential parameters ε_{ij}, $(r_m)_{ij}$ and γ_{ij} are
determined from the simultaneous solution of the following three
equations:

$$2\varepsilon_{ij}(r_m)_{ij}^{11}\left(\frac{3 + \gamma_{ij}}{5}\right) = \left[\frac{1}{2}\left(A_{ii}^{\frac{1}{12}} + A_{jj}^{\frac{1}{12}}\right)\right]^{12}, \tag{10}$$

$$\varepsilon_{ij}(r_m)_{ij}^{6}\left(\frac{11 - 3\gamma_{ij}}{5}\right) = 2\left[\left(\frac{\alpha_{ii}}{\alpha_{jj}}\right)\frac{1}{B_{ii}} + \left(\frac{\alpha_{jj}}{\alpha_{ii}}\right)\frac{1}{B_{jj}}\right]^{-1}, \tag{11}$$

and

$$(2/3)\varepsilon_{ij}(r_m)_{ij}^{8}\gamma_{ij} = C_{ii}\left[\left(\frac{\alpha_{ii}}{\alpha_{jj}}\right) + 2\left(\frac{\alpha_{jj}}{\alpha_{ii}}\right)\frac{B_{ii}}{B_{jj}}\right]^{-1}$$

$$+ C_{jj}\left[2\left(\frac{\alpha_{ii}}{\alpha_{jj}}\right)\frac{B_{jj}}{B_{ii}} + \left(\frac{\alpha_{jj}}{\alpha_{ii}}\right)\right]^{-1}. \tag{12}$$

The derivation of these relations is given by Shashkov et al. [21]. Computed values of these parameters for the neon-argon systems are given in Table 4.

Mason and Saxena [22] found that the thermal conductivity of a binary system, k_{mix}, can be represented by

$$k_{mix} = \frac{k_1}{1 + \phi_{12}(x_2/x_1)} + \frac{k_2}{1 + \phi_{21}(x_1/x_2)}, \tag{13}$$

where

$$\phi_{12} = \frac{1}{2\sqrt{2}}\left(1 + \frac{M_1}{M_2}\right)^{-\frac{1}{2}}\left[1 + \left(\frac{k_1}{k_2}\right)^{\frac{1}{2}}\left(\frac{M_1}{M_2}\right)^{\frac{1}{4}}\right]^2. \tag{14}$$

ϕ_{21} is obtained from ϕ_{12} by interchanging the subscripts 1 and 2. M is the molecular weight of the gas. Based on equation (14) is the relation [23]:

$$\frac{\phi_{12}}{\phi_{21}} = \frac{k_1}{k_2}. \tag{15}$$

Equations (13) and (14) permit the calculation of k_{mix} if the pure thermal conductivity values are known. This method referred to here as SWPI is employed to correlate the mixture conductivity values as a function of composition. Further, equation (15) and one mixture conductivity value will enable the explicit determination of ϕ_{12} and ϕ_{21}. This method of calculating thermal conductivity of mixtures is designated as SWPII. Values of ϕ_{ij} for the two procedures are given in Table 5 as a function of temperature. The weak dependence of ϕ_{ij} on temperature and the usefulness of equation (13) for extrapolating the mixture conductivity values to high temperatures is to be noted.

The comparison of the present conductivity data on pure gases and their four mixtures as a function of temperature with the predictions of the theory based on Lennard-Jones (12-6), exp-six, and (11-6-8) intermolecular potentials is made in Figures 2 through 6. In each case the error bands associated with the experimental data are also shown. For the case of mixtures, Figures 3 through 6, comparison is also included with the values obtained by the procedures SWPI and SWPII. The percentage deviations in all the figures are computed from the following relation:

$$\%\text{Dev.} = 100\{k(\text{exptl}) - k(\text{calc})\}/k(\text{exptl}). \tag{16}$$

k(exptl) values in the above relation are based on the polynomial fits of the actual data points. The representation of actual data points by cubic polynomials is considered adequate as these figures reveal the scatter between the two to be well within the experimental uncertainties.

The predictions of the (12-6) are the worst of all and the theory underestimates the conductivity values almost over the entire temperature range. Figure 2 suggests that the redetermination of pure gas parameters may be in order. Calculations revealed that the agreement for mixtures improves a bit when the parameters for unlike interactions are changed viz., $\varepsilon_{12}/k = 60.18$ K and $\sigma_{12} = 3.10$ A. For both the pure gases exp-six and (11-6-8) potential based values are in fair agreement with the experiments with a slight preference for the latter. This conclusion is consistent with the findings of Nain et al. [19]. The present neon conductivity values may be a few tenths of a percent larger than the actual values due to impurity of 0.1% helium.

The theoretical predictions are relatively inferior to mixture conductivity values in comparison to pure gas values. The computed values for exp-6 and (11-6-8) potentials for most cases are smaller than the measured values as in the case for pure gases. This general trend needs to be evaluated and will be examined in detail with the availability of mixture data on other systems.

The SWPII predicted values are in much better agreement with the experimental data than the SWPI values. We suggest the former procedure (SWPII) for correlation and extrapolation of mixture conductivity data.

Table 1: The constants of the conductivity cell.

Inner diameter of the molybdenum tube	5.700 ± 0.005 mm
Outer diameter of the molybdenum tube	6.300 ± 0.003 mm
Diameter of the molybdenum alloy (MR-50) wire	0.100 ± 0.001 mm
Length of the longer segment of the wire	102.605 ± 0.008 mm
Length of the shorter segment of the wire	42.909 ± 0.008 mm
Eccentricity of the wire mounting	0.250 ± 0.005 mm
Resistance of the longer segment of the wire at 293 K	2.5658 ± 0.0004 Ohm
Resistance of the shorter segment of the wire at 293 K	1.0730 ± 0.0002 Ohm
Effective length of the wire	59.696 ± 0.012 mm

Table 2: Experimental thermal conductivity, $k(T)$, in $mWm^{-1}K^{-1}$ of neon, argon, and their mixtures as a function of temperature and composition.

T,K	k(T)	T,K	k(T)	T,K	k(T)	T,K	k(T)	T,K	k(T)	T,K	k(T)
X(Ar) = 0.0		X(Ar) = 0.2		X(Ar) = 0.4		X(Ar) = 0.6		X(Ar) = 0.8		X(Ar) = 1.0	
373	56.9	399	48.9	394	40.0	398	33.2	392	26.6	370	21.2
475	67.5	496	58.0	474	45.5	490	38.9	505	32.7	648	32.0
580	77.5	592	63.8	567	51.4	576	42.8	597	36.3	807	38.2
693	87.3	643	65.9	647	57.2	654	48.4	658	39.0	949	42.4
781	95.1	805	79.9	779	65.6	753	52.1	816	44.9	1082	44.7
875	102.6	910	88.8	877	70.3	859	58.0	896	49.2	1231	48.5
958	110.0	1030	95.4	998	77.1	978	64.2	991	53.2	1374	54.5
1113	121.0	1101	99.0	1108	82.2	1049	65.8	1071	54.1	1516	58.2
1215	127.0	1166	103	1232	88.0	1145	70.3	1122	57.3	--	--
1341	136.0	1247	108	1347	92.4	1259	73.4	1227	59.6	--	--
1422	141.0	1369	112	1499	96.0	1376	78.2	1320	63.0	--	--
1503	146.0	1517	119	--	--	1051	82.0	1474	67.7	--	--

Table 3: The constants C_i of equation (8)

C_i \\ %Ar	0.00	20	40	60	80	100
$C_1 (mWm^{-1}K^{-1})$	13.53	19.36	10.44	5.766	4.147	-2.688
$10^2 C_2 (mWm^{-1}K^{-2})$	12.88	6.96	7.81	7.63	6.49	8.13
$10^5 C_3 (mWm^{-1}K^{-3})$	-3.534	1.792	-0.6202	-1.958	-2.015	-5.198
$10^8 C_4 (mWm^{-1}K^{-4})$	0.5496	-1.367	-0.5159	0.1696	0.3600	1.648
10^2 Std. Dev.	0.33	0.94	0.33	0.53	0.54	0.62

Table 4: Intermolecular potential parameter

Potential Function	Neon-Neon			Argon-Argon			Neon-Argon			Reference
	ε/k (K)	r_m or σ (A)	α or γ	ε/k (K)	r_m or σ (A)	α or γ	ε/k (K)	r_m or σ (A)	α or γ	
L-J(12-6)	35.7	2.789	--	124.0	3.418	--	66.5	3.103	--	[17]
Exp-six	38.0	3.147	14.5	123.2	3.866	14.0	73.70	3.443	14.17	[18]
11-6-8	43.4	2.710	3	152.8	3.297	3	63.17*	3.094*	3*	[19]

*Present work.

Table 5: The Sutherland-Wassiljewa ϕ_{ij} coefficients as a function of temperature

	SWPI		SWPII	
T(K)	ϕ_{12}	ϕ_{21}	ϕ_{12}	ϕ_{21}
400	1.622	0.861	1.613	0.609
600	1.592	0.873	1.665	0.649
800	1.598	0.871	1.625	0.630
1000	1.608	0.866	1.583	0.606
1100	1.611	0.866	1.575	0.601
1200	1.610	0.866	1.579	0.604
1300	1.604	0.868	1.600	0.616
1400	1.592	0.873	1.639	0.639
1500	1.575	0.880	1.701	0.676

REFERENCES

[1] Touloukian, Y.S., Liley, P.E. and Saxena, S.C., Thermal Conduc-
     tivity: Nonmetallic Liquids and Gases, IFI/Plenum, New York,
     1970.

[2] Marchenkov, E.I. and Shashkov, A.G., J. Eng. Phys., 26 (6),
     762-768, 1974.

[3] Marchenkov, E.I. and Shashkov, A.G., J. Eng. Phys., 28 (6),
     725-731, 1975.

[4] Srivastava, B.N, and Madan, M.P., Proc. Natl. Inst. Sci. (India)
     20A (5), 587-597, 1954.

[5] Srivastava, B.N. and Saxena, S.C., Proc. Phys. Soc. (London)
     70B, 369-378, 1957.

[6] Thornton, E. and Baker, W.A.D., Proc. Phys. Soc. (London) 80,
     1171-1175, 1962.

[7] von Ubisch, H., Arkiv Fysik, 16, 93-100, 1959.

[8] Mathur, S., Tondon, P.K. and Saxena, S.C., Mol. Phys., 12, 569-
     579, 1967.

[9] Saxena, V.K. and Saxena, S.C., J. Chem. Phys., 48 (12), 5662-
     5667, 1968.

[10] Springer, G.S. and Wingeier, E.W., J. Chem. Phys., 59, 2747-
     2750, 1973

[11] Jody, B.J. and Saxena, S.C., Phys. Fluids, 18 (1), 20-27, 1975.

[12] Saxena, V.K. and Saxena, S.C. Chem. Phys. Letters, 2 (1), 44-
     46, 1968.

[13] Faubert, F.M. and Springer, G.S., J. Chem. Phys., 57, 2333-2340,
     1972.

[14] Chen, S.H.P. and Saxena, S.C., Mol. Phys., 29, 455-466, 1975.

[15] Stefanov, B., J. Chem. Phys., 63, 2258, 1975.

[16] Muckenfuss, C. and Curtiss, C.F., J. Chem. Phys., 29, 1273-1277,
     1958.

[17] Hirschfelder, J.O., Curtiss, C.F. and Bird, R.B., Molecular
     Theory of Gases and Liquids, John Wiley, New York, 1964.

[18] Mason, E.A., J. Chem. Phys., 23, 49-56, 1955.

[19] Nain, V.P.S., Azia, R.A., Jain, P.C. and Saxena, S.C., J. Chem.
     Phys., 65, 3242-3249, 1976.

[20] Hanley, H.J.M. and Klein, M., J. Phys. Chem., 76, 1743-1751,
     1972.

[21] Shashkov, A.G., Abramenko, T.N., Nesterov, N.A., Joshi, R.K.,
     Afshar, R. and Saxena, S.C., Thermal Conductivity of Argon,
     Krypton and Their Mixtures at Low Temperatures (90-270 K), to
     be published in Chem. Phys.

[22] Mason, E.A. and Saxena, S.C., Phys. Fluids, 1, 361-369, 1958.

[23] Saxena, S.C. and Gandhi, J.M., Revs. Modern Phys., 35, 1022-
     1032, 1963.

EXPERIMENTAL ASPECTS OF THE TRANSIENT HOT-WIRE TECHNIQUE FOR THERMAL CONDUCTIVITY MEASUREMENTS

C.A. Nieto de Castro and W.A. Wakeham

Department of Chemical Engineering, Imperial College

Imperial College, Prince Consort Road
London, SW7 2BY, England

ABSTRACT

The performance of two transient hot-wire instruments for the measurement of the thermal conductivity of fluids is examined. Specifically, it is shown that the instruments are entirely free of debilitating, convective heat transfer, and that the effects due to the finite length of the wires employed are completely eliminated by a suitable experimental technique. It is demonstrated that the accuracy of measurements made with one instrument, designed for use in the gas phase, is ± 0.2%, whereas for the other apparatus employed in the liquid phase, the accuracy of the results is estimated as one of ± 0.6%.

INTRODUCTION

It has been recognized (1,2) that the transient hot-wire technique offers the opportunity to eliminate spurious convective heat transfer from fluid thermal conductivity measurements. The most recent theory of instruments of this type (2) has established rigorous criteria whereby their successful operation may be judged. In this work we demonstrate that our two instruments (3,4), one used in the liquid phase, the other in the gas phase, satisfy these criteria.

BASIC WORKING EQUATIONS

The thermal conductivity of a fluid, $\lambda(T,\rho)$, at a temperature

T_r and density ρ_r may be obtained from the measured, transient temperature rise, ΔT_m, of a thin, infinitely long vertical wire immersed
in the fluid following initiation of a heat generation, q, within it,
by application of the equations (1,2)

$$\Delta T_c = \frac{q}{4\pi\lambda(T_r,\rho_r)} \ln\left[\left(\frac{4K_o}{a^2C}\right)t\right] \quad , \quad (1)$$

where $K_o = \lambda(T_o,\rho_o)/\rho_o C_{po}$,

$$T_r = T_o + (\Delta T_c(t_1) + \Delta T_c(t_2))/2 \quad , \quad (2)$$

and

$$\Delta T_c = \Delta T_m + \delta T_1 + \delta T_2 \quad , \quad (3)$$

Here, T_o, ρ_o and C_{po} represent the temperature, density and heat
capacity of the fluid in its equilibrium state prior to initiation
of the temperature rise. In addition, a denotes the radius of the
wire, t the time (measured from the initiation of the temperature
rise) and t_1 and t_2 the initial and final times of measurement.
The correction terms δT_1 and δT_2 incorporate the effects arising
from the finite heat capacity of the wire and the finite radial
extent of the fluid. In general, these corrections amount to no
more than 0.5% of the wire temperature rise in properly designed
instruments, and the expressions given by Healy *et al.* (2) permit
their estimation to within a few percent. Other corrections detailed by Healy *et al.* (2) are rendered negligible by a suitable choice
of design parameters and operating conditions.

The linearity between ΔT_c and ln t embodied in Eq. (1) allows
$\lambda(T_r, \rho_r)$ to be evaluated by means of a straightforward regression
analysis of experimental data. Furthermore, the same equation provides a stringent test of the operation of the instrument since if,
and only if, the apparatus conforms to the above mathematical description of it will the expected linearity be observed.

EXPERIMENTAL

Equation (1) is valid only for a wire of infinite length or
any finite segment of an infinite wire. In practice the wires
employed for measurements are necessarily of finite length and must
be rigidly mounted at either end. Temperature non-uniformities
therefore exist along the axis of the wire in the neighbourhood of
these mounting points, so that Eq.(1) is not immediately applicable
to measurements made with such wires. Furthermore, the axial tempera-

ture gradient in the wire gives rise eventually to convective fluid motion (2,5). Before Eq.(1) is used for the interpretation of measurements it is essential that any experimental data correspond to the situation which it describes.

Considering the end effects in finite wires first, Haarman (1,6) has shown that the principal contribution to the axial heat conduction occurs in the fluid for wires of very small cross-sectional area. The average temperature rise of a wire of finite length l_L and radius a_L in a transient hot-wire experiment, using a heat generation per unit length q_L, may therefore be written (1,6)

$$\overline{\Delta T}_L = \frac{q_L}{4\pi\,\lambda(T_r,\,\rho_r)}\,\ln\left[\left(\frac{4K_o}{a_L^2\,C}\right)t\right] - \frac{q_L}{l_L}\,f(K_o,\,t)\,. \qquad (4)$$

Within the second term, which constitutes the end effect, the exact form of $f(K_o,t)$ is unknown, but it is certainly smallest for low thermal diffusivities and short times (1,6). The resistance change of this wire from its unperturbed value, R_L^o, is thus,

$$R_L - R_L^o = \alpha R_L^o\,\overline{\Delta T}_L \qquad (5)$$

where α represents a temperature coefficient of resistance for the wire material assumed constant for simplicity.

Equations similar to (4) and (5) may be written for another wire (denoted by a subscript 'S') of the same material but of a different radius and length. Straightforward algebra then leads to the result that

$$\frac{(R_L - R_L^o) - (R_S - R_S^o)}{\alpha(R_L^o - R_S^o)} = \Delta T_c\left[\frac{R_S^o}{R_L^o - R_S^o}\left[1 - \frac{q_L}{q_S}\,h(t)\right] + 1\right]$$

$$- \frac{R_L^o}{R_L^o - R_S^o}\left[\frac{q_L}{l_L} - \frac{R_S^o}{R_L^o}\frac{q_S}{q_L}\right]f(K_o,t)\,, \qquad (6)$$

where $h(t) = \ln(4K_o\,t/a_S^2 C)/\ln(4K_o\,t/a_L^2\,C)$,

If circumstances can be arranged so that

$$q_L = q_S \quad \text{and} \quad R_L^o/l_L = R_S^o/l_S \qquad , \quad (7)$$

which implies that $a_L = a_S$, then eq.(6) reduces to

$$\frac{(R_L - R_L^o) - (R_S - R_S^o)}{\alpha(R_L^o - R_S^o)} = \Delta T_c \qquad (8)$$

Thus, if the resistance difference between two wires, identical apart from length, is measured as a function of time, when both are immersed in the same fluid and subject to the same heat dissipation per unit length, Eq. (8) may be employed to provide the necessary data for the application of Eq. (1). Even if the conditions (7) cannot be met exactly, so that the cancellation of end effects is not complete, the residual, small correction ($\sim$ 0.1% in the worst case) is easily estimated with sufficient accuracy.

Figure 1 shows a schematic diagram of a device intended to create the situation envisaged in obtaining Eq. (8). Two hot-wire cells containing platinum wires of different length, each immersed in the test fluid are connected in an automatic Wheatstone bridge designed to measure the time, after initiation of a current flow in the wires, at which their resistance difference attains several pre-set values. The design ensures that the heat generation per unit length in each wire is the same. Thus this device provides exactly the time dependence of the numerator of the left hand side of Eq. (8) from which $\Delta T_c(t)$ is easily deduced.

The problem of the onset of convection in a transient hot wire cell has not been completely solved theoretically, however an approximate analysis (2,5) suggests that a finite time will elapse in an experiment before significant (> 0.1% effects on the wire temperature rise will be observed. These effects will take the form of a rapid increase in the rate of heat transfer from the wire and a consequent

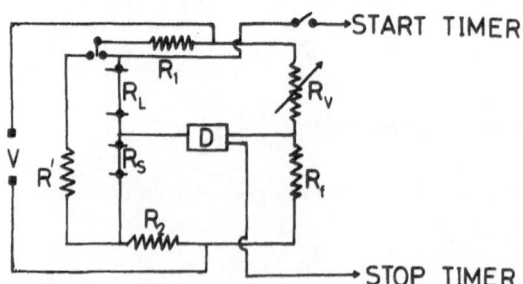

Figure 1   The Automatic Wheatstone Bridge

non linearity in the line of ΔT_c against ln t. That is the apparent
thermal conductivity deduced from the slope of this line will increase. Thus in order to ensure that a particular thermal conductivity
measurement is free of convection it is sufficient to establish that
the temperature increase of the wire is linear in ln t over a finite
and useful time interval.

RESULTS

We have used our thermal conductivity instruments (3) to carry
out measurements designed to demonstrate that they satisfactorily
eliminate effects due to the finite length of the wires employed and
natural convection. The instruments, which are based upon the principle given in Figure 1, have an accuracy in the measurement of the
wire temperature rise of better than ± 0.1%, and a time resolution
of 10 μs. The liquid phase apparatus uses two platinum wires of 10
μm nominal diameter and lengths 157.54 mm and 46.57 mm, whereas in
the gas phase apparatus wires of 5μm diameter are employed of lengths
134.06 mm and 65.50 mm. Figure 2a illustrates the magnitude of the
end effects in the liquid phase measurements. The diagram refers to
measurements made in toluene at 17.54°C under its saturation vapour
pressure. The deviations of the measured temperature rise of the
single short wire suitably corrected with the aid of Eq.(3),from that
calculated on the basis of Eq.(1) are plotted as a function of time
The systematic negative deviation increases with time to a maximum
of -0.2%. Figure 2b contains a similar deviation plot for a two wire
measurement under identical conditions. There is now no evidence of
a systematic deviation and the scatter of the experimental points is
within the estimated precision of the measurements.

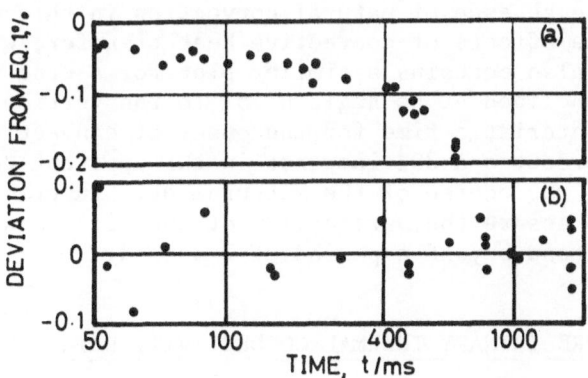

Figure 2   End effects in the liquid phase apparatus

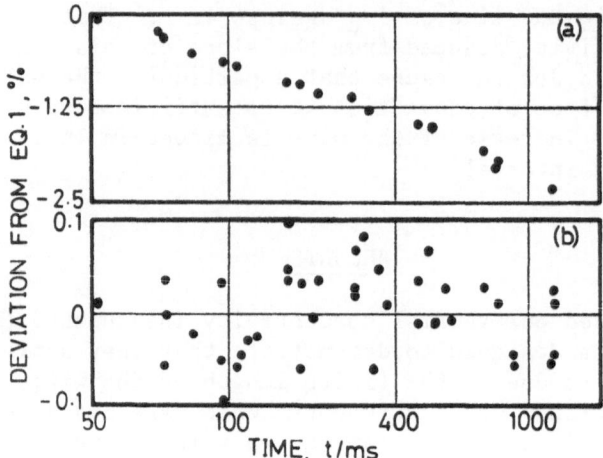

Figure 3   End effects in the gaseous phase apparatus

Figure 3a and 3b show a similar comparison for a measurement in gaseous argon at a temperature of 24.82°C and a pressure of 1.449 MPa. Although the magnitude of the end effects in the short wire is greater than for the liquid phase case, Figure 3b reveals that they are entirely eliminated when a two wire measurement is performed.

In order to demonstrate that our apparatus is free of natural convection, Figure 4 contains a plot of the apparent thermal conductivity (obtained from the local slope of the ΔT_c vs ln t line) as a function of time for the two wire measurement system in n-heptane at 20.1°C. The measurement extends over a period of 4 seconds, and there is no indication of a changing thermal conductivity. This confirms the absence of natural convection in this measurement. To illustrate the effects of convective heat transfer, when they occur, Figure 4 also contains a similar plot for a case where the hot-wires were inclined at an angle of 5° to the vertical so as to reduce the characteristic time for the onset of convection. Here, the convection produces a 30% increase in the apparent thermal conductivity during the course of the measurement. In practice great care is taken to ensure the verticality of the wires and the measurement does not extend beyond a period of 1 second.

PRELIMINARY THERMAL CONDUCTIVITY DATA

Figure 5 shows some preliminary experimental results for the thermal conductivity of argon at 28.75°C as a function of density.

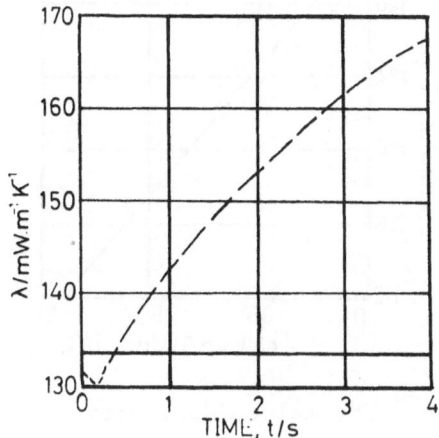

Figure 4   Convective effects

———— Vertical wires   ----- Inclined wires

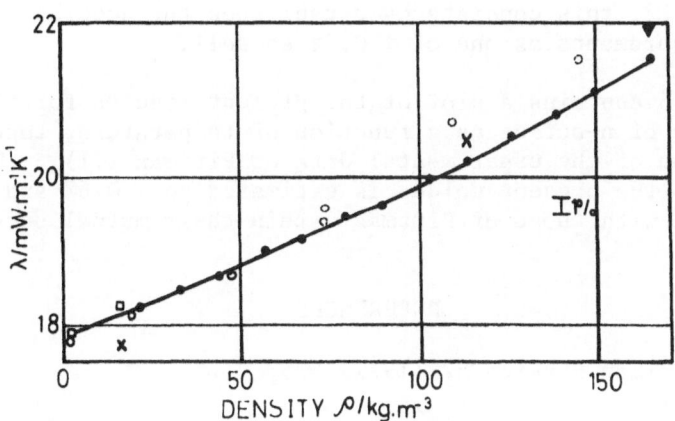

Figure 5   Thermal conductivity of gaseous argon at 28.75°C

● Present work   ○ Sengers (7)
X Bailey and Kellner (8)   □ Tufeu (9)

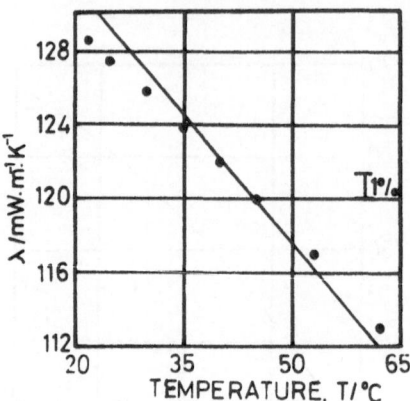

Figure 6   Thermal conductivity of n-octane

● Present work   ——— Pittman (11)

The zero density value, obtained by extrapolation is

$\lambda(T,0)$ = 17.83 mw/m K.

This value is consistent with the best available viscosity data for argon (10) to within ± 0.3%. Since the uncertainty of the viscosity data is ± 0.2%, this consistency establishes the accuracy of the present measurements as one of ± 0.2% as well.

Figure 6 contains a plot of the present results for the thermal conductivity of n-octane as a function of temperature, together with a correlation of the experimental data of Pittman (11).   The uncertainty in the present values is estimated as ± 0.6% and they are in agreement with those of Pittman within their mutual uncertainty.

REFERENCES

1.   J.W. Haarman Physica 52 (1971) 605

2.   J.J. Healy, J.J. de Groot and J. Kestin Physica 82C (1976) 392

3.   C.A.N. de Castro, J.C.G. Calado and W.A. Wakeham J. Phys. E.: Scientific Instruments 9 (1976) 1073

4.   C.A.N. de Castro, J.C.G. Calado and W.A. Wakeham Proceedings of the Seventh Symposium on Thermophysical Properties 1977 (to be published)

5. R.J. Goldstein and D.G. Briggs Heat Transfer, Trans. ASME 86C
    (1964) 490

6. J.W. Haarman Ph.D. Thesis (1969) Technische Hogeschool, Delft

7. J.V. Sengers Ph.D. Thesis (1962) Amsterdam

8. B.J. Bailey and K. Kellner Physica 39 (1968) 444

9. R. Tufeu Ph.D. Thesis (1971) Paris IV

10. J. Kestin, S.T. Ro and W.A. Wakeham J. Chem. Phys. 56 (1972)
    4119

11. R. Kandiyoti, E. McLaughlin and J.F.T. Pittman J. Chem. Soc.:
    Faraday Transactions I 68 (1972) 860

THE THERMAL CONDUCTIVITY OF LIQUIDS - A CRITICAL SURVEY[†]

G. Latini and M. Pacetti

Istituto di Fisica Tecnica-Università di Ancona-Italy

Via della Montagnola 30 - 60100 Ancona - Italy

The present paper intends to suggest thirty empirical, semi-empirical and theoretical formulas appearing in the literature up to 1976, which are very simple and involve easy-to-find physical properties; all the reported correlations are fitted for the estimate of thermal conductivity λ at standard conditions, i.e., at the temperature of 20°C and at atmospheric pressure; nevertheless some formulas may be also utilized at different temperatures. The physical quantities are in the S.I. units. The presented formulas have been tested, with the same input data, for a group of organic liquids and their mixtures belonging to the most important families as shown in Tables I and II. The comparison of experimental and predicted data of λ is developed using the experimental data collected in Ref.[1] which classify the values of λ in three groups labelled A, B, C. Group A collects values with an estimated error within ±2%, group B with an error within ±5%, group C values less accurate than ±5%. Since at a fixed temperature several values of λ exist, in this work the average of the best reliable available data is assumed. The investigated formulas are grouped on the basis of their affinity.

PURE LIQUIDS

The correlation proposed by Weber [2,3,4] and based on experimental work is shown in Eq.(1). Smith [5] changes the value of the constant and gives Eq.(2). Successively Smith [5] splits up the factors and proposes Eq.(3).

[†]Work supported by Consiglio Nazionale delle Ricerche-Italy

$$\lambda = 3.5808 \cdot 10^{-8} C_p \varrho (\varrho/M)^{1/3} \tag{1}$$

$$\lambda = 4.3 \cdot 10^{-8} C_p \varrho (\varrho/M)^{1/3} \tag{2}$$

$$\lambda = 4.1868 \cdot 10^{-4} \left[11 + 8.7885 \cdot 10^{-8} (C_p - 1884.06)^3 + 125(\varrho/M)^{1/3} + 464.16(\mu/\varrho)^{1/9} \right] \tag{3}$$

Palmer [6] introduces in Eq.(2) a multiplying factor depending on liquid degree of association. The Palmer's correlation is shown in Eq.(4). Also Vargaftik [7,8] introduces a corrective factor which may be taken as $\alpha = (M \Delta H_{vb})/(87922.8 \cdot T_b)$, similar to Palmer, at 30°C for most liquids or at $T/2$ for low boiling fluids. Calculated values of $\alpha < 1$ should be taken as unity, and at the temperatures other than 30°C, or $T_c/2$, α may be assumed to vary linearly between the value $(M \cdot \Delta H_{vb})/(87922.8 \cdot T_b)$ at 30°C and 1.0 at T_c. Vargaftik's correlation is given in Eq.(5). Robbins and Kingrea [9,10] propose two corrective factors: the Everett's [11] constant $\Delta S^x = (M \cdot \Delta H_{vb})/T_b + R \cdot 1g(273/T_b)$, an entropy of vaporization term which describes the degree of symmetry of the molecule, and the reduced temperature term T_r; their correlation is shown in Eq.(6); H depends on the molecular structure and N on the density.

$$\lambda = 3.9642 \cdot 10^{-3} C_p \varrho (\varrho/M)^{1/3} \cdot T_b /(M \cdot \Delta H_{vb}) \tag{4}$$

$$\lambda = 4.2713 \cdot 10^{-8} C_p \varrho (\varrho/M)^{1/3} \cdot \bar{\alpha}^1 \tag{5}$$

$$\lambda = 4.1868 \cdot 10^{-5} (88.0 - 4.94 H)/[M \cdot \Delta H_{vb}/T_b + R \cdot 1g(273/T_b)] \cdot (0.55/T_r)^N \cdot C_p \varrho \cdot (\varrho/M)^{1/3} \tag{6}$$

Eq.(7), based on theoretical considerations, is due to Bridgmann [12,13,14]; can be expressed by $\sqrt[3]{M/N_A \rho}$ and w can be obtained by Rao's [15] formula. The equations: (8), due to Osida [16]; (9), due to Hirschfelder, Curtiss and Bird [17]; (10), due to Viswanath [18], are obtained by theoretical considerations more sophisticated than Bridgmann's ones. In Eq.(10) C is a particular "packing factor" depending on molecular structure. The correlation due to Kardos [19] is shown in Eq.(11) and is obtained by theoretical considerations, but the distance ce L between the surface of adjacent molecules substitutes the distance ℓ (utilized by Bridgmann) of their centers. Kardos suggests that L should be assumed approximately constant and equal to $9.5 \cdot 10^{-11}$m.

$$\lambda = 3 \, K w/\ell^2 \tag{7} \qquad\qquad \lambda = 1.2979 \cdot 10^{-2} \cdot T_m /(M^{1/2} V_m^{2/3}) \tag{8}$$

$$\lambda = 2.80 K \cdot \bar{\gamma}^{1/2} \cdot \bar{v}^{-2/3} w \tag{9}$$

$$\lambda = 3.6 \cdot 10^{7} M^{1/2} \Delta H_v \cdot C \bar{v}_m^{-2/3} \bar{T}^{1/2} \tag{10} \qquad\qquad \lambda = L \, w \varrho \, C_p \tag{11}$$

Sakiadis and Coates [20] start from Eq.(11), but consider that L varies with the nature of the liquid and with the temperature and they give Eq.(12) where Δ is an additive quantity calculated using the method of group contributions. Badea [21] also starts from

Kardos correlation and proposes Eq.(13).

$$\lambda = \Delta\left\{1 - \left[\varrho(T_c - T_f)/(\varrho_i T_c - 2\varrho_e T_f)\right]^{1/2}\right\}\cdot w\cdot\varrho\cdot C_p \tag{12}$$

$$\lambda = 1.0798\cdot10^7\cdot C_p^{3/2}\varrho^{2/3}T^{1/2}M^{1/3}/(MC_p - 8314.32)^{-1/2} \tag{13}$$

Narasimhan, Swamy and Narayana [22] connect the Weber and Kardos hypothesis and give Eq.(14). The Eq.(15), due to Smith [23], is essentially empiric and recommended at 30°C. The Eq.(16), due to Denbigh [24], results from the Prandtl's number definition and from the dimensionless equation $LogP_r = (0.2M\cdot\Delta H_{vb})/(8314.321)$. The Eq.(17), due to Scheffy and Johnson [25,26], is obtained by accepting the parameters used by Oside [16] and by extrapolating the Sakiadis and Coates [20] collected data.

$$\lambda = 2.8379\cdot10^7 C_p^{\cdot}\varrho^{0.83}(293/T)^{0.38} \tag{14} \qquad\qquad \lambda = 1.283\cdot10^{13}\varrho^{2.15}C_p^{1.55}M^{0.192}\mu^{0.12} \tag{15}$$

$$\lambda = C_p^{\cdot}\mu/Log^{-1}\left[2.4055\cdot10^5(\Delta H_{vb}/T) - 1.8\right] \tag{16}$$

$$\lambda = 1.951\left[1 - 0.00126(T - T_m)\right]/(T_m^{0.216}M^{0.300}) \tag{17}$$

Missenard [27] proposes Eq.(18) obtained starting from the Debye's [28] equation valid for dielectric solids $\lambda \propto \varrho w C_v \ell$; in such equation he substitutes $w \propto \mu^{v-15}T_b^{1/2}/N^{1/4}\cdot\rho^{1/2}$ and $C_v\cdot\ell \propto C_p/\mu^{1/12}$. Badea [21] gives Eq.(19) obtained by dimensional analysis in the function: $\lambda = f(\rho,m,C_p,\Delta H_v)$. The equations (20), (21), (22), (23), (24) are due to Pachaiyappan and coworkers [29,30,31,23,33]. The Eq.(20) is recommended at 20°C; in Eq.(21) C and n vary with homologous groups; in Eq.(23) C_p and ρ are estimated at 20°C; in all the equations ΔH_v and V_m are estimated at the normal boiling point. Vaidyanathan and Velayutham [34] start from the Eq.(22) and, taking in account that $V_m\Delta H_v = const.$ for members of a series (except for substitutes aliphatics), propose Eq.(25).

$$\lambda = 2.846\cdot10^7(T_b\varrho)^{1/2}C_p^{\cdot}N^{-1/4} \tag{18}$$

$$\lambda = 3.3062\cdot10^{10}M^{1/3}\varrho^{2/3}C_p^{\cdot}\Delta H_v^{1/2} \tag{19}$$

$$\lambda = 8.3455\cdot10^{11}C_p^{\cdot}\Delta H_v^{1/2}M^{1.30}V_m^{-2/3} \tag{20}$$

$$\lambda = 0.41868\cdot C\cdot M^{n}/V_m \tag{21}$$

$$\lambda = 3.0090\cdot10^{10}C_p^{\cdot}\varrho\Delta H_{vb}^{1/2}V_m^{1/3} \tag{22}$$

$$\lambda = 2.27956\cdot10^{10}C_p^{\cdot}\varrho\cdot\Delta H_{vb}^{1/2}V_m^{1/3}(T_c/T)^{1/2} \tag{23}$$

$$\lambda = 8.84\cdot10^8 C_p^{\cdot}\varrho \tag{24}$$

$$\lambda = 7.5\cdot10^8 C_p^{\cdot}\varrho(T_b/T)^{1/2} \tag{25}$$

Sakiadis and Coates [35] developed a new method of correlating λ
of liquids based on a modified statement of the theory of corresponding states. According to this method the ratio of λ of two
members of a given homologous series, at a given temperature, is
the same for the corresponding two members of any other homologous
series at the same reduced temperature. The reduced pressure is
omitted from the correlation. Reid and Sherwood [36] summarize
this method in Eq.(26) where $\lambda_1 = \lambda_{0.6} + A$; A is an additive constant for functional groups; λ_2 is a constant for a given type of
compound; $\lambda_{0.6}$ is at $T_r = 0.6$.

$$\lambda = 2.26 \cdot \lambda_1 - 1.26 \cdot \lambda_2 - 2.10 \cdot (\lambda_1 - \lambda_2) \cdot T_r \tag{26}$$

LIQUID MIXTURES

Comparatively little progress has been made for estimating λ
of liquid mixtures and the data are rather scanty. The simplest
method is shown in Eq.(27), investigated by Tsederberg [37], where
λ_m, λ_1, λ_2 are the thermal conductivities of the mixture and of
the components; m_1, m_2 are the mass fractions. Filippov and Novoselova [38] propose Eq.(28) where the index 2 is referred to the
component with the higher conductivity.

$$\lambda_m = \lambda_1 m_1 + \lambda_2 \cdot m_2 \tag{27}$$

$$\lambda_m = \lambda_2 m_2 + \lambda_1 (1 - m_2) - 0.72 (\lambda_2 - \lambda_1) m_2 (1 - m_2) \tag{28}$$

Bates, Hazzard and Palmer [39] give Eq.(29) valid for water-ethanol
and water-methanol systems. ϕ is a constant characteristic of a
given pair of components of the mixture; the function sinhϕ is
defined as sinhϕ = $0.5(e^\phi - e^{-\phi})$ and its value is available in
standard tables. Following Bondi [40], Riedel recommends Eq.(30)
where x_1 and x_2 are the mole fractions; the function f is given by
$(E_1^0/M_1) + (E_2^0/M_2)$; M_1 and M_2 are molecular weights; E_1^0 and E_2^0 are
standard energies of varporization; the constant b is equal to
$7.0 \cdot 10^{-9}$ when E^0 is expressed in erg/mole.

$$\lambda_m \sinh \phi = \lambda_1 \sinh(m_1 \phi) + \lambda_2 \sinh(m_2 \phi) \tag{29}$$

$$\lambda_m = x_1 \lambda_1 + x_2 \lambda_2 - f b x_1 x_2 \tag{30}$$

CONCLUSIONS

For pure liquids the prediction method and Table II indicate
that the correlations based on a simplified theory are less accurate than the merely empirical ones. Several formulas do not explicitly contain temperature and this limits their application
because it is necessary to estimate all the parameters at various
temperatures. Nevertheless in some correlations corrective temp-

erature factors appear as in Eq.(14), Eq.(23) and Eq.(25). In the cited formulas the parameters are estimated at 20°C. The Eq. (17) is recommended for simplicity because it requires only the values of T_m and M. The Eq.(17) is valid only at $T_r < 0.7$ and it is not suitable for highly polar or inorganic molecules. The Eq. (6), where $0.9 > T_r > 0.4$, is recommended for simplicity and accuracy because all the mechanisms actually affecting λ are present. The correlation summarized in Eq.(26) seems to be very accurate and simple to use, but is not applicable to all organic liquids. Other formulas enough correct are Eq.(5) and the group of Pachaiyappan and coworkers. For liquid mixtures Tsederberg states that the simple law of additivity, Eq.(27), may be applied when both components are non-polar or one component is non-polar and the other slightly polar. If the dipole moments of the molecules are large, the deviations can be considerable. The effect of the temperature upon λ of a liquid mixture depends on whether the conductivities of the pure components increase or decrease with the rise of temperature. The Eq.(28) may be used even when the components are strongly polar. In Eq.(29) the constant φ for a given pair of components can be calculated if λ for at least one mixture is known. In Eq.(30 volume fractions should be employed instead of mole fractions x_1 and x_2 if one of the mixture components is a compound with a large (flexible chain) molecule. In this work the calculations are presented only for the Eq.(27) and Eq.(28).

TABLE 1. Liquid Mixtures – Deviation Δ from Experimental Values at 20°C

Mixture	Mass fractions	λ_1 and λ_2 mW $\bar{m}^1 \bar{K}^1$	λ_m exp. mW $\bar{m}^1 K^1$	λ_m Eq.(27) mW $\bar{m}^1 K^1$	λ_m Eq.(28) mW $\bar{m}^1 K^1$
Benzene	m_1 =0.72	146.8 (A)	149.5	150.8	148.7
Acetone	m_2 =0.28	161.0 (A)	Riedel 48	Δ=0.8	Δ=-0.5
Chloroform	m_1 =0.55	116.4 (A)	124.7	130.1	124.7
Benzene	m_2 =0.45	146.8 (A)	Filippov55	Δ=4.3	Δ=0.0
Carbon Tetrachloride	m_1 =0.25	103.8 (A)	130.1	136.1	130.2
Benzene	m_2=0.75	146.8 (A)	Riedel 48	Δ=4.6	Δ=0.1
Ethyl Al.	m_1=0.25	165.8 (A)	201.0	192.2	187.4
Methyl Al.	m_2=0.75	201.0 (A)	Lees 1898	Δ=-4.4	Δ=-6.7

TABLE 2. Pure Liquids – General Results with Deviation Data ΔZ of the Estimated Values of λ from the Experimental Values at 20°C (the values of λ are in mW m⁻¹ K⁻¹)

	Eq.(1)	Eq.(2)	Eq.(3)	Eq.(4)	Eq.(5)	Eq.(6)	Eq.(7)	Eq.(8)	Eq.(9)	Eq.(10)	Eq.(11)	Eq.(12)	Eq.(13)	Eq.(14)	Eq.(15)	Eq.(16)	Eq.(17)	Eq.(18)	Eq.(19)	Eq.(20)	Eq.(21)	Eq.(22)	Eq.(23)	Eq.(24)	Eq.(25)	Eq.(26)	λexp value
ACETONE	148.4 / −7.8	178.2 / +10.7	168.5 / +4.7	185.9 / +15.5	176.1 / +9.4	167.9 / +4.3	191.7 / +19.1	129.4 / −19.6	159.6 / −1.5	237.0 / +47.2	192.8 / +19.8	158.2 / −1.7	182.5 / −13.3	158.5 / −1.5	188.7 / +17.2	182.4 / +13.3	161.0 / 0	179.4 / +11.4	176.3 / +9.5	140.3 / −12.9	158.5 / −1.6	157.7 / −2.0	157.4 / −2.2	153.5 / −4.7	138.5 / −14.2	138.1 / ---	161.0 A
BENZENE	118.9 / −19.0	142.8 / −2.7	162.1 / +10.4	151.1 / +2.9	141.9 / −3.3	151.9 / +3.5	191.8 / +30.6	123.1 / −16.1	158.8 / +8.2	191.1 / +30.5	186.9 / +27.3	62.7 / −57.3	142.9 / −2.7	132.9 / −9.5	153.2 / +4.4	206.1 / +40.4	153.6 / +4.6	143.7 / −2.1	144.1 / −1.8	121.3 / −17.4	147.2 / +0.3	128.4 / −12.5	134.8 / −8.2	131.0 / −10.8	122.1 / −16.8	---	146.8 A
TOLUENE	111.1 / −17.6	133.4 / −1.1	155.0 / +14.9	142.2 / +5.4	132.5 / −1.8	143.7 / +6.5	168.9 / +25.2	80.4 / −40.4	139.9 / +3.7	172.7 / +28.0	183.9 / +36.3	141.1 / +4.6	137.8 / +2.1	132.1 / −2.1	156.7 / +16.2	118.4 / −12.2	140.2 / +3.9	141.4 / +4.8	148.0 / +9.7	126.0 / −6.6	141.4 / +4.8	130.5 / −3.3	140.7 / +4.3	129.9 / −3.7	126.2 / −6.5	---	134.9 A
CARBON TETRACHL.	106.3 / +2.4	127.6 / +23.0	118.1 / +13.8	137.4 / +32.4	126.8 / +22.2	108.8 / +4.8	120.3 / +15.9	78.7 / −24.2	99.7 / −3.9	98.1 / −5.5	113.8 / +9.6		96.2 / −7.3	110.3 / +6.3	208.4 / +100.8	181.5 / +74.9	123.6 / +19.1	121.4 / +16.9	95.4 / −8.1	100.4 / −3.3		84.5 / −18.6	88.2 / −15.0	120.4 / −16.0	111.5 / +7.4	---	103.8 A
CHLOROFORM	119.0 / +2.2	142.9 / +22.8	135.2 / +16.2	154.0 / +32.3	141.9 / +21.9	121.4 / +4.3	147.7 / +26.9	92.3 / −20.7	57.2 / −50.9	149.2 / +28.2	130.4 / +12.0		108.8 / −6.5	117.6 / +1.0	15.76 / −88.4	15.76 / −35.4	131.1 / +12.6	129.7 / +11.4	104.0 / −10.7	99.8 / −14.3		94.2 / −19.1	96.6 / −17.0	126.8 / +8.9	115.0 / −1.2	---	116.4 A
ETHYL AL.	174.4 / +5.2	209.4 / +26.3	188.3 / +13.6	175.9 / +6.1	166.6 / +0.5	170.5 / +2.8	224.7 / +35.5	158.6 / −4.3	186.1 / −12.2	166.8 / +0.6	210.9 / +27.2	152.1 / −8.2	207.6 / −25.2	172.5 / +4.0	175.7 / +6.0	122.8 / −25.9	171.9 / +3.7	207.3 / +25.0	235.1 / +41.8	171.7 / +3.6	172.5 / +4.0	204.8 / +23.5	205.9 / +24.2	167.0 / +0.7	155.0 / −6.5	173.2 / +4.4	165.8 B
METHYL AL.	206.3 / +2.6	247.7 / +23.2	206.2 / +2.6	218.9 / +8.9	207.3 / +3.1	210.4 / +4.7	288.9 / +43.7	260.4 / +29.6	239.3 / +19.1	591.7 / +194.4	222.4 / +10.6	153.8 / −23.5	234.3 / +16.6	180.5 / −10.2	191.5 / −4.7	119.2 / −40.7	192.6 / −4.2	235.1 / +17.0	246.2 / +22.5	163.1 / −18.9	173.1 / −13.9	217.9 / +8.4	218.5 / +8.7	174.9 / −13.0	159.4 / −20.7	197.3 / −1.8	201.0 B
n-PENTANE	106.9 / −10.0	128.4 / +8.6	152.8 / +29.2	141.9 / +20.0	127.5 / +7.8	117.7 / −0.5	121.0 / +2.3	77.3 / −34.6	100.2 / −15.3	137.2 / +16.0	137.4 / +16.2	131.5 / +6.1	157.8 / +33.4	137.9 / +16.6	134.5 / +13.7	268.0 / +126.6	150.1 / +26.9	143.1 / +21.0	142.3 / +20.3	124.9 / +5.6	127.0 / +7.4	128.4 / +8.6	123.1 / +4.1	128.4 / +8.6	111.9 / −5.4	122.6 / +3.7	118.2 B
n-HEXANE	105.5 / −14.9	126.7 / +2.3	148.5 / +19.9	138.4 / +11.7	125.9 / +1.6	125.3 / +1.1	120.4 / −2.8	72.5 / −41.5	99.7 / −9.5	131.7 / +6.3	152.9 / +23.4	131.5 / +6.1	154.3 / +24.5	140.7 / +13.6	144.7 / +16.8	198.6 / +60.3	143.1 / +15.5	144.8 / +16.9	152.5 / +23.1	132.3 / +6.8	129.8 / +4.9	135.6 / +9.4	135.2 / +9.1	132.2 / +6.7	121.2 / −2.2	128.1 / +3.4	123.9 B
n-OCTANE	101.1 / −23.5	121.4 / −8.1	142.3 / −7.7	129.7 / −1.8	120.6 / −8.7	133.6 / +1.1	113.2 / −14.3	60.0 / −54.6	93.8 / −29.0	123.5 / −6.5	171.5 / +29.8	123.0 / −6.9	148.0 / +12.0	143.5 / +8.6	156.5 / +18.5	111.8 / −15.4	133.2 / +0.8	146.4 / +10.8	167.4 / +26.7	144.4 / +9.3	132.8 / +0.5	146.2 / +10.7	154.3 / +16.8	136.2 / +3.1	134.9 / +2.1	138.1 / +4.5	132.1 B
REFRIG. 11	103.7 / +11.1	124.5 / +33.4	119.9 / +28.5	137.4 / +47.3	123.6 / +32.5	99.9 / +7.1	134.3 / +43.9	69.0 / −26.0	111.3 / +19.3	109.7 / +17.6	118.5 / +27.0		96.6 / +3.5	107.2 / +14.9	200.5 / +114.9	243.0 / +160.5	124.0 / +32.9	111.2 / +19.2	83.4 / −10.6	91.7 / −1.7		75.6 / −19.0	72.6 / −22.2	115.7 / +24.0	98.8 / +5.9	---	93.3 B
REFRIG. 113	107.2 / +41.1	128.7 / +69.3	117.9 / +55.1	138.3 / +82.0	127.8 / +68.2	86.6 / +13.9	165.7 / +118.0	60.5 / −20.4	137.3 / +80.7	144.7 / +90.4	194.6 / +156.0		100.1 / +31.7	119.4 / +57.1	250.5 / +229.6	240.3 / +216.2	115.9 / +52.5	112.3 / +47.8	94.4 / +24.2	108.3 / +42.5		84.5 / +11.2	82.6 / +8.7	130.1 / +71.2	115.5 / +52.0	---	76.0 B
mean Δ	13.1	19.3	18.1	22.2	15.1	4.6	31.5	27.7	21.9	39.3	32.9	15.5	14.9	12.1	52.6	68.5	14.7	17.0	17.4	11.9	4.7	12.2	11.7	14.3	11.7	3.6	
mean deviation of rating A	10.5	12.9	13.3	13.7	8.0	3.4	21.6	28.6	16.3	36.3	21.2	15.5	14.4	7.3	28.7	44.5	9.1	13.7	17.4	9.9	4.7	11.6	11.0	7.6	8.3	3.6	

NOMENCLATURE

C_p = spec. heat at const. press.; C_v = spec. heat at const. vol.; K = Boltzmann const.; M = molec. weight; N = number of atoms in the molec. – N_A = Avogadro const.; R = gas const.; T = abs. temp.; T_b = norm. boil. point; T_c = critical temp.; T_f = freez. point; T_m = melt. point.; T_r = reduced temp.; T_{rb} = T_b/T_c; V_m = molal vol.; m = mass of molec.; v = vol. of molec.; w = sound velocity; γ = ratio of spec. heats; λ = thermal conduc.; μ = dynamic visc.; ρ = density; ρ_c – critical density; ρ_f = dens. at freez. point; ΔH_V = latent heat of vap.; ΔH_{Vb} = latent heat of vaporiz. at norm. boil. point.

REFERENCES

1. Jamieson, D.T. et alii, "Liquid therm. conduct., a data survey to 1973", Her Maj. Stat. Off., Edinburgh, 1975.

2. Weber, H.F., Ann. Physik 10 (3), 103, 1880.

3. Weber, H.F., Sitzber. Koniglich Pr. Akad. wiss, Berlin, 2 Halband, June–Dec., 809, 1885.

4. Malkov, M.P., Pavlov, K.F., Spravochnikpo glubokomu okhlazhdeniyu", Gosyekhizdat, 1947.

5. Smith, J.F.D., Trans. ASME 58, 719, 1936.

6. Palmer, G., Ind. Eng. Chem. 40, 89, 1948.

7. Vargaftik, N.B., "Thermoph. prop. of subst.", F–TS–9537/V, 1959.

8. Vargaftik, N.B., Izv. Vses. Teplot, Inst., 18 (8), 6, 1959.

9. Robbins, L.A., Kingrea, C.L.; Hydr. Process. & Petrol. Ref. 41 (5), 133, 1962.

10. Robbins, L.A., Kingrea, C.L.; Proc. Am. Petrol. Inst. Sect. III, 42, 52, 1962.

11. Everett, D.H.; J. Chem. Soc., 2566, 1960.

12. Bridgmann, P.W.; Proc. Am. Acad., 57, 75, 1922.

13. Bridgmann, P.W., Proc. Am. Acad., 59, 141, 1923.

14. Bridgmann, P.W.; Amer. J. Sci., 7, 81, 1924.

15. Rao, M.R., J. Chem. Phys., 9, 682, 1941.

16. Osida, I., Proc. Phys., Math. Soc. Japan, 21, 353, 1939.

17. Hirschfelder, J.O. et alii, "Molecular theory of gases and liquids", Wiley N.Y., 1964.

18. Viswanath, D.S., Am. Inst. Chem. Eng. J., 13 (5), 850, 1967.

19. Kardos, A., Forsch. Gebiete Ing. 5B, 14, 1934.

20. Sakiadis, B.C., Coates, J., Am. Inst. Chem. Eng. J., 1 (3), 275, 1955.

21. Badea, L., Rev. de Chemie, 26 (3), 209, 1975.

22. Narasimhan, K.S. et alii, Chem. Eng., 82 (8), 83, 1975.

23. Smith, J.F.D., Ind. Eng. Chem., 22, 1246, 1930.

24. Denbigh, K.J., J. Soc. Chem. Ind., 65, 61, 1946.

25. Scheffy, W.J., Tech. Rep. PR85R, Princeton Univ., N.J., 1-99, 1958 (AD 204 891).

26. Scheffy, W.J., Jhonson, E.F., J. Chem. Eng. Data, 6 (2), 245, 1961.

27. Missenard, F.A., "Therm. conduct. of solids, liquids, gases and their mixtures", Ed. Eyrol les, Paris 1965.

28. Debye, P., "Kinetische theorie der materie", 1914.

29. Pachaiyappan, V. et alii, J. Chem. Eng. Data, 11 (1), 73, 1966.

30. Pachaiyappan, V. et alii, Chem. Eng., 74, 140, 1967.

31. Ibrahim, S.H., Kuloor, N.R., Chem. Eng., 6, 147, 1966.

32. Pachaiyappan, V., Brit. Chem. Eng., 16, 382, 1971.

33. Pachaiyappan, V., Vaidyanathan, K.R., Proc. Symp. Thermoph. Prop., 6th, 1973, Liley, P.E. ASME, p. 15.

34. Vaidynathan, K.R., Valayutham, M., Chem. Eng. World IX (5), 69, 1974, Sect. I.

35. Sakiadis, B.C., Coates, J., Inst. Chem. Eng. J., 1 (3), 121, 1957.

36. Reid, R.C., Sherwood, R.K., "The prop. of gases and liquides -
    Their estim. and correl.", McGraw-Hill Book Co., Inc., N.Y.,
    1st ed. 1958.

37. Tsederberg, N.V., Teploenerg., 3 (9), 42, 1956

38. Filippov, L.P., Novoselova, N.S., Vestnik Moskov. Gos. Univ.,
    10 (3), 37, 1955.

39. Bates, O.K. et alli, Ind. Eng. Chem. Anal. Ed., 10, 314, 1938.

40. Bondi, A., A.I.Ch.E.J., 8, 610, 1962.

MEASUREMENT OF THERMAL PROPERTIES OF LIQUIDS

BY THE METHOD OF RADIAL TEMPERATURE WAVES

A. A. Varchenko

Institute of Inorganic Chemistry, Academy of

Sciences of USSR, 630090 Novosibirsk, USSR

A method for the simultaneous determination of thermal conductivity and thermal diffusivity of liquids has been developed based on the measurement of the amplitude and phase of temperature oscillations in a thin wire immersed in a liquid and heated by direct current with a superimposed ac component. A comparison of the obtained results with published data suggests that only the conductive part of thermal conductivity is measured under the conditions of the experiment.

INTRODUCTION

An analysis of the heat conductance equation for a medium propagated by radial temperature waves indicates that under these conditions it is possible to determine thermal conductivity, thermal diffusivity, and volumetric specific heat of the medium in a single experiment [1]. To achieve this, it is necessary to measure amplitudes and phases of small temperature oscillations either with thermocouples [2] or using a known temperature dependence of the resistance of the heater generating temperature waves.

In the present work we have used one of the variations of the second technique – an equivalent impedance method [3]. The method is based on the fact that when a thin wire is heated by direct current with a superimposed ac component the oscillations of the wire resistance, caused by the temperature oscillations, are pha-

se-lagged with respect to the ac current. This results
in wire resistance acquiring the properties of an impedance $Z=R_0(1+A-jB)$ where R_0 is the dc resistance of the
wire. In general case the value of the impedance is determined both by parameters of the wire and by the heat
transfer to a surrounding medium.

So far the equivalent impedance method has been employed in experiments in which the object of the study
was the wire itself. The heat exchange with the surroundings was either made negligibly small [3] or taken into account [4]. In the present case a reversed problem
is solved: the known wire impedance and its other parameters are used to determine thermal properties of an
external medium.

METHOD

The equivalent electrical circuit of a dc-heated
thin wire has been investigated by a number of authors
in connection with the bolometer theory. It can be shown
[5] that the impedance of such a wire is

$$Z= R_0(1+A-jB)=R_0\left[1+2I^2R'/(jmc\omega +hS-I^2R')\right], \qquad (1)$$

where I is the direct current, $R'=dR_0/dT$ is the temperature derivative of the wire resistance; m, c, S are, respectively, the mass, specific heat, and lateral surface
area of the wire; h is the heat-transfer coefficient.
In the case of a purely conductive heat exchange h is a
complex quantity which can be written as [1]:

$$h=\frac{\lambda}{r}x\sqrt{-j}\ H_1(x\sqrt{-j})/H_0(x\sqrt{-j})=\frac{\lambda}{r}\left[ReH(x)+j\ ImH(x)\right], \qquad (2)$$

$$x=r\sqrt{\omega/a}, \qquad (3)$$

where λ and a are thermal conductivity and thermal diffusivity of the medium, respectively; r is the wire radius, H_0 and H_1 are the Hankel functions, $ReH(x)$ and
$ImH(x)$ are the real and imaginary parts of the function
$x\sqrt{-j}\ H_1(x\sqrt{-j})/H_0(x\sqrt{-j})$. By substituting (2) into (1)
and separating real and imaginary parts we find:

$$tg\,\phi =ImH(x)/ReH(x)=\left[2B-\frac{mc\omega}{I^2R'}(A^2+B^2)\right]/(A^2+B^2+2A) \qquad (4)$$

and

$$\lambda S\ ReH(x)/r=I^2R'(A^2+B^2+2A)/(A^2+B^2). \qquad (5)$$

Both x and $ReH(x)$ are known functions of $tg\,\phi$ and can be either plotted or approximated by polynoms using the tables of mathematical functions [6]. After substituting x and ReH(x) into (3) and (5), respectively, we obtain the expressions for calculating a and λ:

$$a = \omega r^2/x^2, \qquad (6)$$

$$\lambda = I^2 R'(A^2+B^2+2A)/\left[2\pi L(A^2+B^2)\,ReH(x)\right], \qquad (7)$$

where L is the wire length. As can be seen from Fig. 1a the error in the determination of x and ReH(x) from the measured tgϕ grows rapidly with increasing x. Acceptable values of x should be regarded those lying within the .1 – 1 range. For organic liquids and the wire diameter of about 20 microns this corresponds to frequencies from 3 to 300 Hz.

Equations (4) and (7) involve the quantity A which is a difference between the real part of the impedance and the dc wire resistance. For its direct determination a bridge circuit was used (Fig. 1b) which could be simultaneously balanced for both the ac (by means of R_p and C) and the dc (by means of R) currents. If C_p is sufficiently large so that its reactive resistance at a working frequency can be neglected then the bridge balance conditions are:

$$\frac{R_p R_0(1+A-jB)}{R_0(1+A-jB)+R_p} = \frac{R}{1+j\omega RC} \qquad \text{and} \quad R_0 = R,$$

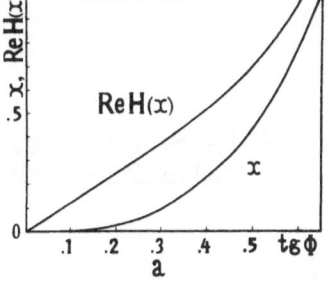 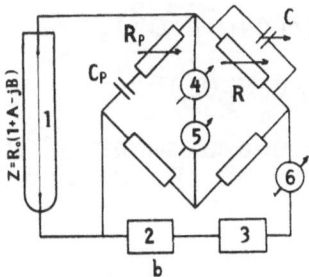

Fig. 1. a) Plot of $x = r\sqrt{\omega/a}$ and ReH(x) versus tgϕ. b) Block diagram of the measuring bridge circuit. 1 – sensor, 2 – dc source, 3 – audio oscillator, 4 – dc zero-indicator, 5 – ac zero-indicator, 6 – dc amperemeter.

Table I. Thermal properties of water calculated from the impedance of the wire at different frequencies

Freq. [Hz]	$A \cdot 10^2$	$B \cdot 10^3$	$a \cdot 10^9$ [m²/s]	λ [W/m K]	$\rho c \cdot 10^{-4}$ [J/m³K]
3	1.40	4.60	148	.619	418
4	1.32	4.56	148	.618	419
5	1.25	4.52	149	.620	415
6	1.20	4.48	150	.620	414
7	1.16	4.46	149	.620	415
8	1.12	4.42	149	.620	415
9	1.08	4.40	148	.619	418
10	1.05	4.37	148	.620	417
15	.94	4.26	149	.620	415
20	.85	4.16	147	.618	422
30	.75	4.00	148	.619	419
40	.67	3.86	150	.622	415
60	.57	3.63	150	.622	415
68	.54	3.56	151	.623	413
80	.50	3.46	150	.623	415
116	.42	3.21	149	.621	417

from which we find

$$B = \omega RC / \omega^2 R^2 C^2 + (1 - R/R_p)^2, \tag{8}$$

$$A = (R/R_p - \omega RCB)/(1 - R/R_p). \tag{9}$$

At lower frequencies when the influence of C_p can not be neglected, C and R_p in Eqs. (8) and (9) should be replaced, respectively, by $C - \Delta C$ and $k \cdot R_p$, where

$$\Delta C = C_p/(1 + \omega^2 R_p^2 C_p^2),$$

$$k = 1 + 1/(\omega R_p C_p)^2.$$

Thus, the measurement of a and λ is reduced to the determination of R, C, and R_p by balancing the bridge circuit.

EXPERIMENT AND DISCUSSION

A Pt wire .02 mm thick and 20 cm long welded to Pt leads was used in the experiment as a sensor. Its resistance as a function of temperature was determined experimentally and used for calculating the wire tempera-

Table II. Thermal conductivity of some organic liquids calculated from the impedance measurements

Liquid	λ [W/m K]	Published data
Benzene	.138	.1367 [7]
Toluene	.128	.1278 [7];.1281 [8];.1285 [12]
m-Xylene	.127	.1263 [7]
Carbon tetra-chloride	.0964	.0950 [7];.0972 [12]

ture and the temperature derivative R´. The ac balance of the bridge was indicated with a tuned narrow-band amplifier and the dc balance - with a photocompensation amplifier. The impedance was measured at frequencies from 20 to 160 Hz (in a few cases from 3 to 300 Hz) and at the sensor temperature near 30°C. The overheat of the sensor with respect to a surrounding liquid was about 5°C. Table I shows the results of calculating a,λ and ρc (volumetric specific heat) for water along with the impedance parameters A and B at different frequencies. Table II shows the results of calculating thermal conductivity of benzene, toluene, m-xylene, and carbon tetrachloride together with the results of some other authors.

In recent years a number of works [7,8,9] appeared in which the contribution of the radiative component into the effective thermal conductivity of liquids was studied. It was found that the effect of radiation decreases with decreasing measuring gap. At room temperatures the difference between λ values obtained by extrapolating to the zero-width measuring gap and those obtained with a 4 - 5 mm gap can amount, depending on a liquid, to 4 - 7 %. The radiation effect can be also strongly reduced if a thin wire is used as a heater [10, 11]. In our case both these requirements are met since the temperature waves propagating in a liquid attenuate across the distances of the order of .1 mm which is equivalent, in a way, to the use of a narrow measuring gap.

The above considerations, as well as a good agreement with the data either calculated for the zero-gap [7,8] or obtained with a gap of .5 mm [12], suggest

that only the conductive part of thermal conductivity
is measured by the present technique.

REFERENCES

1. L. P. Filippov, Investigation of thermal conductivity of liquids, Moscow University, Moscow, 1970
   (in Russian).
2. P. Andersson and G. Bäckström, J. Appl. Phys. 44,
   2601 (1973).
3. Ya. A. Kraftmakher, Zh. Prikl. Mekhan. Tekhn. Fis.
   №5, 176 (1962).
4. A. A. Varchenko and Ya. A. Kraftmakher, Phys. Stat.
   Sol. (a), 20, 387 (1973).
5. A. Van der Ziel, Solid state physical electronics,
   Prentice-Hall, Inc., N.Y. 1957, p. 472.
6. E. Jahnke and F. Emde, Tables of functions with formulae and curves, N.Y., Dover Publications, 1945.
7. H. Poltz and R. Jugel, Int. J. Heat Mass Transfer,
   10, 1075 (1967).
8. V. Z. Geller, I. A. Paramonov and G. A. Tatevosov.
   In: Thermophysical properties of liquids. Nauka,
   Moscow, 1973, p. 93 (in Russian).
9. T. V. Gurenkova, P. A. Norden and A. G. Usmanov,
   ibid., p. 97.
10. A. A. Men', Teplofizika Vysok. Temper. 11, 762 (1973)
11. K. L. Wray and T. J. Connolly, J. Appl. Phys. 30,
    1702 (1959).
12. D. R. Tree and W. Leidenfrost. In: Thermal conducti-
    vity. Proceedings of the 8-th Conference. Plenum
    Press, N.Y., 1969.

SESSION G   ROCKS AND SOILS

Session Chairmen: C.J. Shirtliffe
                     National Research Council
                     Ottawa, Ontario

                     W.P. Shimmel, Jr.
                     Sandia Lab.
                     Albuquerque, New Mexico

THERMAL CONDUCTIVITY OF S.E. NEW MEXICO ROCKSALT AND ANHYDRITE*

R. U. Acton

Sandia Laboratories

Albuquerque, NM 87115

Abstract

The thermal conductivity of several rocksalt materials has been determined. Some of the materials were core samples from well AEC 8, Carlsbad, New Mexico. These samples ranged from nearly pure halite (NaCl) to nearly pure anhydrite ($CaSO_4$). Core sample crystallite size ranged from about 3 centimeters to essentially packed salt sand (≈0.5mm). The samples exhibited thermal conductivities from ≈1.5 to 7.5 W/mK which depended upon purity and grain size.

A one meter cube of rocksalt from the Mississippi Chemical Company's S.E. New Mexico potash mine was obtained for other experiments. The thermal conductivity of one sample from each of the orthogonal directions of the cube was measured. This material had a high conductivity of ≈8.5 W/mK and was very isotrophic. A core of rocksalt from the Morton Salt Company, Paynesville, Ohio had a thermal conductivity of 6 W/mK, which is in the upper band of the results on cores from well AEC 8.

Finally, a concrete made with salt sand and rocksalt aggregate was determined to have a conductivity of ≈2 W/mK.

A longitudinal heat flow apparatus was used to determine the thermal conductivity. An analysis of the experiment gave an accuracy within $\pm$ 15% on geological samples and within $\pm$ 10% on 304 stainless steel.

*This work supported by the Department Of Energy

Introduction

Salt beds are geological formations that have lain undisturbed by seismic activity for millions of years.
Because of this fact, these salt beds are attractive
sites for the disposal of the radioactive wastes that
would be produced by the large scale production of
electricity by atomic reactors. The construction of
a waste disposal facility requires a complete characterization of the salt formation and an analysis of the
interactions of the waste and waste containers with the
salt surroundings. These thermal conductivity measurements form a small part of that characterization and
analysis.

Materials

Well AEC 8

Generalized descriptions of the core sections from well
AEC 8 near Carlsbad, New Mexico, are given in Table 1.
These descriptions are based on visual examination as
well as some x-ray diffraction analysis. Core compositions varied from about 95% halite (NaCl) to 95%
anhydrite ($CaSO_4$). Some cores contained other foreign
materials such as calcite, clay and sandstone. The
color of the cores ranged from white and opaque for core
527; through dark gray (smokey) and translucent for core
837; to reddish brown with both opaque and translucent
grains for core 281. The crystallite sizes varied between 0.5mm to 3 cm. Core densities averaged $2170 kg/m^3$
with a 15% spread.

Paynesville, Ohio

One core section (10cm long) from Paynesville, Ohio was
measured. No information is available about its composition or location. It is very similar in appearance to
AEC 8 core 837.

Mississippi Formation

Several cubes of rocksalt were obtained from the
Mississippi Chemical Company's S.E. New Mexico potash
mine. These cubes, 1 meter on a side, were used in
bench heating experiments. Three samples, corresponding
to the orthogonal directions, were taken from one cube
and labeled 'X', 'Y' and 'Z'. This material is salmon
in color. The three samples are indistinguishable from

one another. The densities for samples X, Y and Z are
2156, 2090 and 2157 kilogram per cubic meter, respectively.

The blocks are composed of halitic rock with 1-2%
sylvite (KCl). The halite crystal size varies from 2-3
mm to 1 cm with some crystals as large as 2 cm. Crude
layers are evident based on crystal size and varying
degrees of translucency. The layers are approximately
one centimeter thick and sometimes contain facturing
parallel to the layers. The sylvite is slightly milky
and is commonly outlined by a halo of orange iron oxide.
A few small (0.5 mm) fluid inclusions were noted--but
are not abundant. The halite does not show any well
defined foliation. The halite crystal cross-sections
are polygonal on faces normal to the bedding.

Saltcrete (concrete)

The composition of the concrete made with salt sand and
rocksalt aggregate is given in Table II.

TABLE II
Saltcrete
1560 Kg/m^3

Composition in Percent by Volume	
Portland Cement Type 3	13.93
Saltsand	31.10
Coarse Salt:	
#8 Sieve to 1/4 inch	16.65
1/4 to 1/2 inch	15.37
Brine	3.66
Water	19.29
	100.00

Concretes of this nature are of interest for several
reasons. They may be used for stemming bore holes,
grouting instrument packages in place of backfilling
around construction or cannisters. The composition of
the concrete is varied in an attempt to match the
properties of the surrounding strata.

TABLE I.   WELL AEC 8 CORE DESCRIPTIONS

Depth M	Density Kg/m^3	
281	2180	Roughly equigranular (up to 8 mm) reddish halite crystals (50% of rock) are set in a fine grained reddish sandstone consisting mainly of quartz and feldspar. Crude layering of sandstone and halite exists. Minor anhydrite is present.
504	2300	Calcite banded anhydrite (95%) which is finely crystalline (>.5 mm). Clear neomorphic halite (5%) is concentrated in small lenticular areas where banding has been contorted.
527	---	Massive appearing, finely crystalline anhydrite (95%) of rock with some poorly developed bedding. Halite (>5%) is dispersed through the rock.
560	2220	Halite (95%) consisting of large (up to 3 cm) crystals which vary from clear to light reddish color. Fluid inclusions (1-3 mm) are common in this halite. Some interstices are filled with a pink-colored anhydrite. Some minor polyhalite and unidentified clay minerals (especially along crudely horizontal surfaces).
628	2180	Large (1-2 cm) halite crystals with common small fluid inclusions (up to 2-3 mm) for 85% of rock. Anhydrite and calcite comprise about 15% of the rock, fill the interstices, and also form some crude bedding planes through the rock.

TABLE I.   WELL AEC 8 CORE DESCRIPTIONS
(continued)

Depth M	Density Kg/m^3	
646	2180	Finely crystalline anhydrite and calcite, without apparent microscopic structure, forms about 75% of rock; coarsely crystalline (up to 1 cm) neomorphic halite has formed generally vertical masses through the anhydrite, and is about 25% of the rock. Halite has a texture where individual masses appear to crudely radiate about dispersed centers or upward from laminations in the anhydrite.
777	2120	Very similar to 504m'; 90% calcite banded anhydrite with about 10% halite.
837	2000	Medium (2-4 mm) to coarsely crystalline (1-2 cm) clear halite (95% of rock). The halite is foliated at an angle of about 30° from the horizontal. About 5% of the rock consists of anhydrite in disseminated and lenticular forms.
?	2210	Morton Co., Paynesville, Ohio, similar to core 837 meters, AEC 8.

Experimental Method

A longitudinal heat flow apparatus was used to determine
the thermal conductivities of the materials of this report. The apparatus was designed specifically for use
with geological core sections. One feature of design
is that cores of different diameters and length may be
measured without effecting a change in the equipment.

The core sample rests upon a nichrome wound heater constructed of zirconia tiles. Thermocouples are placed in
drilled holes along the sample length and into the
centerline. A heat flux transducer is placed on the top
flat end of the sample. A piece of asbestos paper
cushions the transducer from the upper cold plate. The
upper cold plate may be used to apply pressure to the
stack, to insure proper interface contact, by a screw
arrangement. A schematic of the apparatus is shown in
Figure 1. Not shown in Figure 1 is lead foil tape
usually' affixed to the top flat sample face to insure
intimate contact between the sample and the heat flux
transducer. The entire sample stack is surrounded by a
powered silica insulation.

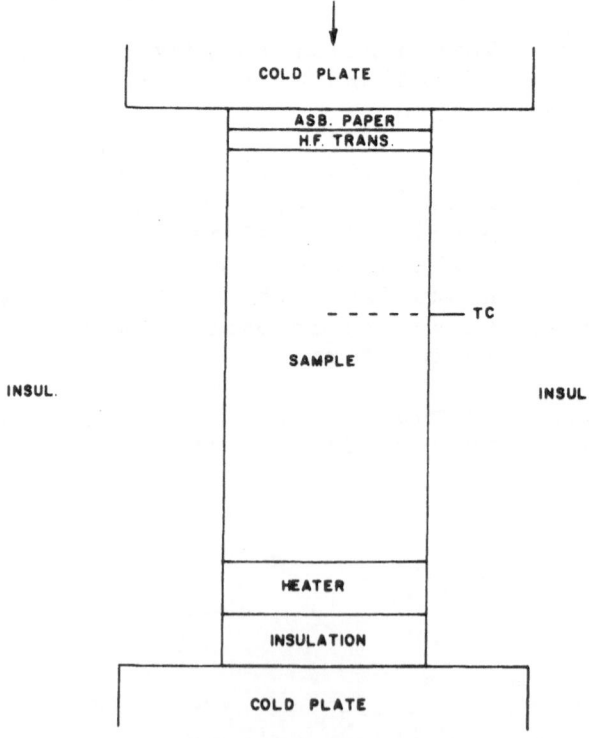

Figure 1.   Schematic of Thermal Conductivity Apparatus

The measurement procedure consists of applying constant power to the heater and allowing time for thermal equilibrium to be established. Once equilibrium is established, the heat flux transducer and thermocouple outputs are recorded. The thermal conductivity is then calculated from the equation

$$\lambda = \frac{Q/A}{dT/dX} \tag{1}$$

where

λ = thermal conductivity
Q/A = heat flux (transducer output)
dT/dX = temperature gradient (thermocouples output and location).

Analysis

The experiment has been analyzed using the finite differencing heat transfer code CINDA.[1] An auxiliary code, HEATMESH-71,[2] was used to generate node and conductor data required by CINDA. In the analysis, the experimentally measured thermal conductivity was input to the code as was either (i) the experimental sample hot face temperature or (ii) the transducer determined flux. The temperature gradient produced by the code was then compared to the experimental temperature gradient. The agreement between analysis and experiment has varied from 2 to 14% for geological materials. Part of this variation is believed to be due to the nonuniformity of some of the materials. Experiments on a sample of 304 stainless steel have yielded results that are within $\pm$ 5% ($300 \leq T (K) \leq 400$) and $\pm$ 10% ($400 \leq T (K) \leq 500$) of the recommended values[3] for that material.

CINDA requires re-coding for each change in sample dimensions. In addition to this, there was a desire to be able to check each data point, as it was taken, on the time-sharing terminal in the laboratory. For this purpose, an analytical solution was used. The experiment may be described as a pin protruding from a plane surface heated through its base and losing heat out of its curved surface and other end. The differential equation governing this configuration is

$$\frac{d^2\theta}{dx^2} - m^2\theta = 0 \tag{2}$$

where

$$\Theta = T - T_s$$

$$m = \sqrt{\frac{hC}{\lambda A}}$$

T = temperatures at x

T_s = temperature of surroundings

h = heat transfer coefficient at surface

C = circumference

A = cross sectional area

λ = thermal conductivity

Chapman[4] gives the general solution to this equation as

$$\Theta = Be^{-mx} + De^{mx} \tag{3}$$

where B and D are arbitrary constants evaluated (or specified) by the boundary conditions. Eq. (3) is more conveniently expressed in terms of the hyperbolic functions:

$$\frac{\Theta}{\Theta_o} = \frac{T - T_s}{T_o - T_s} = \frac{\cosh m(L - x) + H \sinh m (L - x)}{\cosh mL + H \sinh mL} \tag{4}$$

where

T_o = temperature of base

L = sample length

$$H = \frac{h_e}{\lambda m}$$

h_e = heat transfer coefficient at sample end.

All of the heat dissipated by the sample was conducted through the base ($x = 0$), Thus the heat flow rate, Q is given by

$$Q = - \lambda A \ (dT/dX)_{x = 0} \tag{5}$$

Introduction of the temperature distribution from Eq. (4) yields

$$Q = \lambda m A (T_o - T_s) \frac{\sinh mL + H \cosh mL}{\cosh mL + H \sinh mL} \qquad (6)$$

If h_e is allowed to equal zero, then H becomes zero and Eq. (6) reduces to the heat flow rate out of the curved surface only

$$Q = \lambda m A (T_o - T_s) \tanh mL \qquad (7)$$

Thus by using equation (4), (6) and (7), we may calculate the temperature gradient and the hot and cold face heat fluxes. Estimates of h and h_e are determined from a knowledge of the thermal conductivity of the powdered insulation and of the heat flux transducer. The experimental hot and cold face temperatures and the experimental thermal conductivity are used in these equations to yield an analytical temperature gradient and heat fluxes which are compared to their experimental counterparts.

The above technique does not actually give an accuracy figure but it does establish whether or not the data is self-consistent - a valuable aid to the technician who must decide whether to accept that data point and proceed or to look for trouble.

Results

The thermal conductivities of the cores that contained mostly salt are shown in Figure 2; those of mostly anhydrite in Figure 3 and the saltcrete in Figure 4. Data for the mutually perpendicular samples from the one meter cube of Mississippi rocksalt are shown in Figure 5.

Two of the cores from well AEC 8 and one of the Mississippi samples contained brine inclusions. These inclusions caused the samples to decrepitate - violently in two cases. The decrepitations were accompanied by a loud report and powdered insulation was thrown into the air. Core AEC 8, 527 was crumbled when the experiment was disassembled. When the sample crumbled is not known, therefore, all of the data for this sample is suspect. The Mississippi 'Y' sample decrepitated within moments after the power was turned off after taking the last data point. The sample cracked nearly in two about one

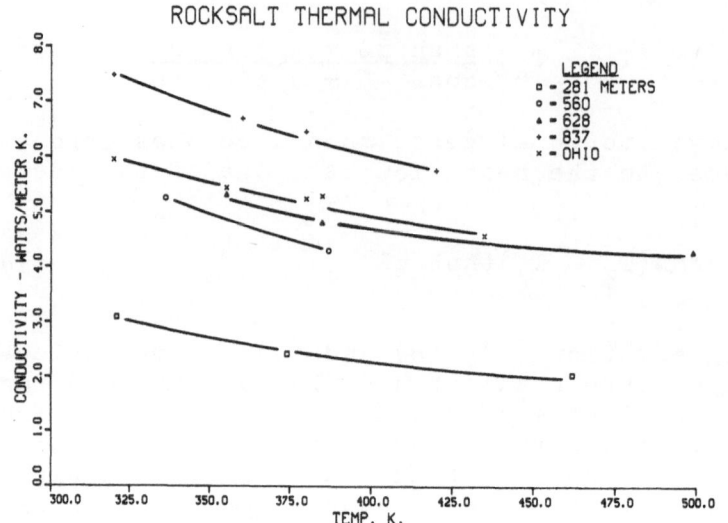

Figure 2.   Thermal Conductivity of Well AEC8
             Cores Containing 50% or More Halite.

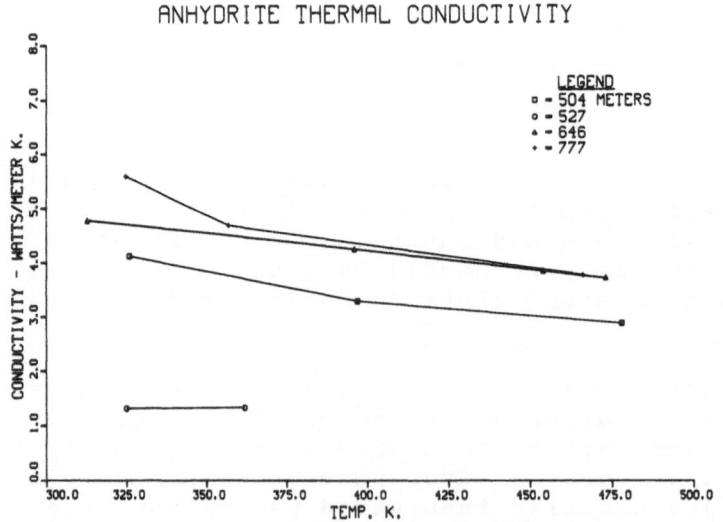

Figure 3.   Thermal Conductivity of Well AEC8
             Cores Containing 50% or More Anhydrite.

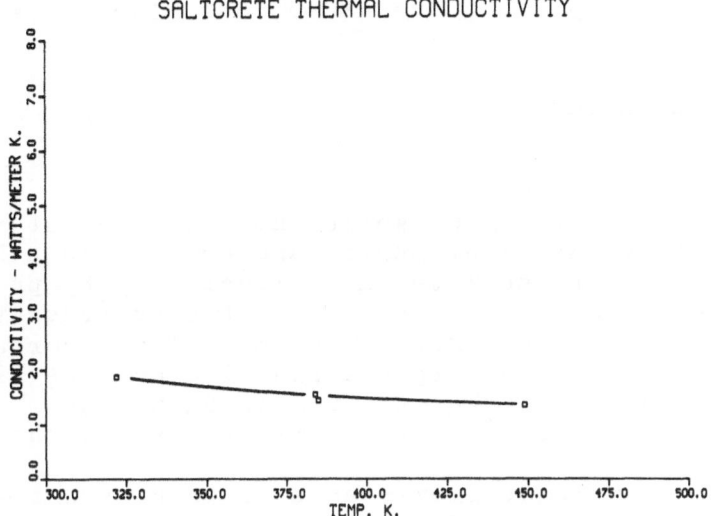

Figure 4.   Thermal Conductivity of A Concrete
Made with Saltsand and Rocksalt
Aggregate.

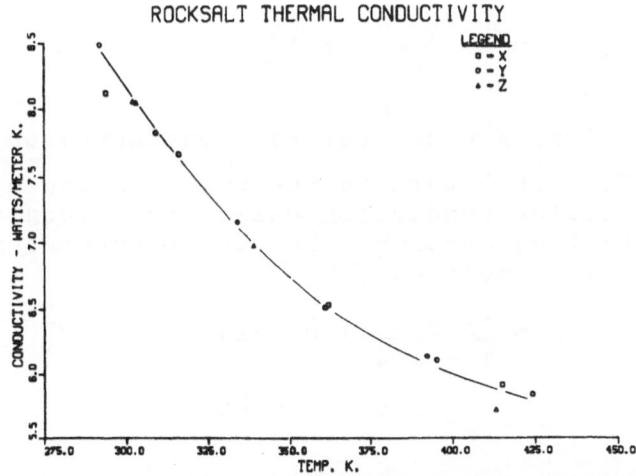

Figure 5.   Thermal Conductivity of a One Meter
Cube of Rocksalt From Mississippi
Chemical Co. S.E. New Mexico Potash
Mine. Measured in the Orthogonal
Directions.

inch from the hot face. The temperature at that location
was approximately 525K. Core AEC 8, 560 also decrepitated.

Discussion of Results

Salts

The thermal conductivity of solid, nonmetallic materials
is influenced primarily by purity and grain size. Heat
is transported in these materials principally by phonons
(lattice vibrations). Phonons are scattered (resisted)
by grain boundaries and impurity atoms. Those cores,
therefore, that had the larger grain size and the higher
purity exhibited the higher thermal conductivity. Phonon
transport is characterized by a $\frac{1}{T}$ relationship. Halite

and anhydrite can range from opaque to transparent –
again depending on grain size and purity. At higher
temperatures, heat may be transported through transparent materials by radiation. Radiant conduction is
characterized by a T^3 relationship. The three orthogonal
samples had the same thermal conductivity and yielded
enough data to test the above relationships. A nonlinear least squares routine[5] was used to fit the data
to an equation of the form

$$\lambda = \frac{b_1}{T - b_2} + b_3 T^3$$

where b_1, b_2 and b_3 are parameters (constants) to be

determined. The first term on the right in the equation
provides the lattice conduction while the second term
gives the radiant conduction. The least squares routine
gave the following equation:

$$\lambda = \frac{1549.9}{T - 100} + 0.13199 \times 10^{-7} T^3 \qquad (8)$$

$$290 < T < 450$$

where

λ = thermal conductivity, Watts per meter Kelvin

T = temperature, Kelvin.

Thirteen of the experimental data points within one percent of the curve generated by this equation. The other two data points are within 2.2% and 2.8%, respectively. Stephens and Maimoni[6] present an equation for Carey mine rocksalt from Winfield, La., where b_1 is 1423.2 and b_3 is 0.35581 x 10^{-8}.

A vertical traverse of a salt bed would go from dirty, impure salt at the edges to clean salt at the center. Herrin and Clark[7] have investigated the temperature gradients in the salt section of the Permian Salado formation of West Texas and Eastern New Mexico and have concluded that the effective thermal conductivity of the formation is 5.44 W/mK at 300 K. The average of the range of values for the current study is 5.75 W/mK at room temperature.

Saltcrete

The behavior of the saltcrete is typical of concretes in general. The value of the conductivity is a function of the density which in turn is a function of the particular aggregate and sand used. J. P. Moore[8] et al., show an increase by a factor of 12 in the thermal conductivity of concretes as the density varies from 400 to 2400 Kg/m^3. They also show a steady decline in the conductivity as a function of drying time -- reaching a constant value at about 200 days. The decrease was 21%. Other than the effect caused by water loss, the thermal conductivity of concretes is insensitive to temperature over the range of normal usage.

Conclusions

The thermal conductivity of a number of core sections of rocksalts has been investigated. Room temperature values for the thermal conductivity ranged from 3 to 8.5 W/mK with the purer materials having the higher values. No anisotropy was found in the thermal conductivity of purer rocksalt when mutually perpendicular samples were measured.

The dirty rocksalts are not being considered for radioactive waste isolation and the data was collected as base line information only. For the cleaner rocksalts, the thermal conductivity should be investigated to somewhat higher temperatures because this may influence the value of the thermal radiation conduction parameter b_3 in the equation (8).

REFERENCES

1 Chrysler Improved Numerical Differencing Analyzer, CINDA-3G, TN-AP-67-287, Space Div., Chrysler Corp., New Orleans, LA, 1967.

2 HEATMESH-71: A Computer Code for Generating Geometrical Data Required for Studies of Heat Transfer in Axisymmetric Structures, SCL-DR-72004, V. K. Gabrielson, Sandia Laboratories, Albuquerque, NM, 1972.

3 Thermal Conductivity, Metallic Elements and Alloys, Thermo-Physical Properties of Matter. Vol. 1, the TPRC Data Series, Y. S. Touloukian et al., IFI Plenum, NY, 1970, p. 1175.

4 Heat Transfer, 2nd ed., Alan A. Chapman, p. 59, MacMillan Co., NY, 1967.

5 TJMARI - a Fortran Subroutine for Nonlinear Least Squares Parameter Estimation, T. H. Jefferson, SLL-73-0305, Sandia Laboratories,

6 Thermal Conductivity of Rocksalt, D. R. Stephens and A. Maimoni, UCRL-6894 Rev. II, Lawrence Radiation Laboratory, California, 1964.

7 Heat Flow in West Texas and Eastern New Mexico, E. Herrin and S. Clark, Jr., Geophysics, Vol. XXI, No. 4, p. 1087, 1956.

8 Some Thermal Transport Properties of a Limestone Concrete, J. P. Moore et al., ORNL-TM-2644, Oak Ridge National Laboratory, 1969.

Acknowledgement

The author wishes to thank Dennis Powers, R. E. Stinebaugh and P. L. Nelson for furnishing the cores and their descriptions and A. J. Anaya for sample preparation and set-up.

THERMAL RESPONSES IN UNDERGROUND EXPERIMENTS IN A DOME SALT FORMATION*

G. H. Llewellyn

Oak Ridge National Laboratory

Oak Ridge, TN 37830

ABSTRACT

Identification of suitable sites and construction of radio-active waste repositories is the goal of the National Waste Terminal Storage (NWTS) Program. To provide design information for a repository in dome salt, in-situ experiments with nonradio-active heat sources are planned.

Three such experiments using electrical heat sources are scheduled to be carried out in a salt dome. The purpose of these experiments is to acquire rock mechanics data to ascertain the structural deformation due to the thermal load imposed, to study brine migration and corrosion, and to provide thermal data. A data acquisition system is provided with these experiments to monitor temperatures, heat fluxes, stresses, and ground displacement.

A thermal analysis was made on models of each of these experiments. The objective of the analysis is to verify the capability of making accurate transient temperature predictions by the use of computer modeling techniques. Another purpose is to measure in-situ thermal conductivity and compare the results with measurements taken from core samples.

*Research sponsored by the Division of Waste Management, U.S. Department of Energy under contract W-7405-eng-26 with the Union Carbide Corporation, Nuclear Division. By acceptance of this article, the publisher or recipient acknowledges the U.S. Government's right to retain a nonexclusive, royalty-free license in and to any copyright covering this article.

The HEATING5 computer program was used to predict transient
temperatures around the experiments for periods up to 2 years using
two-dimensional and three-dimensional heat transfer models. The results of analysis are presented with the associated boundary conditions used in the individual models.

INTRODUCTION

The U.S. Department of Energy (formerly U.S. Energy Research
and Development Administration) is considering alternatives for the
permanent disposal of radioactive waste. It appears that permanent
disposal can be most effectively accomplished by converting the
waste to a highly concentrated form and burying it in geological
formations. Heat generation in the high-level waste (HLW) and the
resulting major thermal problems must be accommodated in the design
of the waste repository. These considerations limit the amount of
waste material that can be stored in a given repository area.

TEMPERATURE LIMITS IN REPOSITORY DESIGN

Many geological formations are under consideration by the
National Terminal Waste Storage (NTWS) Program for the establishment of federal repositories. Dome salt, bedded salt, shale, and
granite are presently being investigated. The Office of Waste
Isolation (OWI) is currently in the process of formulating a con-
ceptual design for a repository in salt formations. There are
many factors that must be considered in establishing proper heat
loads for such a repository: (1) structural integrity of the rooms
and formation, (2) stability of the solidified waste, (3) brine
migration, and (4) pertinent environmental considerations.

The design involves detailed heat transfer and rock mechanics
studies. The thermal expansion of the salt induces strain and
causes deformation that could ultimately result in collapse of the
room if the temperature is not controlled. It is also necessary
to control the temperature to limit upheaval and eventual subsidence
on the surface that might in time lead to the formation of a lake
or sever a protective sheath of shale beneath an aquifer. It may
also be desirable to limit temperatures at specific depths due to
environmental considerations.

IN-SITU EXPERIMENTS

In order to provide reliable design information, in-situ
experiments have been made in Lyons, KS, which have been documented
in Project Salt Vault[1] using both radioactive waste and electrical

heaters for waste simulation. At present, experiments involving only simulated waste have been planned for installation in a domed salt formation by OWI to validate the previous work done in the salt vault experiments. The present experiments in the domed salt formation do not involve any radioactive material. It is these experiments which will be discussed.

The objectives of the thermal analysis involved in the experiments are: (1) ascertain strategic location for thermocouples and instrumentation in the salt bed, (2) exhibit the feasibility of using a computer program to accurately predict the transient temperature distribution anywhere in the experiment, (3) determine method of obtaining in-situ thermal conductivity as a function of temperature from regression analysis of the temperature data, and (4) compare the thermal conductivity obtained from the in-situ calculations with laboratory measurements made on the core samples.

CONFIGURATION OF DOMED SALT EXPERIMENTS

Three experiments have been designed for installation in an existing salt mine in a domed salt formation at Avery Island, LA, as shown in Fig. 1. These experiments will be used to compare the effect of sand and salt backfill on the pressure exerted on the protective sleeve in addition to the objectives previously mentioned. The experiments are strategically located between the salt pillars. Experiments A and B are monitored within a 50-ft radius, while experiment C is confined within a 40-ft radius. The experiments are separated by at least 110 ft, which in 2 years should cause no thermal interference with each other.

Experiment A is provided with a 5 kW continuously operating heater that is centrally located in the allocated area. Experiments B and C are similarly equipped with 4 kW heaters. All of the experiments contain stress meters, extensometers, and thermocouples. Experiment A is primarily concerned with temperature distribution, and is therefore more heavily instrumented with thermocouples, while experiment C contains more strain gages for determining pressures encountered on the protective sleeve due to the thermal expansion of the salt.

The purpose of these experiments is the acquisition of thermal, rock mechanics, brine migration, and corrosion data. These data will be used in the design of a federal repository similar to the one shown in the artist's conception (see Fig. 2).

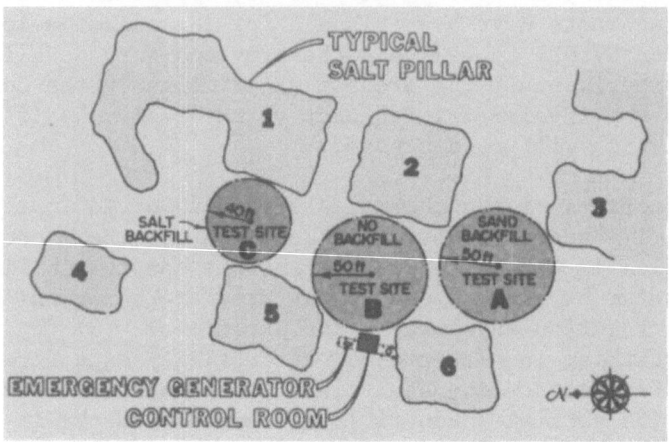

Fig. 1. Experiment Locations at a Depth of 550 Ft in a
         Domed Formation at Avery Island, LA.

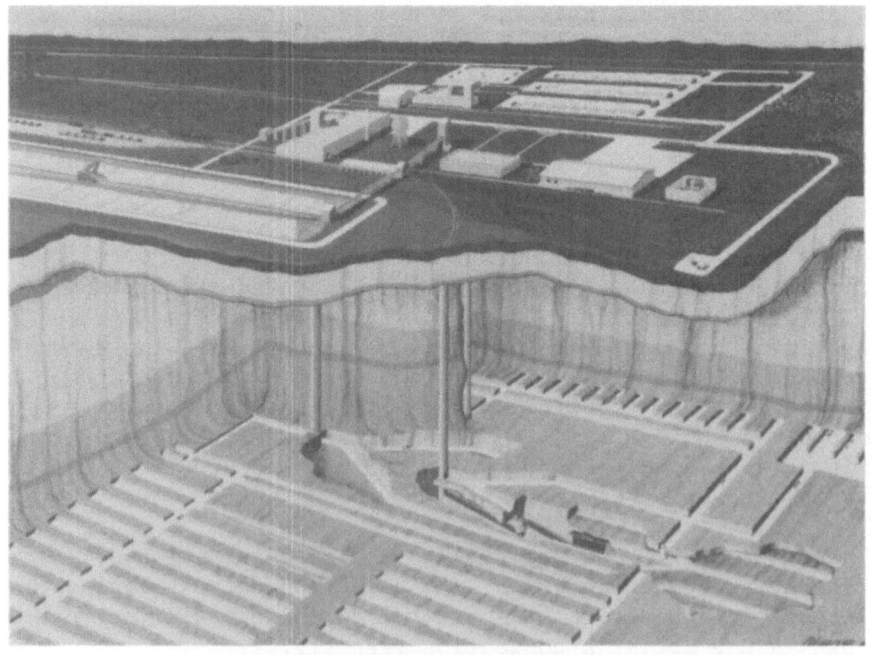

Fig. 2. Artist's Conception of a Waste Repository.

CALCULATIONS OF TEMPERATURE DISTRIBUTION

Analytical solutions have been developed[2] to obtain transient temperature distributions in salt similar to those derived by Carslaw and Jaeger[3] for point, finite line, and disk sources in an infinite media. The thermal conductivity of halite varies by a factor of 2.6 to 1 over the temperature range of 20°-250°C. For this reason, the solution will have to be obtained from a nonlinear analysis.

HEATING5[4] and TRUMP[5] computer codes have been successfully employed at Oak Ridge National Laboratory (ORNL) for determining the temperature distribution in near- and far-field models in conjunction with the NTWS program. The HEATING5 program has been used for most of these analyses and offers the capability of implicit solution techniques which yields a high ratio of real time to computer time in a transient analysis. The models employed in these analyses contain from 500 to 2500 nodes and require from 12 to 90 minutes of CPU time on our IBM 370-195 computer.

A plan and section view of experiment A is shown in Fig. 3. A 12-in. diameter sleeve, 1/4-in. thick, protects the heater canister from the pressure of the salt. Sand is used as backfill material. Thermocouple probes extend down to 25 ft and provide up to seven thermocouples in the vertical location. The plan view indicates the layout of the in-situ instrumentation. Note that heat flux monitors have been installed at strategic locations to check the heat flow in both the axial and radial directions.

Experiments B and C contain 8 peripheral heaters on a 6-ft diameter circle around the central heater and have a power rate that is a function of time. These heaters are 14 ft long and extend from 7 ft to 21 ft below the floor of the room. The central heaters used in each of the experiments will be operated on a continuous basis at a power level of 4 kW. The heat zone, filled with cast aluminum, has a diameter of 8 in. with an active length of 8 ft that is heated with 1/2-in. tubular heaters.

Experiment A contains sand as a backfill. Experiment B contains no backfill between the undisturbed salt and the protective sleeve, whereas experiment C contains crushed salt. It is desired to determine the effect of the backfill material on the pressure exerted on the protective sleeves in experiments B and C. There are 239 thermocouples, 8 resistance thermometers and 84 thermisters, 5 heat flux meters, and 11 stress gages in these experiments.

A data acquisition system is used to log and store most of the experimental data. The data will be stored on magnetic tape during

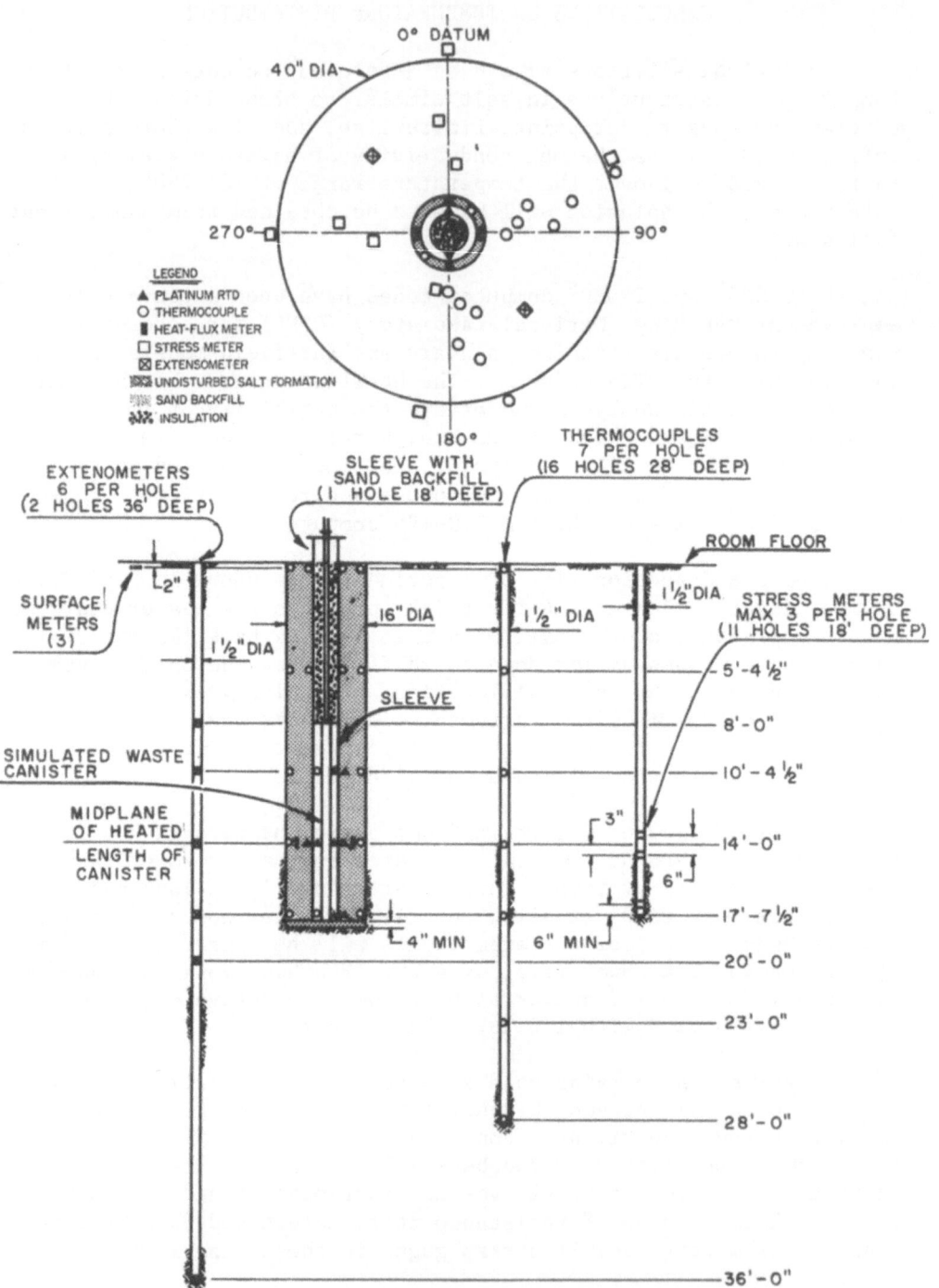

Fig. 3.   Plan and Profile of Instrumentation Employed in
Experiment C with a Sand Backfilled Hole.

the 1 1/2-year experiment. In addition, unmonitored extensometer data will be manually recorded to obtain deformation in the salt.

PHYSICAL PROPERTIES OF DOMED SALT

The halite formation is essentially pure sodium chloride containing less than 0.1% moisture and found to exhibit practically no anisotrophy in work performed by D. D. Smith[6] using a new laser technique. The thermal diffusivity and specific heat obtained from his analysis are shown in Fig. 4. A density of 2.16 gm/cc was used for the salt in these determinations. It is presently planned to take biaxial samples on the cores from the experimental holes and use DYNATECH analyzers for obtaining the measurements.

ERRORS DUE TO THERMOCOUPLE BACKFILL MATERIAL

Several problems were encountered in the emplacement of the thermocouples in 1 1/2-in. holes drilled in the floor of the room. The undisturbed salt has a thermal conductivity about 10 times that of crushed salt or sand which were prime candidates for backfill material. A two-dimensional nonlinear heat transfer model was used to compare the error by backfilling the thermocouple hole with a variety of materials. Errors as high as 8°F were found as shown in Fig. 5. By using compounds with a constant thermal conductivity of 2.0 Btu/hr·ft·°F, the maximum error could be reduced to about 1.5°F. It was finally decided to use crushed salt for the backfill material due to the fact that it will reconstitute itself into the solid form in the hotter regions and in the cooler regions it would be of less significance due to the higher conductivity.

HEAT TRANSFER MODEL FOR EXPERIMENT A

The predictions of the temperature in these experiments have been obtained from transient analysis using the HEATING5 program. A two-dimensional model of experiment A is shown in Fig. 6. The boundary conditions assume natural convection and radiation off of the floor to 70°F ambient air. The initial temperature was assumed to be 70°F throughout the models, although probably a gradient of about 1°F per 100 ft could exist in the actual situation. It should be pointed out at this point that thermal conductivity measurements within plus or minus 10% are deemed more than adequate for these studies.

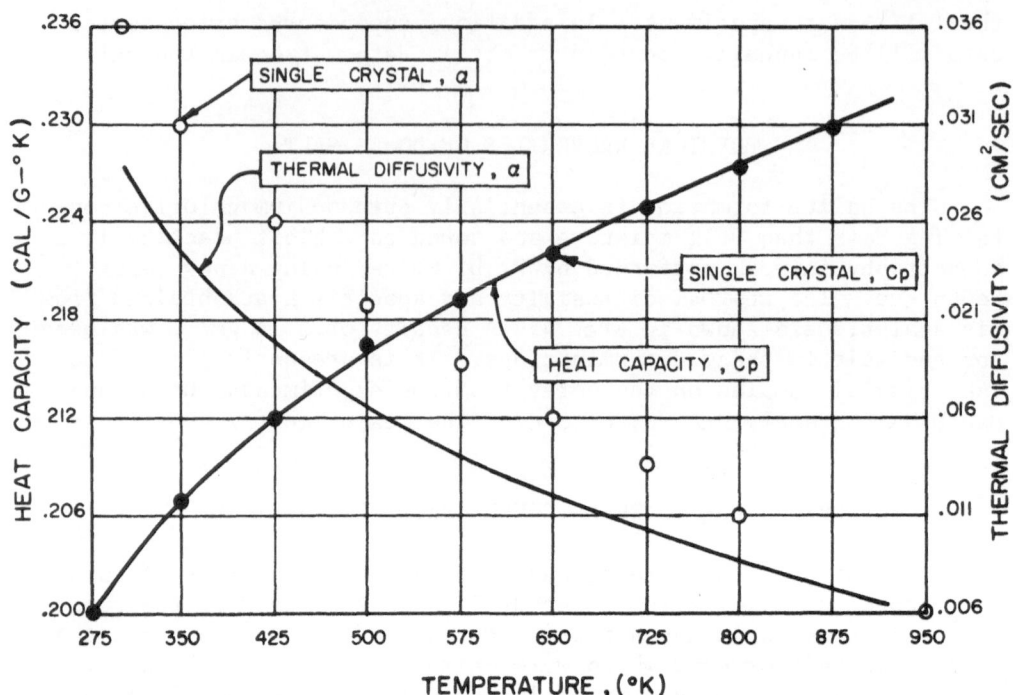

Fig. 4. Specific Heat and Thermal Diffusivity as Functions of
Temperature Obtained from Domed Salt Samples Taken at Avery Island.

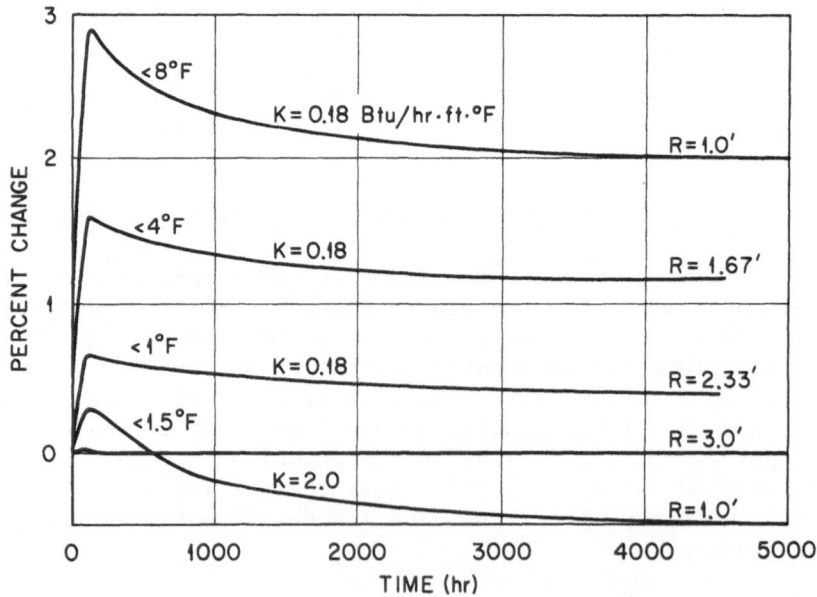

Fig. 5. Temperature Errors as Functions of Time for Backfill
Material with Different Thermal Conductivities.

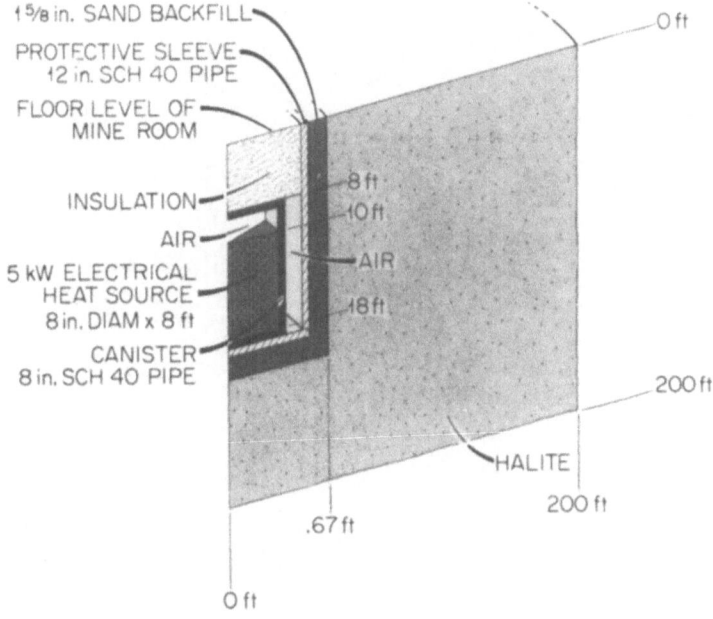

Fig. 6.   Two-Dimensional Heat Transfer Model for Experiment A.

SIMULATION OF AN ARRAY OF CANISTERS FOR EXPERIMENT C

A special two-dimensional nonlinear computer code was developed to determine the power transient in the peripheral heaters to produce a maximum temperature of 400°F in the salt in 1 year with the central heater at a constant power level of 4 kW throughout the transient. This program utilizes the principle of superposition to simulate the effect of heat flow from a matrix of canisters of high-level waste.

The temperature distributions obtained from these calculations are accurate only within the radius of the peripheral heater. To verify these results, a three-dimensional model using cylindrical coordinates was formulated as shown in Fig. 7. The results of the three-dimensional computer analysis produced an identical temperature distribution within the radius of the peripheral heaters and confirmed the reliability of the special two-dimensional computer code. The three-dimensional analysis gives accurate temperatures throughout the model.

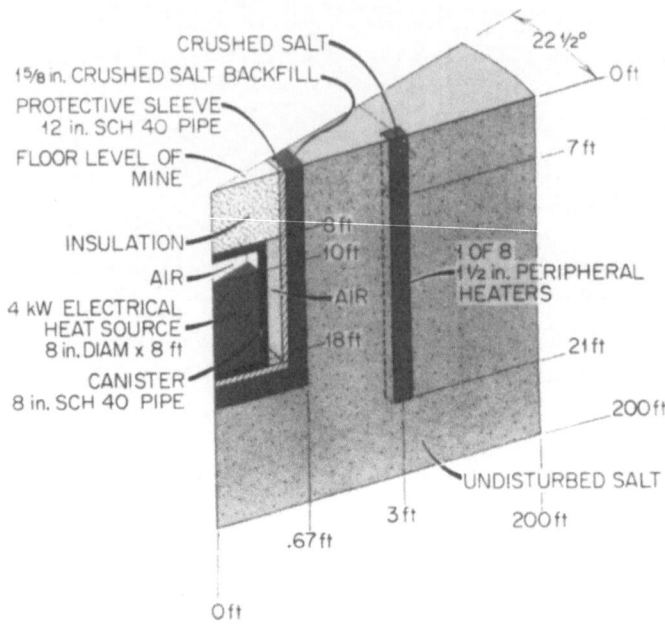

Fig. 7.   Three-Dimensional Heat Transfer Model for Experiment C.

RESULTS FROM THE ANALYSIS OF CASE A

The temperature distribution is shown in contours generated from the computer output after 6 months, 12 months, and 18 months in Fig. 8.   The heat source in these transients is the central heater which is producing heat at a constant rate of 5 kW through-out the transient.

SUMMARY

In-situ experiments have been designed to obtain pertinent physical and mechanical properties in a domed salt formation to gather data that will be used in the design of a federal repository for the terminal storage of radioactive waste.   In order to strate-gically locate the heat sources and thermocouples in these experi-ments, it was necessary to thermally model the experimental site. The results of these experiments will be used to evaluate the in-situ properties of the salt, establish brine migration rates, determine corrosive effects of the salt, and evaluate the short-term effects of heat on the adjacent salt.   The comparison of

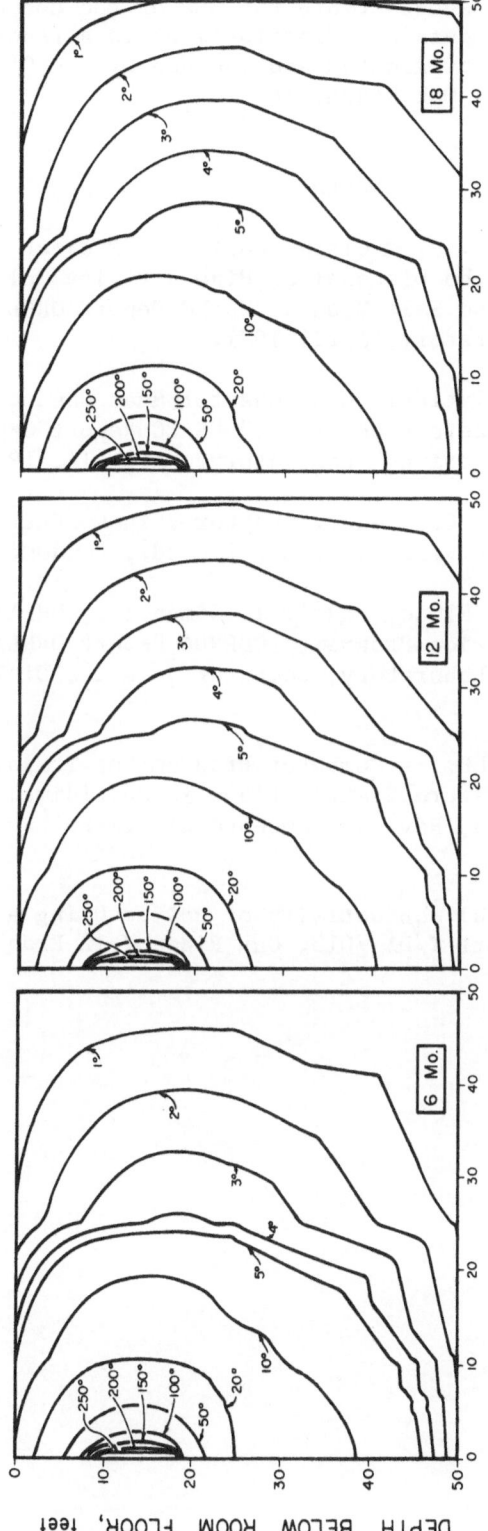

RADIAL DISTANCE, feet

DEPTH BELOW ROOM FLOOR, feet

Fig. 8. Contours Showing Increase in Salt Temperature in Experiment A After 6 Months, 12 Months, and 18 Months.

results will be used to substantiate the use of the computer in predicting transient temperature distributions in a repository. These efforts have been originated and managed by the Office of Waste Isolation, UCC-ND, Oak Ridge, TN.

REFERENCES

1. R. L. Bradshaw and W. C. McClain, editors, "Project Salt Vault: A Demonstration of the Disposal of High-Activity Solidified Wastes in Underground Salt Mines," USAEC Report ORNL-4555, Oak Ridge National Laboratory, April 1971.

2. J. P. Nichols, "Analytical Solutions for Heat Conduction in the Proposed National Waste Repository," USAEC Report ORNL CF 71-4-24, Oak Ridge National Laboratory, March 14, 1971.

3. H. S. Carslaw and J. C. Jaeger, Chapter 2 in Conduction of Heat in Solids, Oxford University Press, 2nd ed., London, 1959.

4. W. D. Turner, D. C. Elrod, and I. I. Siman-Tov, "HEATING5 - An IBM 360 Heat Conduction Program," USERDA Report ORNL/CSD/TM-15, Oak Ridge National Laboratory, Computer Sciences Division, March 1977.

5. A. L. Edwards, "TRUMP: A Computer Program for Transient and Steady-State Temperature Distributions in Multidimensional Systems," UCRL-14754, Rev. 3, Lawrence Livermore Laboratory, September 1, 1972.

6. D. D. Smith, "Thermal Conductivity of Halite Using a Pulsed Laser," USERDA Report Y/DA-7013, Oak Ridge Y-12 Plant, December 13, 1976.

SPACE-TIME-TEMPERATURE PROFILES FOR CONDUCTIVE HEAT FLOW IN COUNTRY ROCK SURROUNDING HOT INTRUSIVE BODIES[*]

M.S. Spergel[+] and P.W. Levy[#]

[+]York College, CUNY, Jamaica, N.Y. 11451

[#]Brookhaven National Lab., Upton, N.Y. 11973

ABSTRACT

Temperature vs distances profiles during geological time have been calculated for a variety of realistically shaped magmatic bodies intruded into country rock. The calculations were made with a computerized finite difference formulation of the classical time dependent conductive heat-flow equation. The formulation can be adopted to numerous situations. These include:emplacement at different rates, variations in the shape of the intrusion, e.g. domed cylinders, cones, or even arbitrary shapes as determined by field measurements. The intrusions may remain in contact with or become isolated from the magma pool. Also, the thermodynamic properties of the intrusion may be adjusted, e.g. to include latent heat of fractional crystallization. The surrounding rock may be homogeneous or strata of different kinds of rock. The surface above the intrusion may assume any reasonable shape. Faulting in the surrounding rock is being included and magma convection will be added in the future.

Completed calculations of the surface heat-flow gradients are consistent with those found in regions of known geothermal activity. This calculational procedure, in conjunction with local surface and subsurface temperature and heat-flow measurements appears to provide a useful approach to evaluating geothermal energy sources and new techniques for geothermal exploration and characterization.

INTRODUCTION

The increasing interest in geothermal energy
sources has been the motivation for a number experimental and theoretical investigations of hot magmatic
intrusions into the upper levels of the continental
crusts. Described below is a procedure for calculating
thermal histories for points both within and surrounding arbitary shaped intrusions. These intrusions may
vary in content and temperature and may be intruded into
stratigraphically complex regions. Many features of
this procedure are similiar to those described in
recently appearing publications. Other feature are
novel, especially those which have been introduced to
facilitate computation for shaped bodies in a strati-
graphic situation containing (mathematical) discontin-
uities corresponding to physical features such as
faults.
     Calculations which can furnish a realistic thermal
history of a intrusion and the surrounding region are
useful for many purposes. They can provide estimates,
especially when abetted by physical measurements (e.g.
knowledge of local geology, heat flow, seismic data etc),
of the size, shape, temperature, total thermal content,
time of emplacement and other physical properties of a
given intrusion. They can also be used to evaluate
measurement procedures such as fluid inclusion or
thermoluminescence paleotemperature techniques.

BACKGROUND

The United States has a large number of localites
which appear to contain thermal intrusions suitable for
development into geothermal energy sources. (McGetchin,
1975; White and Williams 1975; Reynolds, 1976). The
literature contains descriptions of several analytical
methods for calculating temperature profiles in and
around geothermal bodies. These include, for example,
the early papers of Lovering (1935,1936,1955), the
classic papers of Carslaw and Jaeger (1959) and Jaeger
1964. The last mentioned paper includes calculation of
temperature as a function of time for several geologi-
cally realistic situations. It has recently been adopted
to geothermal calculations of the Long Valley, California Caldera (Lachenbruch et.al,1976; this paper
contains numerous pertinent references.) Included in the
analytical methods are the recent steady state calculations of Horai (1976a,1976b).

The analytical methods, such as those mentioned
above, lack flexibility and are difficult to apply to
the space and time dependent conditions encountered in
geothermal calculations. However, numerical finite
difference methods are particularly well suited to problems involving time dependent conditions and to dis-
continuities. For example, finite difference methods
are readily adopted to handle the discontinuity in
thermal conductivity at the interface between the
intrusion and country rock, expansion or injections of
the intrusion, etc. Most importantly, finite difference
methods are readily programmed for the currently available large computers. In recent years, they have been
utilized by McGetchin (1975) and Witsan (1976). In
particular, McGetchin (1977) has used finite difference
methods for calculations on the thermal intrusion at
Fenton Hill on the edge of the Valles Caldera in the
Jemez Mountains of New Mexico.

The numerical finite difference procedure outlined
below has been formulated to describe realistic magmatic
intrusions in a variety of cases. The body may have any
realistic shape, e.g. cylinder, cone, sphere, spherically
capped cylinder or an arbitary shape. The intrusion may
remain in contact with the magma pool, and fed rapidly
or slowly by a pipe. Also the contact may be broken or
reestablished at will. The surrounding rock can be
arbitarly stratified, faulted, and each segment(layer)
given a set of thermal parameters. The thermal properties of the magma can be specified. Lastly, it is planned
to include additional features such as convection in the
magma body and the presence of water in the surrounding
medium.

THEORY

First, the classical time dependent conductive
heat flow equation is rewritten in finite difference
form. It is then applied to each cell in a finite three
dimensional net which includes the intrusion and the
surrounding rock. The properties of each cell are
established before the calculations is initiated. During
the calculation they vary in accord with the appropriate
physical conditions. Discontinuous phenomena such as
latent heat are included by using Kronecker-like delta
functions to account for the magnitude and direction of
latent heat flow into or out of a cell.

During the calculations, both the cell size and
temperature increments can be adjusted to remove the

instabilities inherent in finite difference calculations.
Also, to optimize the computer usage, small cells and
time steps may be used when the parameters, e.g.temper-
ature, are varying rapidly and larger increments and
cells used when the parameters change slowly.

With the present formulation, in each case it is
necessary to determine if adequate boundary conditions
have been established at the interface of the grid and
the surrounding material. The upper surface, which
represents the air-ground interface, does not pose a
problem. The surface temperature is fixed at 0°C. It
is assumed that the material outside of the grid other
than at the air-ground interface is the same as the
material at the grid surfaces. A number of different
boundary conditions at the grid surface have been tested.
For example it can be assumed that the second derivative of the temperature with heat conductive equation,
i.e. the rate of heat flow across the boundary, is
continuous. This assumption has been tested with a
fixed size intrusion in grids of different size. Usually
grids with edges that are 2 or 3 times the diameter of
the intrusion appear to be adequate, larger grids do
not produce appreciably different time temperature
profiles.

RESULTS

Temperature-distance profiles at various times have
been calculated for a number of physically realistic
yet simply shaped basaltic and granitic intrusions into
a granitic medium. The shapes included are solid cones,
right cylinders and domed cylinders. It was assumed
they were emplaced instantly. Numerical values for
thermal conductivity, specific heat, etc., were obtained
for Clark (1966). Also latent heat effects have been
evaluated.

The geometry of the intrusions is illustrated in
Fig.1. The horizontal surface, corresponding to the
air-ground interface is maintained at zero degrees C.
The radiogenic heating is not included. Its effect can
be included by superimposing an additional temperature
gradient of approximately 25°C/km.

Table I-Typical Measured Surface Thermal Gradient Values

| Location | Heat flow in HFU (=10^{-6} cal/cm^2/sec) | |
	Near Anomally	On Fringe of Anamoly
Radiogenic	1.53	1.53
Valles Caldera	3.7-5	2.8
Socorro	10	2.5

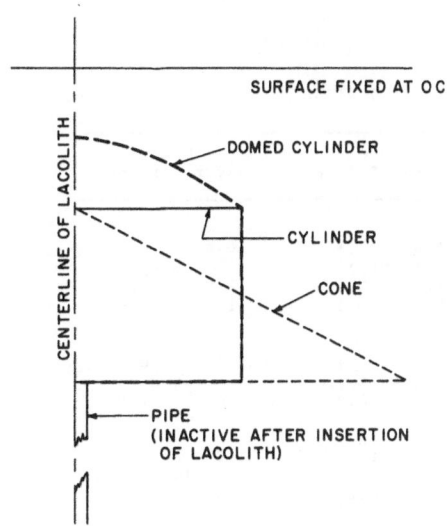

Fig. 1. Lacolith geometries used for computations described in this paper. The bases of all intrusions lie 6km below the surface.

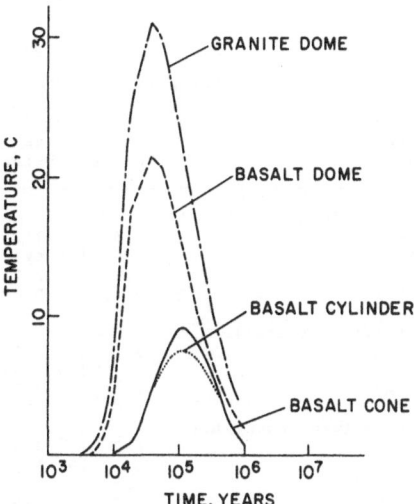

Fig. 2. Temperature vs time after insertion, 100 m below surface and at a point on the lacolith centerline. All bodies were inserted at 1100°C.

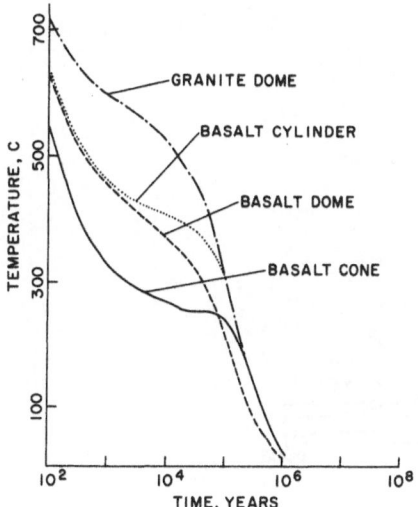

Fig. 3. Temperature vs time after insertion, on the roof of the lacolith and at a point on the center line.

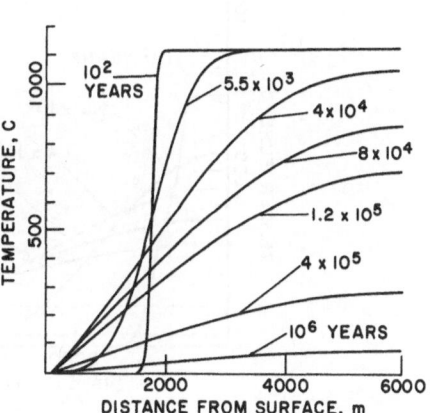

Fig. 4. Temperature vs distance from surface along the centerline of the domed lacolith and at various times after insertion.

TABLE II

CALCULATED THERMAL GRADIENTS AT SURFACE[a]

Intrusive Medium	Cylinder		Cone		Dome	
	Center	Fringe	Center	Fringe	Center	Fringe
Basalt (w/o latent heat)						
Peak	5.8	1.2	4.6	2.8	10.8	0.4
(Time)		(80 Ky)		(140Ky)		(40 Ky)
400 Ky	2.2	2.2	2.3	2.4	2.8	2.7
1000 Ky	0.45	0.5	0.5	0.5	1	0.8
Basalt (latent heat)						
Peak	6	1.2	4.6	2.4	11.2	0.3
		(100 Ky)		(140 Ky)		(40 Ky)
400 Ky	2	2.1	2.2	2.3	2.8	2.8
Granite (w/o latent heat)						
Peak	8.3	2.5	not		15.8	0.6
		(100 Ky)				(40 Ky)
400 Ky	3.7	3.4	computed		5	4
1000 Ky	0.9	0.9			1.8	1.4

a. In heat flow units (1 HFU=10^{-6} cal/cm/sec)

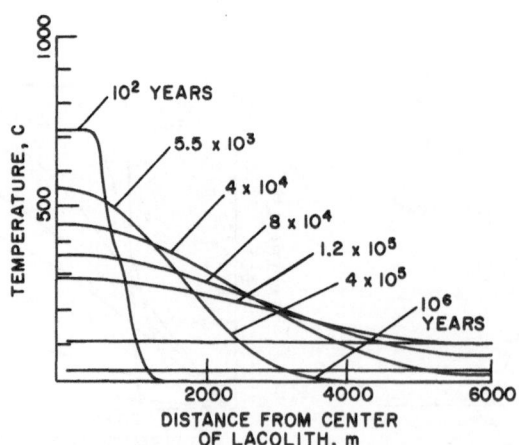

Fig. 5. Temperature vs distance from centerline of the domed lacolith, at the depth of the lacolith roof and at various times after insertion.

Thermal histories at a point 100 meters below the ground surface and centered over the intrusions are shown in Fig. 2. The bases of all bodies considered are 6km below the surface: In general, granitic inclusions produce higher temperatures than basaltic bodies. This is attributable to the greater specific heat of granites. At 100m below the surface, it is apparent that the inclusion of 100 Cal/gm latent heat in the intrusion does not produce a detectable effect. This illustrates an advantage of the numerical approach over an analytical one. To include latent heat effects in analytical calculations one usually employs an adjusted initial temperature or specific heat. The analytical approach tends to over compensate for the latent heat effects.

Thermal histories at the center of the top of various intrusions are shown in Fig. 3. Temperatures near lacoliths, at various times after intrusion, are illustrated in Figs. 4 and 5. The curves behave in the expected manner.

The reliability of these calculations is illustrated by comparing computed gradients with measured values. Typical measured values near or over known thermal anamolies, are given in Table I. Calculated values for cases in Figs 2 to 5 and additional ones are given in Table II. Clearly the measured and compiled values are comparable.

To summarize, the finite difference calculations described above appear to be both useful and practical. Furthermore, although not discussed above, they can be carried out with reasonable computer use; the excutation times never exceeded 9 minutes and utilized only 1/3 of the available storage. Thus, it would appear that this technique can be developed into a useful tool for calculating numerous and varied characteristics of thermal intrusions.

BIBLIOGRAPHY

Carslaw, H.S. and J.C. Jaeger, Conduction of Heat in Solids, 2nd ed., (Oxford University Press, New York, 1959).
Clark,S.P., Handbook of Physical Constants,(The Geological Soc. of Amer. N.Y.,N.Y. 1966).
Horai,K., Heat flow anomaly associated with dike intrusion 1, J.Geophy. Res. 79, 1640,1974.
Horai,K.,Heat flow anomaly associated with dike intrusion 2, J.Geophy. Res. 83, 894, 1976.
Jaeger, J.C.,Thermal Effects in Instrusions, Rev.Geophy, 2, 443,1964.

Kolstad, C.D., and T.R. McKetchin, Thermal Models for the
Valles Caldera with References to a Hot-Dry-Rock Geo-
thermal Experiment, LA-UR-77-33 (Los Alamo's Scientific
Laboratory, N.M. 1977).
Lachenbruck, A.H., J.H. Sass, R.J. Munroe, and T. Moses,
Geothermal Setting and Simple Heat Conduction Models for
the Long Valley Caldera, J. Geophys. Res. 81, 769 1976.
Lovering, T.S., Theory of heat conduction applied to
geological problems, Bull. Geol.Soc.Am.,46,69 1935.
McGetchin, T.R., Complied, Solid Earth Geosciences
Research Activities at LASL, LA-5956 PR, (Los Alamos
Scientific Laboratory, N.M., 1975).
Nitson, U., C. Kolstad and T. McGetchin, Thermal History
of Plutons, in Summaries of Physical Research in
Geosciences, ERDA 76-44, 21, 1976.
Reynolds, G.T., Energy research for physicists, Physics
Today, 29, 34, 1976.
Sanford, A.R., R.P. Mott, Jr., P.J. Shuleski, E.J.
Rinehart, F.J. Caravella, R.M. Ward and T.C. Wallace,
Geophysical Evidence for a Magma Body in the Crust in
the Vicinity of Socorro, New Mexico, ONR-CSM Symposium
on the Nature and Physical Properties of the Earth's
Crust, Vail, Colorado, August 2-6,1976.
White D.E. and D.L. Williams, Ed., Assessment of
Geothermal Resources of the United States - 1975, U.S.
Geological Survey, Circ. 726, 1975.

*Research supported in part by the U.S. Department of
Energy.

THERMAL CONDUCTIVITY OF CERTAIN ROCK TYPES AND ITS RELEVANCE TO

THE STORAGE OF NUCLEAR WASTE

V.V. Mirkovich and J.A. Soles

Canada Centre for Mineral and Energy Technology

Dept. of Energy, Mines and Resources, Ottawa, Canada

Nine rocks selected from the surface of three plutons have been examined petrographically and their thermal conductivities measured in the temperature range of 100° to 500°C. The thermal conductivities of different rock types were found to vary considerably with composition. They also changed with temperature, but at different rates for different rocks. The study indicates that thermal conductivities of rocks and their changes with temperature are important factors when considering their suitability in storing nuclear wastes.

INTRODUCTION

One of the major problems in nuclear energy generation is the disposal of nuclear waste. Several methods for its disposal have been considered, one of which is storage in rock masses. Because certain products of nuclear fission continue to decay, heat is generated and a knowledge of the thermophysical properties of the surrounding rock mass is required to predict changes that may adversely affect the storage site.

Thermal conductivity of a rock is an important property because it establishes the rate at which heat is dissipated (in the steady-state configuration) through that rock mass. Consequently, it provides a sensitive measure by which one can compare the stability of different rock masses where thermal stresses are involved.

Thermal conductivities of many rocks have been reported in the literature (1,2,3,4,5,6,7) and a fairly comprehensive review of the subject was given recently by Horai (8). Unfortunately,

because of often inadequate descriptions and wide variations in
the composition and texture of rocks of the same type, the thermal
conductivity of a particular rock cannot be established with con-
fidence from previously published data. Thermal conductivity
measurements were therefore made as a function of temperature, and
in some cases were related to a visible anisotropy of the rock.

PROCEDURES

(a) Method of Measurement

A comparative-method thermal conductivity apparatus, described
in detail elsewhere (9), was used for this work. Briefly, a 25.4-
mm diameter cylinder of rock 25.4 mm long is placed between two
standards of the same dimensions. The column thus formed is placed
on a heat stabilizer (a cylinder of the same dimensions and usually
made of the same materials as the standards) and held between the
heat source and heat sink. The entire assembly, shown in Figure
1, is surrounded by a cylindrical heat guard.

Temperatures in the cylinders and the heat guard were measured
with 28 B&S guage chromel-alumel thermocouples. Pyroceram 9606*
was used for the standards. All measurements were made at successively increasing temperatures. The temperature gradient within
the samples was between 10° and 20°C.

This apparatus was used because it accommodates rock samples
which are sufficiently large to integrate the contribution of
individual grains to the overall thermal conductivity of the specimen, yet are sufficiently small to facilitate selection of specimens which were free of visible cracks.

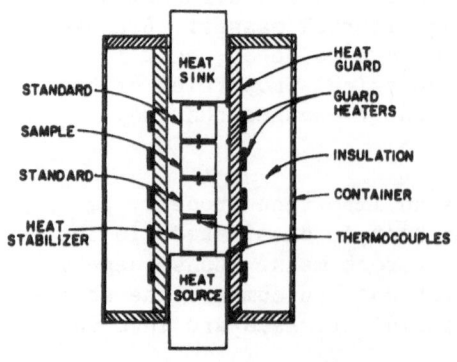

Figure 1. Thermal conductivity apparatus, schematic.

(b) Materials

The rock samples used for
this study were selected from
surface specimens that were
collected from three different
plutons. The three plutons are:
(1) syenitic, (2) gabbroic and
(3) granitic complexes. The
compositions and brief descriptions of the specimens are
given in Table 1.

*Supplied by Corning Glass,
Corning, New York, U.S.A.

TABLE 1

Composition of Rock Specimens from Three Plutons

Sample number	Sample orient	Rock type	Texture; structure	Density g/cm³	Volume per cent of minerals											
					Qz	K	P (An)	Am	Px	Mi	Ca	Ep	Cl	Mg	Sc	Acc
			Syenitic Rocks													
1 2	0° 90°	Syenite	F-gr, hypid.; hom.	2.74	0.5	56	23(15)	12u,h	4	<1b	–	–	<1	3	–	<1
3 4	0° 90°	Monzonite	M-gr, hypid.; hom.	2.80	–	40	28(10)	<1h	14	13b	–	–	–	4	–	1
5 6	0° 90°	Monzonite	M-gr, hypid.; hom.	2.78	0.5	40	35(10)	18h	–	4b	–	–	–	0.5	–	<1
7	–	Alt. Syenite	M-gr, hypid.; hom.	2.66	–	55	20(15)	5u	–	<1b	3	–	10	–	–	<1
8	–	Granite	F-gr, hypid.; hom.	2.58	31	30	35(15)	<1	–	3m	–	–	–	–	–	<1
			Gabbroic Rocks													
9 10	∥ ⊥	Chloritized Amphibolite	M-gr, hypid.; slightly foliated	2.99	–	–	2(10)	35h	19	–	3	<1	38	2	–	<1
11	–	Gabbro	M-gr, hypid.; hom.	2.95	–	–	33(45)	19u	32	↓	<1	↑	15	<1	–	–
12	–	Scapolite "Gabbro"	M-gr, hypid.; hom.	3.04	–	–	5(?)	3u	43	1b	2	2	–	<1	38	<1
			Granitic Rocks													
13 14	∥ ⊥	Granodiorite	M-gr. hypid.; foliated	2.70	18	8	53(25)	6h	–	14b	–	<1	–	–	–	<1

LEGEND: F-gr = fine grain (<3mm); M-gr = medium grain (2-6mm); hypid. = hypidiomorphic; hom. = homogeneous; Qz = quartz; K = potassium feldspar; P = plagioclase: (An) = anorthite content; Am = amphibole: h = hornblende, u = uralite; Px = pyroxene; Mi = mica: b = biotite, m = muscovite; Ca = calcite; Ep = epidote; Cl = chlorite; Mg = magnetite, ilmenite; Sc = scapolite; Acc = accessory (apatite, sphene, sulphides).

TABLE 2

Smoothed Thermal Conductivities for the Different Rock Types, W/cm°C

Temp. °C	Rock Specimen No.													
	1	2	3	4	5	6	7	8	9	10	11	12	13	14
100	-	-	-	-	-	-	-	.0293	.0333	.0285	-	-	.0229	-
125	.0207	.0207	-	-	-	-	.0217	.0283	.0328	.0280	-	-	.0225	-
150	.0203	.0203	.0197	.0210	.0210	.0210	.0213	.0272	.0322	.0275	.0244	.0223	.0221	.0178
200	.0196	.0196	.0192	.0194	.0203	.0203	.0204	.0253	.0312	.0265	.0234	.0216	.0213	.0170
300	.0180	.0181	.0179	.0180	.0191	.0188	.0186	.0220	.0288	.0241	.0214	.0200	.0197	.0155
400	.0167	.0167	.0166	.0166	.0180	.0173	.0168	.0192	.0258	.0218	.0195	.0182	.0176	.0140
500	.0151	.0152	.0149	.0152	.0171	.0159	.0151	.0164	.0221	.0194	.0175	.0163	.0151	.0125

	Syenitic Rocks						Gabbroic Rocks			Granitic Rocks	

The cylindrical specimens were cut from the samples according to the structural anisotropy. For rocks 9-10 and 13-14 the symbols $\parallel$ and $\perp$ in Table 1 indicate that the cylinders were cut parallel or perpendicular to a visible foliation. Specimens 1 to 6 represent three megascopically homogeneous, syenitic types from each of which two mutually perpendicular specimens were cut to demonstrate that there would be no thermal anisotropy in a structurally isotropic rock.

The 25.4 x 25.4-mm cylinders were ground to a tolerance of ±0.025 mm. Two thermocouple holes, 0.75 mm in diameter, were drilled radially to the cylinder axis at a distance of 1.3 mm from each end.

RESULTS AND DISCUSSION

The results of the thermal conductivity measurements are shown in Figures 2 and 3, and the smoothed thermal conductivity values for all rock specimens are summarized in Table 2. The curves were fitted to the data by inspection; each represents 8 to 10 conductivity measurements. The average deviation of the experimental data from the smoothed values is ±1.1%. The standard deviation calculated for all data is ±1.2%.

Figure 2 shows the data for the three specimen pairs of homogeneous rocks from the syenitic pluton. Each pair consists of specimens cut perpendicularly to each other. Curves 1 and 2 represent the thermal conductivities of a fine-grained syenite, and

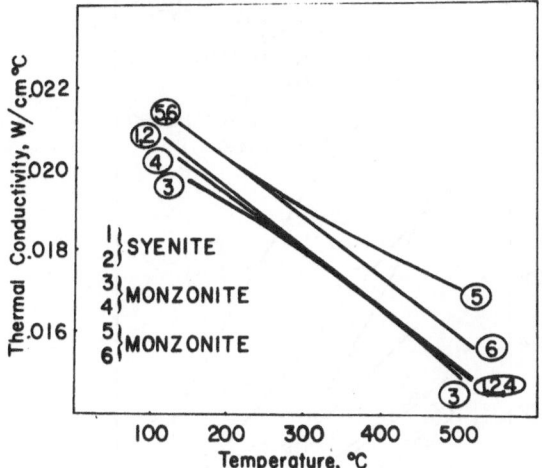

Figure 2.   Thermal conductivity of the syenitic rocks.

curves 3 and 4 the conductivities of a medium-grained monzonite.
The fact that the four curves are virtually superimposed indicates
that there is no difference in thermal conductivity due to either
the difference in the direction of the cut or the difference in
mineralogical composition. Curves 5 and 6 show the conductivities
of a monzonite which is structurally and compositionally similar
to the monzonite samples 3 and 4, but has distinctly, although
slightly, higher conductivities than the latter. Moreover, the
thermal conductivities for the two directions of cut, from samples
5 and 6, diverge pronouncedly at higher temperatures.

The data in Figure 2 show that rocks of similar composition
may have different thermal conductivities whereas rocks of different
compositions may have similar conductivities. They also indicate
the difficulty of accurately predicting the thermal conductivity
of a rock from its mineralogical composition and from conductivity
data obtained previously on similar rock types.

The thermal conductivities of the other rocks are shown in
Figure 3, with the data for the syenitic rocks indicated by the
shaded area. Specimens 7 and 8, an altered syenite and a granite
respectively, are also from the syenitic complex. Not surprisingly,
the thermal conductivity of the altered syenite falls mostly within
the shaded area, but that of the mineralogically distinctive
granite is quite different from all other rock types. At low
temperatures, its conductivity is relatively high; however, with
increasing temperature it decreases rapidly so that at 500°C it
equals that of the syenites.

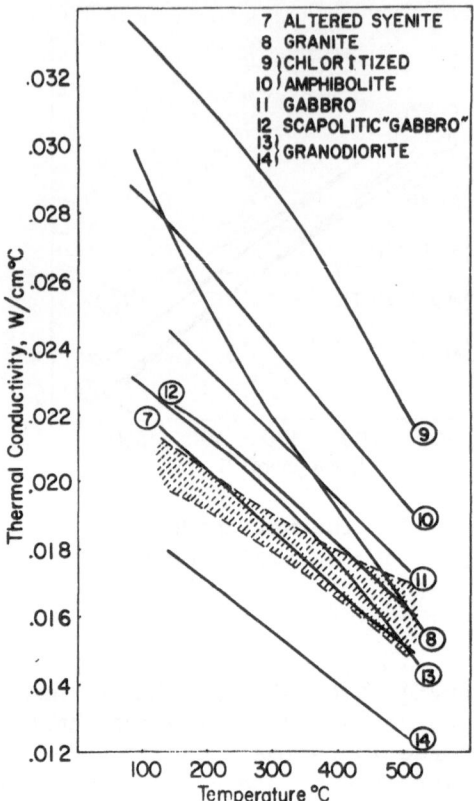

Figure 3. Thermal conductivities of the various plutonic
rocks. The syenitic group is
shaded.

The highest thermal conductivity was obtained with chloritized
amphibolite from the gabbroic pluton. Curve 9 represents the
thermal conductivities measured parallel to a foliation in the
rock, and curve 10 the conductivities measured perpendicular to
the foliation. The foliation is barely visible, yet the difference
between the conductivities is substantial. The high thermal conductivity of this rock can probably be attributed to the high
content of chlorite, a hydrated magnesium-iron aluminosilicate.
Because the foliation is caused by orientation of the chlorite,
there is a large difference in thermal conductivity of the samples
in different directions.

The other two rocks from the gabbroic group show appreciably
lower thermal conductivities. At higher temperatures, the conductivity of the gabbro approaches the syenitic region, and that of
scapolite "gabbro" actually lies within the syenitic region. The
content of chlorite does appear to be the principal factor influen-

cing the thermal conductivity of these rocks. A decreasing quantity of chlorite, from 38% in chloritized amphibolite to 15% in gabbro to 0% in scapolite "gabbro," is reflected in their respective thermal conductivities.

The lowest thermal conductivity for all rocks in this study was obtained on granodiorite from the granitic pluton, specifically on sample 14 which was cut perpendicular to the foliation. It is interesting that the thermal conductivity curves of both foliated rocks (i.e., the chloritized amphibolite and the granodiorite) have similar shapes, in that the thermal conductivity of specimens cut parallel to the foliation (9 and 13) decreases more rapidly with increasing temperature than that of the specimens cut perpendicular to the foliation (10 and 14).

Considering a nuclear waste repository, where heat will be generated for long periods of time, the conductivity of the storage rock will be one of the factors which will determine the heat loads that can be tolerated at that point. Room temperature conductivity would not be useful for selecting the storage rock type because the conductivity may be different at other temperatures. As the temperature of a repository will be appreciably above the ambient temperature, determination of the more favourable rock types should be made on the basis of higher temperature thermal conductivities.

For highest possible heat dissipation one would tend to select rocks of high thermal conductivity. However, this should not be the sole criterion for selection because thermal conductivity of a particular rock may change more rapidly than another with increasing temperature. For example, rock 8 in Figure 3 has a high thermal conductivity at low temperatures, but it drops quickly with increasing temperature to the syenite range at 500°C. In contrast, the thermal conductivity of rock 10 drops more slowly. Consequently, for any given storage temperature above 135°C (at which the two curves intersect) rock 10 would be able to accommodate a larger heat load. Furthermore, should there be an excess generation of heat, the temperature rise necessary for its dissipation in rock 8 would be greater than in rock 10 because the conductivity of rock 8 decreases more rapidly with increasing temperature. One could imagine an extreme case in which the thermal conductivity of a rock decreases so rapidly with increasing temperature that a relatively modest increase of heat load may trigger a melt-down. In this case, the temperature rise caused by the increased heat load substantially decreases the thermal conductivity of the rock, which in turn causes another temperature rise and decreases the conductivity still further, and so on.

In conclusion, the thermal conductivity is an important parameter which should be known over a wide temperature range before judging the suitability of a particular rock mass as a medium for storing nuclear waste.

REFERENCES

1. Bullard, E.C., "Heat flow in S. Africa"; Proceedings of the Royal Society of London; v. 173, pp 474-502, 1939.

2. Mossop, S.C. and Gafner, G., "The thermal constants of some rocks from the Orange Free State"; J. Chem., Met. and Mining Soc. of S. Africa; v. 52, no. 4, pp 61-76; 1951.

3. Carte, A.E., "Heat flow in the Transvaal and Orange Free State"; Proceedings of the Physical Society, London, v. B67, pp 664-72; 1954.

4. Carte, A.E., "Thermal conductivity and mineral composition of some Transvaal rocks"; Am. J. Science; v. 253, no. 8, pp 482-90; 1955.

5. Kunii, D. and Smith, J.M., "Heat transfer characteristics of porous rocks"; J. A.I.Ch.E.; v. 6, no. 1, pp 71-78; 1960.

6. Stephens, D.R., "High temperature thermal conductivity of six rocks"; Univ. of California, Lawrence Radiation Laboratory, UCRL-7605, 1963.

7. Mirkovich, V.V., "Experimental study relating thermal conductivity to thermal piercing of rocks"; Int. J. Rock Mech. Min. Sci.; v. 5, pp 205-18; 1968.

8. Horai, Ki-iti, "Thermal conductivity of rock-forming minerals"; J. Geophys. Res.; v. 76, no. 5, pp 1278-1308; 1971.

9. Mirkovich, V.V., "Comparative method and choice of standards for thermal conductivity determinations"; J. Am. Cer. Soc.; v. 48, no. 8, pp 387091; 1955.

10. Soles, J.A., "Petrography of certain rock types relative to their stability in nuclear waste repositories". In preparation.

THERMAL DIFFUSIVITY AND THERMAL EXPANSION OF SOME ROCK TYPES IN
RELATION TO STORAGE OF NUCLEAR WASTE

V.V. Mirkovich

Canada Centre for Mineral and Energy Technology

Dept. of Energy, Mines and Resources, Ottawa, Canada

Thermal diffusivities between 25 and 600°C and thermal
expansions between 25 and 500°C were determined for nine rocks
from three plutons. These igneous rocks were obtained from the
surfaces of the three rock masses. The thermal diffusivities of
the nine specimens varied considerably and were found to decrease
at different rates with increasing temperature. It was established
that at 500°C, the thermal expansion of the specimens varied by a
factor of 2.5. An empirical relation was used to determine the
high-temperature piercing rate as a possible indicator for the
spallability of these rocks exposed to low-temperature heat loads.

INTRODUCTION

Depending on the type of nuclear waste, the generation of
heat due to continuing decay of the waste material may extend over
many decades. In considering rock masses for repositories of
nuclear waste, the effect of heat (among other geophysical and
physical factors) on the surrounding rock should be examined. To
predict the temperature gradients in a rock subjected to a transient heat load, one must determine its thermal diffusivity.
Also, as the temperature gradients are the cause of differential
thermal expansions, which, because of the confining nature of the
underground nuclear waste repository site, result in stresses,
the thermal elongation of the rocks should be known.

Because of the very nature of rocks, thermal diffusivity
measurements are difficult to perform. Most rocks have relatively
coarse structures, which requires sizable specimens for represent-

ative sampling. Yet their thermal diffusivity is usually low
which, on the other hand, requires small samples so as not to
attenuate the heat pulses to a point where their detection becomes
difficult. An additional complication is the characteristic
irreversible structural change that most rocks undergo at higher
temperatures: once the measurements are made at a higher temperature, they cannot be repeated at a lower level. Either because
of the measuring difficulties or for other reasons, with the
exception of the recent work by Hanley *et al*. (1), virtually no
data on thermal diffusivity of rocks can be found in the literature.

PROCEDURES

a) Methods of Measurement

Thermal diffusivity of the rocks was measured by a novel
method described in detail previously (2). In essence, the method
is based on the principle of an infinite cylinder and radial heat
flow symmetry. The progress of the heat pulses, generated on the
surface of the cylinder either in periodic or transient form, is
monitored inside the cylinder, at the centre, and in the same
radial plane at a point near the surface. For this study only
the transient measuring mode was used. In such a case the surface
of the cylinder is heated at a constant rate. The temperature
differential between the two measuring points is recorded until
it becomes constant, i.e., reaches the maximum. Thermal diffusivity is obtained from the increase of the temperature differential,
time and the geometry of the sample.

Because some rocks release water vapour and other gases on
heating, the measurements were performed in air. The principle
of measurement in air by this method was discussed previously (3).
The sample column was composed of two identical cylindrical rock
samples, each 25.4 mm in diameter and 25.4 mm long. To prevent
axial heat losses, a cylindrical insulator was placed at each end.
The column assembly was held together by two pairs of stainless
steel wires binding the cylindrical insulators. The heat pulses
are generated by a 0.51-mm Nichrome heating wire wound non-inductively around the sample column. To avoid electrical short-
circuiting the wire is covered with a thin layer of refractory
cement. Two thermocouple holes extend through one insulator and
about 18 mm deep into the first sample. The assembly is shown in
Figure 1. For measurements at higher temperatures the sample
assembly is placed in a tube furnace.

A commercial dilatometer* employing fused-quartz media was used for the thermal expansion measurements. The vertical tube-type furnace is closed at the lower end by a plug extending up to the bottom of the central zone of uniform temperature. This plug supports a fused quartz tube, 22 mm OD and 500 mm long and sealed at the bottom end with a fused-quartz plate. The rock specimen, 51 mm long by 8 mm in cross-section, rests on this plate and, in turn, supports a hollow fused-quartz rod, which transmits the elongation of the specimen to a sensitive Ames dial gauge fixed to the external quartz tube. No corrections were applied to compensate for the expansion of the quartz tube relative to the 51-mm length of the specimen, which was considered insignificant. A thermocouple was used to measure the temperature about the rock specimen. The temperature was increased at approximately 9°C/min.

b) Materials

The nine rocks for this study were obtained from three Canadian locations, each representing a particular pluton. A brief description of the samples is given in Table 1. Monzonite Samples 2 and 3 are petrographically similar, but they differ in mineralogical composition, principally in the contents of plagioclase, amphibole, pyroxene and mica. The petrographic compositions are presented in a separate report (4).

For the thermal diffusivity measurements, samples were cut into cylinders, 25.4 mm in diameter and 25.4 mm long. As the measuring method used in this work permits only radial heat flow, the specimens of all foliated rocks were cut so that the schistosity was perpendicular to the principal axis of the cylinder. To form the measuring column, two cylindrical samples of the same rock were used. In one of the cylinders two 0.75-mm diameter thermocouple holes were drilled from one end to a depth of about 18 mm: one in the centre through the principal axis and the other parallel to the principal axis near the cylindrical surface. For thermal expansion measurements rocks were cut in the form of bars, 51 mm long with an 8 x 8 mm rectangular cross-section. In the case of the foliated rocks, specimens were prepared with the schistosity both parallel and perpendicular to the length of the bar.

RESULTS AND DISCUSSION

The thermal diffusivities are shown in Figures 2 and 3 and in Table 2. The curves were fitted to the data by inspection and each curve represents some 30 thermal diffusivity measurements. The

---

*Manufactured by Industrial Engineering Instrument Company of Allentown, Pa., U.S.A.

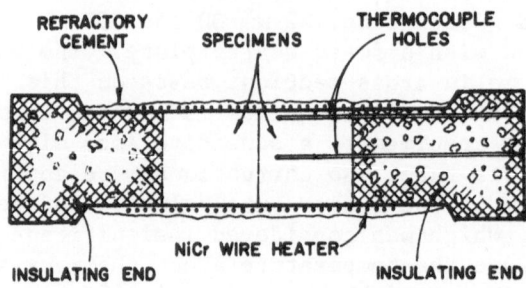

Figure 1.   Sample assembly for measurement of
thermal diffusivity of rocks.

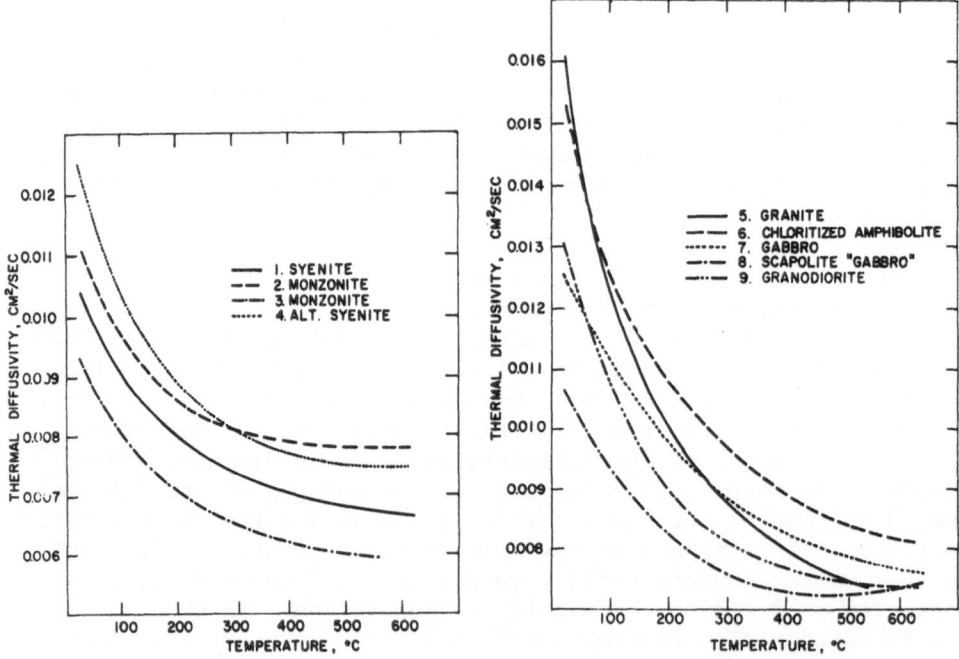

Figure 2.   Thermal diffusivity
of four syenite rocks.

Figure 3.   Thermal diffusivity
of granite, chloritized amphibolite, scapolite "gabbro" and
granodiorite.

TABLE 1

Description of Rock Specimens from Three Canadian Plutons

Sample Number	Rock Type	Texture; Structure	Density g/cm^3
Pluton 1			
1	Syenite	F-gr, hypid.; hom.	2.74
2	Monzonite	M-gr, hypid.; hom.	2.80
3	Monzonite	M-gr, hypid.; hom.	2.78
4	Alt. Syenite	M-gr, hypid.; hom.	2.66
5	Granite	F-gr, hypid.; hom.	2.58
Pluton 2			
6	Chloritized Amphibolite	M-gr, hypid.; slightly foliated	2.99
7	Gabbro	M-gr, hypid.; hom.	2.95
8	Scapolite "Gabbro"	M-gr, hypid.; hom.	3.04
Pluton 3			
9	Granodiorite	M-gr. hypid.; foliated	2.70

Legend: F-gr = fine grain (3 mm); M-gr = medium grain (2-6 mm); hypid. = hypidiomorphic; hom. = homogeneous.

TABLE 2

Thermal Diffusivity of Nine Canadian Rocks from Three Plutons, cm^2/sec

Rock	Temperature, °C							
	25	50	100	200	300	400	500	600
1. Syenite	.0104	.0099	.0091	.0080	.0074	.0070	.0068	.0067
2. Monzonite	.0111	.0116	.0097	.0086	.0081	.0079	.0078	.0078
3. Monzonite	.0093	.0088	.0081	.0071	.0065	.0062	.0060	–
4. Alt. Syenite	.0124	.0116	.0104	.0089	.0081	.0077	.0075	.0075
5. Granite	.0160	.0142	.0122	.0100	.0088	.0080	.0075	–
6. Chloritized Amphibole	.0152	.0141	.0125	.0108	.0097	.0089	.0084	.0081
7. Gabbro	.0125	.0120	.0102	.0098	.0089	.0083	.0079	.0077
8. Scapolite "Gabbro"	.0106	.0102	.0094	.0083	.0076	.0073	.0072	.0074
9. Granodiorite	.0130	.0121	.0108	.0090	.0081	.0077	.0075	.0074

average deviation of the experimental data from the smoothed values
and the standard deviation for all data are 0.00014 cm^2/sec.

Figure 2 shows the thermal diffusivities of the syenitic rocks
from pluton 1. While there is an obvious difference in the thermal
diffusivities of these rocks, it is somewhat unexpected that the
difference between the two monzonites, particularly at higher
temperatures, should be so large. The thermal diffusivity of this
group of rocks decreases with increasing temperature. The decrease
between room temperature and 600°C ranges from 0.0033 to 0.0050
cm^2/sec. The thermal diffusivity of the granite of pluton 1, shown
in Figure 3, is substantially higher than that of the syenitic
rocks, but it decreases rapidly so that at 600°C it falls into the
region of syenitic rocks. With the exception of scapolite "gabbro",
the lowest thermal diffusivities of the rocks of plutons 2 and 3
are appreciably higher than those of the syenitic rocks and generally
remain higher with increasing temperature. The difference in the
thermal diffusivity values between room temperature and 600°C for
plutons 2 and 3 (with the exception of scapolite "grabbro") ranges
from 0.0050 to 0.0085 cm^2/sec.

For storage of heat-generating materials, rocks with higher
thermal diffusivity would appear to be more desirable because, for
equal heat loads, the thermal gradient in such rocks will be lower
than that for those of lower thermal diffusivity. However, it
should be taken into consideration that a steep drop in thermal
diffusivity with increasing temperature may be a cause of thermal
instability.

The thermal diffusivities of chloritized amphibolite and
granodiorite were measured parallel to the schistosity, since,
presumably the cavity of the repository would be constructed perpendicularly to the schistosity of the rock so as to provide a path
of least resistance for heat transmission.

Figure 4 compares the thermal diffusivities of granodiorite
rocks obtained in this investigation and that by Hanley *et al.* (1),
which were measured with the laser-flash technique (5). There is
no difference of any substance for the data from room temperature
to 250-300°C. Above that temperature, however, the results diverge
so that at 600°C the present data are higher by some 25%. Of
course, measurements on rocks of the same category as, for example,
Monzonite 3 can give significantly different results. However,
the other possibility is that the relatively large quantity of heat
repeatedly deposited by the laser beam and the subsequent large
temperature rise of the back face of the sample eventually alters
the structure of the rock.

The thermal expansion of the rocks of all three plutons is
shown in Figure 5. The continuity of the curves indicates that no

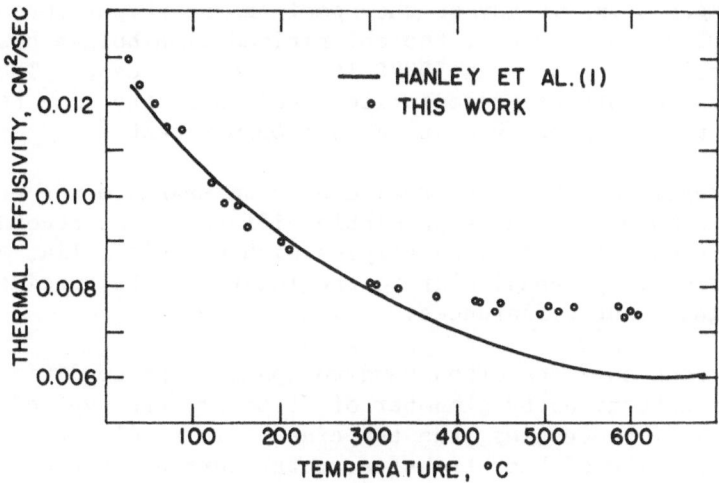

Figure 4. Comparison of thermal diffusivities of
granodiorite rocks measured by two
different methods.

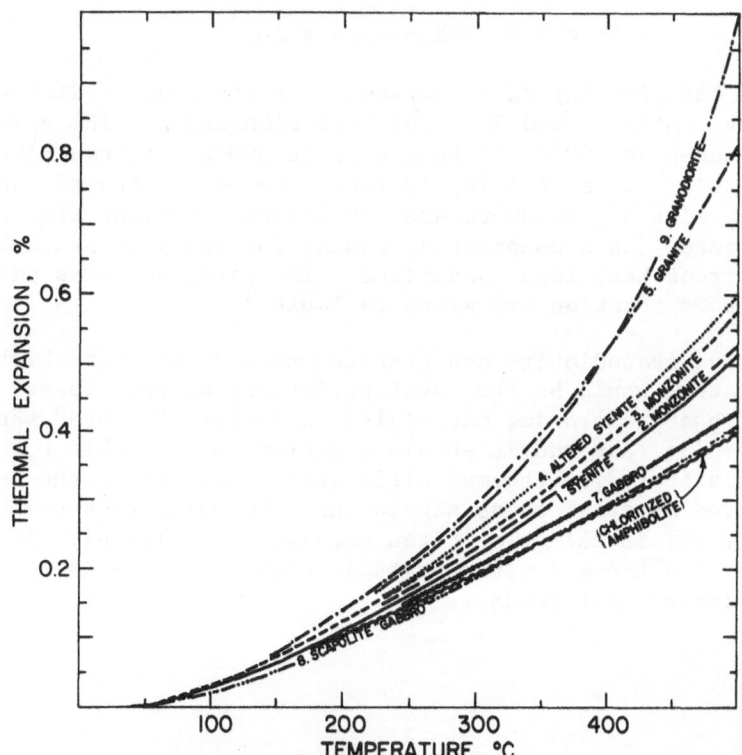

Figure 5. Thermal expansion of nine rocks from
three plutons.

drastic changes occur in any of the specimens at temperatures up
to 500°C. Of the nine rocks, the chloritized amphibolite has the
lowest thermal expansion: at 500°C it is 0.40 per cent. The
thermal expansion of the granodiorite specimen at the same temperature is 2.5 times higher, amounting to 1.00 per cent.

For several rock types thermal expansion measurements were
made on two samples, but no appreciable difference was recorded.
Also, measurements performed on samples with visible folia, oriented
either parallel or perpendicular to the length of the specimen, did
not show significant differences.

Heated rock surfaces often tend to spall. The conditions for
spalling are determined by a number of factors: the type of rock
and its physical properties, the temperature level of the heat
source and the rate of heat transfer at the surface, the geometry
of the rock (or rock cavity) and finally the proportion of heated
to non-heated surface. Unfortunately no relation exists which
incorporates all these factors. However, an empirical equation,
which gives the high-temperature piercing rate of rocks, was
obtained in a previous study (6). The piercing rate is:

$$R = 21.3\alpha E - 1.75 \times 10^{-2}$$

where R is the piercing rate, cm/sec, α is the thermal diffusivity
of the rock, cm^2/sec, and E is per cent elongation. The values
for α are taken at 200°C and that of E at 500°C. Although the
validity of the above relation is likely to be limited to conditions
of piercing with high-temperature jet flames, it might also serve
as an indicator (on a comparative basis) for spalling of rocks
under different heat load conditions. The piercing rates calculated
from the above equation are given in Table 3.

Clearly, granodiorite and granite, because of their high
spalling rates, would be the worst performers as rock masses for
storage of heat-generating materials. Scapolite "Gabbro" shows the
lowest piercing rate and it should therefore be a stable rock mass,
but its relatively low thermal diffusivity would limit the quantity
of the stored material so as not to cause too-high rock temperatures.
The best of all is the chloritized amphibolite. Its piercing rate
is low, yet its thermal diffusivity is high, enabling it therefore
to accept larger heat loads.

TABLE 3

Piercing Rates for Rocks Exposed to High
Heat Transfer at the Surface

Rock	$\alpha_{200°C}$ cm^2/sec	$E_{500°C}$ %	R cm/sec
1. Syenite	0.0080	0.51	0.07
2. Monzonite	0.0086	0.53	0.08
3. Monzonite	0.0071	0.58	0.07
4. Altered Syenite	0.0090	0.60	0.10
5. Granite	0.0100	0.82	0.15
6. Chloritized Amphibolite	0.0108	0.40	0.07
7. Gabbro	0.0098	0.43	0.07
8. Scapolite "Gabbro"	0.0083	0.41	0.05
9. Granodiorite	0.0090	1.00	0.17

REFERENCES

1. Hanley, E.J., DeWitt, D.P. and Taylor, R.E. "The thermal transport properties at normal and elevated temperature of eight representative rocks"; Presented at the Seventh Symposium on Thermophysical Properties, May 10-12, 1977, Gaithersburg (NBS), Md., U.S.A.

2. Mirkovich, V.V. "Thermal diffusivity measurements of Armco iron by a novel method"; Rev. Sci. Instr.; v 48, no. 5, pp 560-65; 1977.

3. Mirkovich, V.V. "An apparatus for measurement of thermal diffusivity in air"; CANMET Report 77-21; Canada Centre for Mineral and Energy Technology, Department of Energy, Mines and Resources, Ottawa, Canada; December, 1976.

4. Mirkovich, V.V. and Soles, J.A. "Thermal conductivity of certain rock types and its relevance to the storage of nuclear waste"; to be published in the Proceedings of the 15th Int'l Thermal Conductivity Conference, 1977.

5. Parker, W.J. et al. "Flash method of determining thermal diffusivity, heat capacity and thermal conductivity"; J. Appl. Physics; v 32, no. 9, pp 1679-84; 1961.

6. Mirkovich, V.V. "Experimental study relating thermal conductivity to thermal piercing of rocks"; Int. J. Rock Mech. Min. Sci.; v 5, pp 205-18; 1968.

SESSION H   INSULATION

Session Chairmen: R.P. Tye
                    Dynatec R/D Co.
                    Cambridge, Massachusetts

                    P. Bondi
                    University of Bari
                    Bari, Italy

THERMAL CHARACTERISTICS OF GAS-FILLED FIBER POWDER INSULATION SYSTEMS

Decatur B. Rogers[+] and John W. Williamson[#]

[+]Prairie View A & M University
Prairie View, Texas 77445

[#]Vanderbilt University, Nashville, Tennessee

ABSTRACT

This investigation was initiated to study gas-filled fiber-powder insulations. It has been experimentally shown that these systems may exhibit a thermal conductivity as low as one-half that of air. Selected gas-filled, fiber-powder systems were measured and the important role that powder particle size plays was demonstrated. The powders which were investigated included carbon black, perlite, flame-prepared silica, and silica-carbon black mixtures.

The existing theoretical model of Luikov, Shashkov, Vasiliev, and Fraiman was used for gas-filled powders and extended to the case of gas-filled, fiber-powder systems by adding Bankvall's theoretical model for fibrous materials. These models are useful in evaluating candidate powders and gases and they can be used to establish conductivity limits of gas-filled powder systems.

INTRODUCTION

This paper presents the theoretical and experimental results of thermal properties of various gas-filled powders, or fiber-powder insulations. The research was directed at modeling the heat transport mechanisms and the comparison of these predictions with experimentally determined results. In particular, an attempt to identify powder systems which would exhibit thermal conductivities in the range of 0.006 to 0.015 Btu-in/hr-ft^2°F has been made.

From past experience, it is evident that thermal conductivities in the selected range can be achieved with evacuated systems. In a powder system, gas conduction is the predominant mechanism for heat transport at moderate pressures, and this mode continues to be the primary mechanism as gas pressure is reduced. When the gas pressure approaches the value at which the mean free path of the molecule is comparable with the pore dimensions formed by the powder, there is a marked reduction in the effective thermal conductivity. To achieve this same result with non-evacuated powders it is necessary to obtain free molecular flow by decreasing the characteristic dimension of the pores, and this can be accomplished by using powders with very small diameters. Other researchers have come to the same conclusion, but there are few data on the thermal conductivity of finely divided powders filled with gas at normal pressures and temperatures.

ANALYTICAL MODEL

Many mathematical models have been developed, and for a chronological listing and discussion of these the reader is directed to the work of Rogers (1). In this study, the term "effective" thermal conductivity (ke) is used to define the conductivity of the powder system. The term "apparent" thermal conductivity is used to define the specific transport mechanism which contributes to the effective thermal conductivity of the powder system. Consequently, the "effective mean thermal conductivity" is made up of contributions from gas conduction, radiant transfer, and solid conduction.

The theoretical model used in this study is that of an open pore system seeded with finely divided powders. The powder model is after Luikov, Shashkov, Vasiliev, and Fraiman (2) with some alterations, and the open pore system is a fiber system, modeled after Bankvall (3). While Bankvall's model is quite adequate for predicting heat transfer through fiber, only his expression for conductance through the solid fiber was used. A computer program was written for both the powder-gas case, as well as the powder-fiber-gas situation.

While the theoretical models seem to be valid from a physical standpoint, three parameters exist in the model for which there is little reliable data in the literature. These parameters are the thermal conductivity of finely divided powder, contact resistance between powder particles, and the accommodation coefficient. Approximate values of these parameters can be chosen and used as constants within the theoretical models for further investigation. This method was used to examine the correlation between theoretical values and published data.

Data by Smith and Wilkes, (4), and Verschoor and Greebler (5) were used for this purpose. In 1944 Smith and Wilkes measured the

effective thermal conductivity of three samples of commercial
carbon black. Their results appear in Table 1. The comparison of
results produced by the fiber-powder model with these data are also
shown. The correlation is excellent, yielding differences of + 1
percent. The table shows the theoretical prediction of the effect
produced by adding a loose fiber structure. The model predicts
values for the effective thermal conductivity to be slightly lower
than that resulting for just the powder system alone.

It is important to note that the data of Table 1 show the effective thermal conductivity to be less than that of air. Both
sets of data show a decrease of approximately 27%. Three other experimental studies also show a reduction in the effective thermal
conductivity. The work by Kistler and by Kistler and Caldwell (6)
in 1934, and the study of Johnson and Hollweger (7) have experimentally demonstrated this fact.

MATERIALS SELECTED FOR TEST

In order to select combinations of powders and gases which may
achieve the goal set for effective thermal conductivity, it was
necessary to first identify those characteristics which are desirable for each of the constituents.

The gas should be non absorbing (thus possess a small value of
the van der Wall constant (a) with a small dipole moment) and exhibit a low thermal conductivity (gas of high molecular weight).
Since air is abundant, accessible at no cost and involves no safety
hazard, it would appear to be the logical choice as the interstitial
gas for most powder insulations.

The powder should have small particle diameter and be opaque.
Powders considered in this study included perlite, fumed silica and
carbon black. Rosita perlite would probably be the best powder
because it is a cheap mineral which remains opaque at the desired
sizes, and it also has low thermal conductivity. Based on the work
of Bankvall and Pelanne (8), a low density glass fiber matrix would
be adequate for containment of the powder.

EXPERIMENTAL METHODS AND APPARATUS

Both the comparative method and the steady flow calorimeter
method were used to experimentally determine the effective thermal
conductivity. For the comparative method a half inch thick layer
of air was used as the known substance. The upper surface of the
test specimen was maintained at a constant temperature by boiling
a refrigerant, R-11. Energy from an electrically heated flat plate

TABLE 1

Comparison of Previous Experimental Data with Results Generated from the Theoretical Model

Type of Powder	Particle Size (μ)	Density (lbm/ft3)	Mean Temperature (°F)	Thermal Conductivity			
				K*** AIR	K*** exp	K*** e*	K*** e**
Smith & Wilkes Data (4)							
Carbon Black							
1. Grade 6	0.028	16.2	133	0.01649	0.01175	0.0118	0.0117
2. Monarch 71	0.015	12.0	133	0.01649	0.01192	0.0090	0.0089
Verschoor & Greebler Data [5]							
1. Carbon Black	0.010	4.5	300	0.0193	0.0142	0.0140	0.0139
2. Carbon Black	0.018	8.1	300	0.0193	0.0167	0.0164	0.0162

*Effective thermal conductivity predicted using only the powder portion of the model.
**Effective thermal conductivity predicted using the entire fiber powder model.
***Units are Btu/hr-ft-°F.

was transmitted in series through the air and then through the test specimen. The temperature drop across the air gap was accurately measured. By knowing the temperature of the isothermal surface produced by the boiling refrigerant and spatial dimensions of the apparatus, the effective thermal conductivity of the test specimen could be determined. The steady flow calorimeter method was used as a check of these results.

DISCUSSION OF RESULTS

A series of ten tests was performed with four different powders, and data from these tests appear in Table 2. The effective thermal conductivity presented in this table was determined using the comparison method. Cab-O-Sil (M-5) was the primary powder of importance because of its size ($\sim 0.014\mu$) and low apparent density (~ 3.0 lbm/ft^3). The large value for the effective thermal conductivity shown in Table 2 may be due in part to the moisture content of the sample and probably to radiant energy transport through the transparent fumed silica (Cab-O-Sil). Fumed silica is transparent to 95% of radiation with a wave length less than 2μ. Several schemes were tried in an effort to reduce the radiant transport through the Cab-O-Sil. These methods included mixtures of Cab-O-Sil with either carbon black or perlite. Carbon black was effective at preventing heat transfer by radiation, and the mixture of 50% Cab-O-Sil and 50% Monarch 1100 had the lowest thermal conductivity. As a consequence, this mixture was used for filling a fiber matrix. This system showed an increase in the thermal conductivity by 25%, but this increase was due in part to the moisture absorbed from the atmosphere of the lab.

CONCLUSIONS

This study has shown that gas-filled powder systems and gas-filled, fiber-powder systems can have effective thermal conductivities which are substantially lower than that of air. The best result occurred with a mixture of 50% Cab-O-Sil and 50% carbon black; this mixture produced a thermal conductivity of approximately 45% that of air. The theoretical models adopted to study these fine powder systems produced fair results; the differences between theoretical and experimental results generally ranging from 10% to 40%. The theoretical models do show trends of the various parameters involved, and should prove to be useful in predicting the behavior of various combinations of powder, gas, and fiber.

The goal of 0.00125 Btu/hr-ft-°F, set by the authors, was not achieved. However, the theoretical results indicate that if Perlite powder of a size equal to 0.01 micron is used this goal might be realized.

D. B. ROGERS AND J. W. WILLIAMSON

TABLE 2

Summary of Experimental and Theoretical Results for Several Powder Configurations

Powder	Particle Size (μ)	Powder Density (lbm/ft3)	Mean temp. °F	Thermal Conductivity			K_{exp}/K_{air}
				K_{exp} Btu/hr-ft-°F	K_{calc} Btu/hr-ft-°F	% Diff*	
Cab-O-Sil (M-5)	0.014	2.85	73.8	0.0115	0.0092	20	0.77
75% Cab-O-Sil (M-5) + 25% Monarch 1100	0.014	6.97	75.1	0.0087	0.0045	48	0.56
50% Cab-O-Sil (M-5) + 50% Monarch 1100	0.014	9.54	71.7	0.0062	0.0055	11	0.44
25% Cab-O-Sil (M-5) + 75% Monarch 1100	0.014	12.77	73.2	0.0115	0.0074	36	0.77
Monarch 1100	0.014	23.15	74.5	0.0123	0.0101	18	0.82
Grace Filler GR-53020	1.000	6.89	75.4	0.0125	0.0112	10	0.83
Chem-Sil No. 3	0.100	14.62	69.5	0.0144	0.0122	15	0.96
50% Cab-O-Sil (M-5) + 50% Grace Filler GR-53020	0.507	5.39	71.8	0.0100	0.0114	14	0.67
Cab-O-Sil (M-5) + Aluminum Sheets	0.014	3.05	65.3	0.0117	--	--	0.79
50% Cab-O-Sil (M-5) + 50% Monarch 1100 + Fiber	0.014	11.34	66.2	0.0083	0.0053	36	0.56

*Diff = $(K_{exp} - K_{calc})/K_{exp}$

REFERENCES

1. D. B. Rogers, Investigation of the Effective Thermal Conductivity of Gas-Filled Fiber-Powder Insulation Systems for Residential and Commercial Structures, Ph.D. Dissertation, Vanderbilt University, Mechanical Engineering Department, 1975.

2. A. V. Luikov, A. G. Shashkov, L. L. Vasiliev, YU. E. Fraiman, Thermal Conductivity of Porous Systems, Int. J. Heat Mass Transfer, Vol. 11, 1968, pp 117-140.

3. C. Bankvall, Heat Transfer in Fibrous Materials, Journal of Testing and Evaluation, Vol. 1, No. 3, May, 1973, pp 235-243.

4. W. R. Smith and G. B. Wilkes, Thermal Conductivity of Carbon Blacks, Industrial and Engineering Chemistry, Vol. 36, No. 12, 1944, pp 1111-1112.

5. J. D. Verschoor and P. G. Greebler, Heat Transfer by Gas Conduction and Radiation in Fibrous Insulations, Transactions of the ASME, August, 1952, pp 961-968.

6. S. S. Kistler and A. G. Caldwell, Thermal Conductivity of Silica Aerogel, Industrial and Engineering Chemistry, Vol. 26, No. 6, June, 1934, pp 658-662.

7. C. L. Johnson and D. J. Hollweger, Some Heat Transfer Considerations in Nonevacuated Cryogenic Powder Insulation, Advances in Cryogenic Engineering, Vol. 2, 1966, pp 77-88.

8. C. M. Pelanne, Experiments on the Separation Heat Transfer Mechanisms in Low-Density Fibrous Insulation, 8th Conference on Thermal Conductivity, Thermal Conductivity Proceedings, Edited by C. Y. Ho and R. E. Taylor, Plenum Press, 1969, pp 897-911.

HEAT TRANSFER IN MICROSPHERE INSULATION IN THE PRESENCE OF A GAS

G.R. Cunnington, Jr.[+] and C.L. Tien[#]

[+]Lockheed Palo Alto Research Laboratory
Palo Alto, California

[#]University of California
Berkeley, California

INTRODUCTION

Evacuated porous materials have seen wide use as cryogenic insulations for many years [1,2] and numerous studies [2-9] have addressed the technology of heat transport in these powder, fiber, foam and multilayer types of insulation. Although they are comprised of a wide variety of materials, the mechanisms of heat transfer by radiation and solid conduction within the media are well understood. However, when a fluid is present in the void spaces between and/or within particles the heat transport is less well defined, particularly at cryogenic temperatures and in the pressure range intermediate to free molecular and continuum regimes [10-12].

The present paper reports the results of an experimental investigation of the effects of a non-condensible interstitial gas on microsphere insulation heat transfer for the cryogenic to room temperature range. Gases are helium and nitrogen at pressures from 10^{-8} to 1 atmosphere. These data are compated with an existing analytical model for combined gaseous and solid conduction, and a good engineering correlation is demonstrated for this particular case.

ANALYTICAL CONSIDERATIONS

The major heat transfer mechanisms in packed spheres or powder insulation under evacuated conditions consist of the surface radiation transfer across the voids and the constricted conduction through the contacting surfaces. If a gas is present in the void space between spheres a second conduction mechanism, as illustrated by the physical model of Fig. 1, is considered. For a medium with coupled conduction and radiation transport, it has been shown that the effective (or apparent) thermal conductivity, k, defined in the Fourier law

$$q = -k(T) \frac{\partial T}{\partial X} \tag{1}$$

can be well approximated under most conditions as the linear summation of the conduction and radiation contributions.

$$k(T) = k_s(T) + k_r(T) \tag{2}$$

where $k_s(T)$ is the combination of constriction and gas phase contributions. For optically thick conditions, such as in most microsphere insulations, this is an excellent approximation. The conduction through the gas is a function of not only the thermal conductivity of the gas but also of the relationship between points on the surfaces of adjacent spheres (δ_r) and the mean-free path of the gas. An apparent thermal conductivity for this mode can be expressed as [12].

$$k_{gc} \sim F'' \, k_g \, f(k_s'/k_g) \tag{3}$$

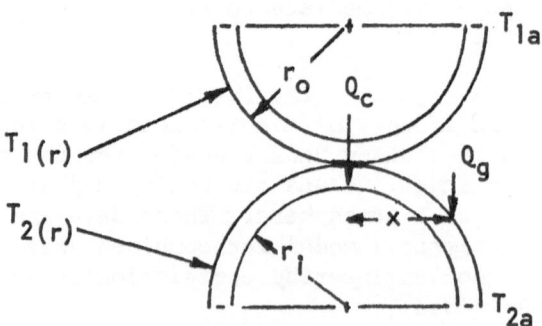

Figure 1.   Heat transfer physical model

where k_g is the conductivity of the gas at atmospheric pressure modified[8] by a relationship of the local spacing to the mean free path, k's is a function of the conductivities of the gas within the granule and the granule material itself, and f" is a geometric factor describing the packing.

In the general case the properties of the solid and gases are temperature dependent, but the integration of the equation is cumbersome considering temperature dependencies of both thermal conductivities and gas mean free paths. In the solution used in this work an approximation was made whereby the properties are evaluated as the average temperature between boundaries for the gas phases and by the expression for the glass as a function of local temperature [13]. At low pressures and temperatures second consideration in calculation of gas phase effects is the definition of a relationship between local pressure in the bed and the pressure measured by a gauge remotely located from the bed. For Knudsen numbers $(L/\delta_a) \gg 1$ a pressure gradient exists with a temperature gradient in the porous media and one must then include the influence of dP/dx in the rigorous solution. For this analysis the simplifying assumption was to evaluate pressure at a mean pressure, defined as

$$\bar{P} = \frac{P_g (T_H^{\frac{1}{2}} + T_C^{\frac{1}{2}})}{2T_g^{\frac{1}{2}}}$$

where P_g is gauge pressure and T_g is temperature of the gauge, and use this to compute the mean free path term for the gas conduction equation.

The equation for the gas conduction contribution to overall thermal conductivity is that suggested for porous granules [11] and is

$$k_{gc} = k_g \left\{ \frac{5.8 \ (1-m)^2}{1-k_g/k_{gr}} \left[\frac{1}{1-k_g/k_{gr}} \quad \ell n(k_{gr}/k_g) - 1 - K/2 \right] + 1 \right\} \quad (4)$$

The gas conductivity term in equation (4) is further defined as

$$k_g = k_{go} / (1 + 2XL/\delta_a)$$

where k_{go} is gas thermal conductivity at atmospheric pressure, L is gas mean free path, $X \equiv [2\gamma/(\gamma+1)] \ (2-\alpha)/(\alpha Pr)$, and $\delta_a \equiv (2md_{gr})/3(1-m)$, where d_{gr} is the mean diameter of the spheres in the bed. For the sphere effective conductivity

$$k_{gr} = k_s \{1+[2m(1-v)/(2v+1)]/1-[m(1-v)/(2v+1)]\}$$

where m' is the void volume of the sphere itself, $v \equiv k_s/k_g$ and k_s is the thermal conductivity of the sphere material. For the microspheres used in this investigation the interior sphere volume was evaluated to a pressure $< 10^{-1}$ pa (by long time vacuum baking) [13] so that $v \to \infty$ and $k_g \approx 2K_s(1-m')/(2+m')$.

PHYSICAL CHARACTERISTICS OF MICROSPHERES

A number of hollow glass spheres are marketed in a range of sizes/sphere wall thicknesses. The spheres examined for loose-fill type cryogenic insulation range in diameters from 15 μm to 150 μm and in nominal wall thicknesses from 0.5 to 2.0 μm. Sphere densities range from 110 to 350 kg/m^3, resulting in typical bulk densities from 60 to 230 kg/m^3. Several formulations of glass can be used to yield various strengths of spheres (see, for instance, Technical Information on 3M Brand Glass Bubbles, Special Enterprises Department, 3M Company, Minneapolis, Minnesota). The 3M Company also has supplied microspheres coated with thin aluminum films.

For the uncoated microspheres (Type B-12-AX, 3M Company) used in this investigation, a low magnification study (~100X) on an optical microscope showed that the particles are spherical, and with size variation as shown in Table I, the mean diameter being about 80 μm. A Scanning Electron Microscope (SEM) examination of the clear glass spheres as well as crushed spheres indicated that the spherical shells are of uniform thickness ranging from 0.3 to 2.0 μm and exhibit a smooth surface even at very high magnification (up to 30,000X).

Table 1.   Size Distribution for Microspheres

Diameter X 10^6 m	%
< 20	4
20-30	5
30-40	6
40-50	7
50-60	8
60-70	9
70-80	15
80-90	20
90-100	13
100-110	6
110-120	4
> 120	3

EXPERIMENTAL RESULTS

Data from three separate heat transfer experiments have
been used to evaluate the effect of interstitial gas conduction
on the apparent thermal conductivity of packed beds of microspheres. In all cases the microspheres are uncoated and of the
size distribution of Table I. Experiments were conducted using
an 18-cm-diameter guarded hot plate apparatus [6] with a 40°K
temperature difference between boundaries, a 40-cm diameter
flat-plate calorimeter [6] with a 220°K temperature difference
between boundaries and a 1.2m diameter spherical cryogen storage vessel filled with liquid nitrogen. These three types of
apparatus covered a range of temperatures and pressures as well
as two distinct sphere packing arrangements (flat plate versus
spherical insulation annulus).

Change in apparent thermal conductivity with gas pressure
for nitrogen at an average temperature of 300°K is shown by
the solid data points of Fig. 2. These data, from the guarded
hot plate apparatus, are compared with data for 80 μm nominal
diameter solid spherical particles [10] in air at an average
temperature of 315°K. The change in apparent thermal conductivity in going from vacuum to atmospheric pressure is greater,
by nearly four times, for the solid spheres than for the thin

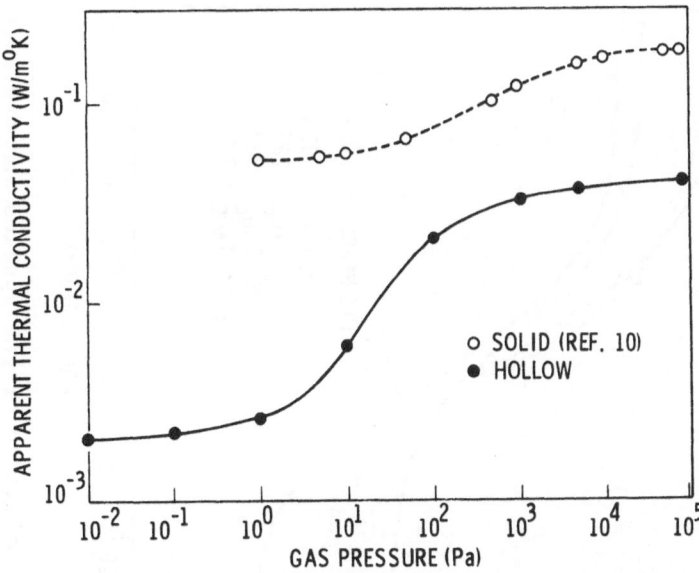

Figure 2.   Thermal conductivity of solid and hollow microspheres

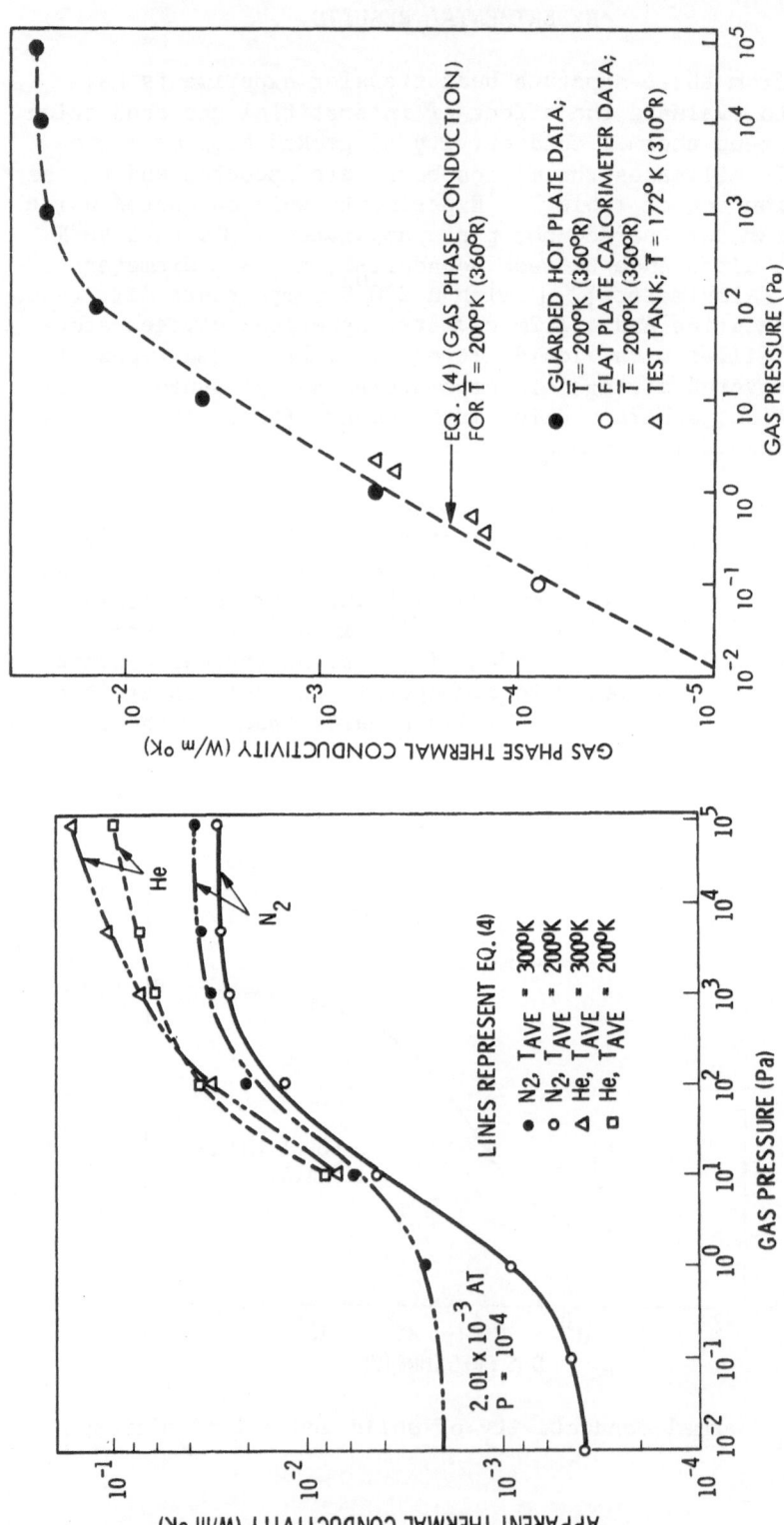

Figure 4.   Gas phase contribution to apparent
              thermal conductivity

Figure 3.   Influence of gas species and pressure
              on apparent thermal conductivity

walled hollow spheres. This illustrates the influence of the
sphere wall thermal resistance on the total resistance. In
this case, the resistance is not dependent only on the constriction and the gas regions (Q_c and Q_g of Fig. 1), but it is also
a function of the sphere wall resistance which varies with distance from the contact area.

The influence of temperature, pressure and gas species on
apparent thermal conductivity are illustrated in Fig. 3. At
atmospheric pressure the temperature dependence of conductivity
is equivalent to that of the interstitial gas, as this is the
dominant heat transfer mechanism at these moderate temperatures.
At lower pressures both temperature and pressure are interre-
lated through the Knudsen number function which governs transport
through the gas phase. This is observed by the cross-over of the
helium data between pressures of 10^2 and 10^3 pa. At a pressure
of 10^2 pa the Knudsen number is greater for 300°K than for 200°K
so the heat transfer is reduced for the former condition. At
a pressure of 10^5 pa the thermal conductivity with helium is
approximately five times that for nitrogen, which is the same
as the ratio of gas thermal conductivities. The solid and
dashed curves of Fig. 3 represent the gas phase conductivity from
equation 5, added to the measured thermal conductivity for high
vacuum conditions for each temperature. A reasonable engineering
correlation of the experimental data is obtained using this equation.

A comparison between data for the three different experiments is shown by Fig. 4 for nitrogen as the interstitial gas.
The data from each experiment is consistent, and the agreement
between the data and (5) is quite good. The data from the 1.2m
diameter tank experiment falls slightly below the prediction,
and this is believed to be the result of either an uncertainty in
calculation of the true packed bed pressure using an ionization
gauge pressure measurement remote from the insulation space or
the difference in packing between flat and spherical insulation
spaces.

CONCLUSION

The model suggested by Kagner [4] for computing the effect
of an interstitial gas on the heat transfer in packed beds of
granular particles having an internal void structure provides a
good correlation with experimental data for hollow-thin walled
microspheres. The adequacy of this model for engineering calculations has been demonstrated for several gases over a range
of temperatures and pressures.

NOTATION

f = function of

F = packing factor

k = thermal conductivity

$K = 1 - k_g/k_{gr}$

L = mean free path

m = void fraction

P = pressure

Pr = Prandtl number

q = heat flux

T = temperature

$v = k_s/k_g$

x = distance

$X = [2\gamma/(\gamma+1)] (2-\alpha)/(\alpha Pr)$

GREEK SYMBOLS

α = accommodation coefficient

γ = ratio of specific heat

δ = local distance between surfaces

SUBSCRIPTS

a = average

C = cold boundary

g = gas or gauge

g_c = gas conduction

g_o = gas at atmospheric pressure

g_r = granule or microsphere

H = hot boundary

r = radiation

s = solid

REFERENCES

1. Scott, R. B., Cryogenic Engineering (Van Nostrand, New York, 1959).

2. Fulk, M. M., Progress in Cryogenics, Vol 1 (1959) 65.

3. Kropschot, R. H., Birmingham, B. W., Mann, D. B., "Technology of Liquid Helium," NBS Monograph 111 (1968) 170.

4. Kaganer, M. G., Thermal Insulation in Cryogenic Engineering (IPST Press, Jerusalem, 1969).

5. Glaser, P. E., et al, Thermal Insulation Systems -- A survey NASA SP-5027 (1967).

6. Tien, C. L., Cunnington, G. R., Advances in Heat Transfer, Vol 9, (Academic Press, New York, 1973) 349-417.

7. Caren, R. P., Cunnington, G. R., Chem Eng Prog Sym Series 64,
    No. 87 (1968) 67.

8. Tien, C. L., Cunnington, G. R., Cryogenics 12 (1972) 419.

9. Cropschot, R. H., Advances in Cryogenic Engineering, Vol. 16
    (1971) 104.

10. Masamune, S. and Smith, J. M., "Thermal Conductivity of
    Spherical Particles," Ind. Eng. Chem. Fwd., 2, (1963).

11. Luikov, A. V., et al, "Thermal Conductivity of Porous Systems,"
    Int. J. Heat Mass Transfer, 11 (1968).

12. Ogniewicz, Y. and Yovanovich, M. M., "Effective Conductivity of
    Regularly Packed Spheres: Basic Cell Model with Constriction",
    Paper 77-188 AIAA 15th Aerospace Sciences Meeting, Los Angeles
    (Jan. 1977).

13. Parmley, R. T. and Cunnington, G. R., "Evacuated Load-Bearing
    High-Performance Insulation Study", Lockheed Missiles and Space
    Company, Inc., NASA CR-135342, (Dec. 1977).

OPACIFIED SILICA REUSABLE SURFACE INSULATION (RSI) FOR THERMAL PROTECTION OF THE SPACE SHUTTLE ORBITER

Howard E. Goldstein, Daniel B. Leiser,* Marnell Smith, and David Stewart

Ames Research Center, NASA

Moffett Field, California 94035

INTRODUCTION

The Space Shuttle Orbiter will be the world's first reusable space vehicle. Basic to its development is a heat shield that can survive multiple reentries at temperatures to 1400° C. For most of the Orbiter's surface, a lightweight material was required that had good insulative properties, and could survive temperatures to 1260° C. It would also have to survive large temperature gradients (>1000° C/cm) and severe thermal shock during reentry.

A family of materials called reusable surface insulation (RSI) was developed for this application in the late 1960s and early 1970s. These low density, rigidized, fibrous ceramic materials have high emittance, water impervious coatings. The state of the art for such materials was evaluated in early 1972 at a Space Shuttle technology conference (1). The evaluation showed RSI materials to be the most cost effective thermal protection system for the Orbiter. After extensive testing, a silica RSI, LI-900, density of 0.144 g/cc (9 lb/ft^3), developed by Lockheed Missiles and Space Company, was adopted in early 1973 for use on most of the Orbiter's surface (2). The two major advantages of the material are its high resistance to thermal shock (3) and its high insulative efficiency (2). The thermal shock resistance of LI-900 is a result of its extremely low thermal expansion coefficient (0.54 μcm/cm-°C) and its high strain to failure (0.1% to 0.2%). Insulative efficiency results from small diameter fibers (1 to 2 μm) and the low bulk density.

---

*Research Associate, Stanford University.

Research on advanced RSI concepts was pursued at Ames Research Center to develop more thermally efficient, stronger RSI materials. As a result, a second generation of improved silica RSI materials was produced. This paper describes those materials, one of which (called LI-2200) was adopted for use on limited areas of the Orbiter's heat shield. LI-2200 is being manufactured by Lockheed Missiles and Space Company.

MATERIAL DESCRIPTION

Figure 1 shows the advanced RSI schematically. It is a rigid sintered fibrous silica material containing about 3% by weight 1200 grit silicon carbide powder. It is made by dispersing the silica fibers using a water-fiber aeration system, partially draining, blending with silicon carbide powder, and mixing in a V-blender. (The mixing step is essential to obtain good dispersion and optimum fiber length.) The slurry is then degassed and molded to form a billet. The billet is sintered at 1300° to 1400° C. Density is controlled by the fiber-to-water ratio, V blending time, molding procedure, and sintering schedule. Strength characteristics and thermal conductivity of the final product can be tailored by controlling the isotropy. For the Shuttle heat shield it was found desirable to have the fibers oriented with their long dimension primarily parallel to the surface (Fig. 1) to minimize through-the-thickness thermal conductance at some sacrifice in strength. This characteristic is obtained by controlling the fiber orientation using the molding parameters. Scanning electron micrographs (SEM) of the fiber before processing, of the tile after molding and drying, and then after sintering, are shown in Fig. 2. The silicon carbide particles are not visible in the SEM. The relatively slight bonding

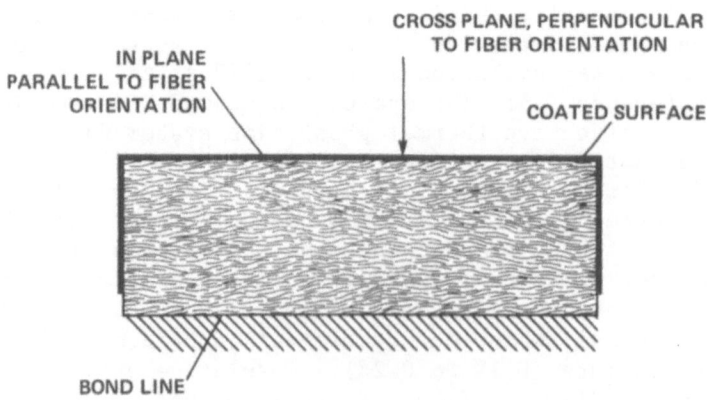

Figure 1. Fiber orientation in silica RSI tile.

50X

500X

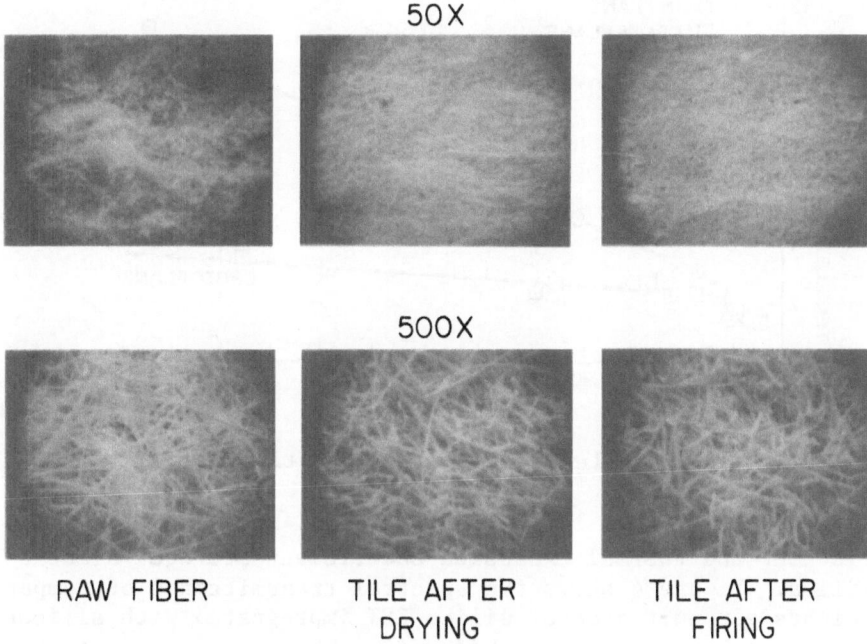

RAW FIBER   TILE AFTER   TILE AFTER
                    DRYING   FIRING

<u>Figure 2.</u> Microstructure of Ames fabricated silica RSI during
processing.

shown and the flexibility of the small diameter glassy fibers
result in an extremely high strain to failure for a ceramic (0.2%
to 0.4%); the strain to failure is nearly independent of density.
However, RSI tensile strength is quite density-dependent (Fig. 3).
The ratio between the weak and strong direction strength is also
density-dependent.

Thermophysical Properties

The thermophysical properties of the material are as follows:

1. Composition: 97% silica fiber (99.7% SiO_2), 1-2 μm fiber
    diameter; 3% silicon carbide (1200 grit)

2. Density: 0.14 to 0.45 g/cm^3

3. Coefficient of thermal expansion: (RT-1100° C) 0.54 μm/m-°C

4. Specific heat: 0.23 cal/g-°C

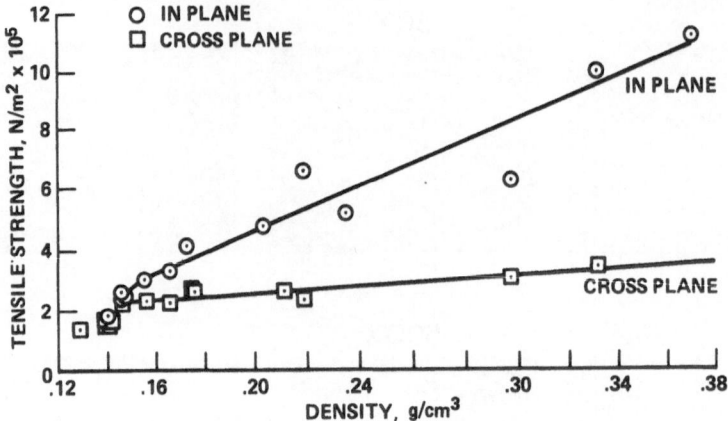

Figure 3. Tensile strength as a function of density.

Specific heat and thermal expansion coefficient are equivalent to pure silica. Figure 4 shows the spectral transmittance of comparable weights per unit area of silica RSI impregnated with silicon carbide and without silicon carbide (4). These data demonstrate that the silicon carbide powder opacifies the RSI, decreasing the radiant transmission significantly. Figure 5 shows the transmission at two densities for equal thickness of opacified material. These data demonstrate that a higher fiber density also decreases transmittance because of light scattering. Figure 6 shows the thermal conductivities at 0.01 atm derived for opacified and unopacified silica RSI (0.24 g/cm^3 density) from the measurements of Ref. 4 and from thermal response test data. This pressure corresponds to that at which peak heating occurs on the Orbiter during atmospheric

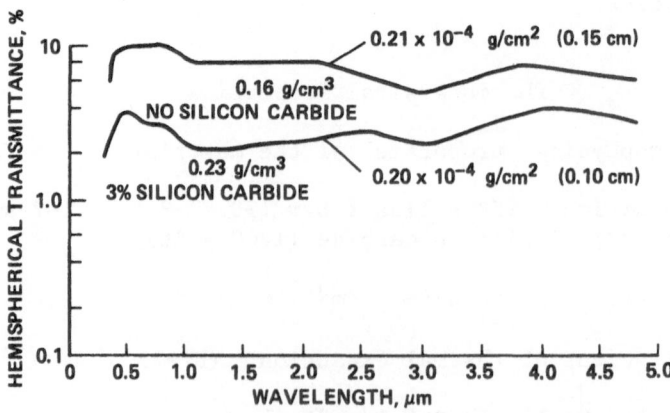

Figure 4. Transmittance of silica RSI equal weights.

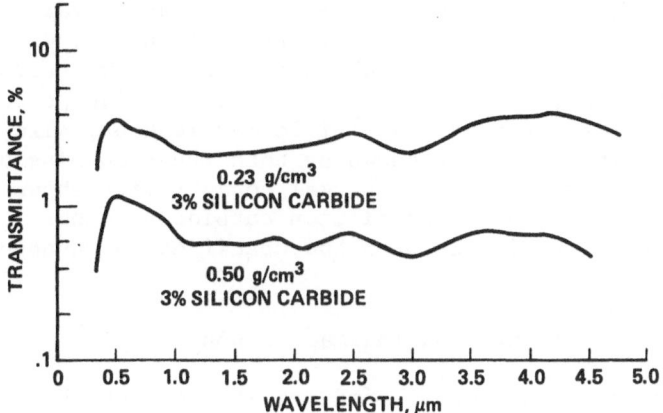

Figure 5. Transmittance of silica RSI equal thickness 0.10 cm.

reentry. At temperatures below about 700° C the conductivities of
opacified and unopacified materials are equivalent. This is
because heat transfer in this temperature and pressure range is
primarily by solid conduction, which is controlled by contact
resistance between the fibers. Gas conduction and convection are
not significant heat transfer mechanisms at low pressure because of
the small pore size of the insulation. Radiation is insignificant
because the peak radiation intensity occurs at relatively long wavelengths (>5 μm) at which the fibers are opaque and also because the
radiation fluxes are low. As the temperature increases above 700° C,

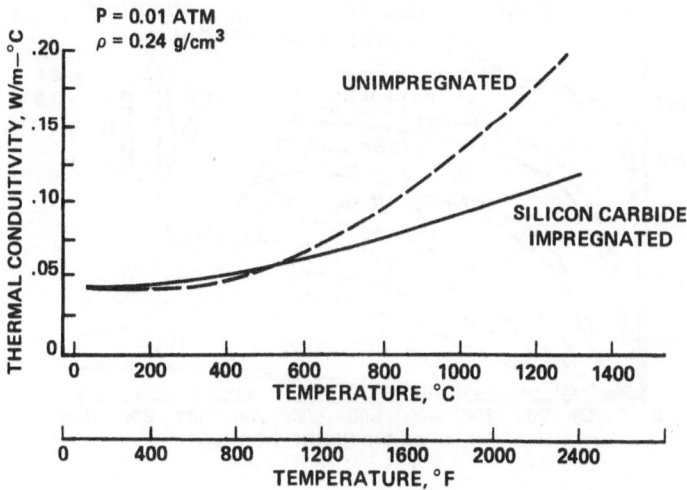

Figure 6. Thermal conductivity of silica RSI.

however, the radiation fluxes increase and the peak radiation is at
shorter wavelengths where the fiber is transparent.   The fibrous
matrix is an effective scatterer which significantly decreases the
conductance; however, addition of the silicon carbide powder, which
absorbs in the near infrared and visible wavelengths, also has a
strong effect.   The effect is shown by both the transmission data in
Figs. 4 and 5 and the effective thermal conductivity shown in
Fig. 6.   This absorption by the silicon carbide accounts for the
difference in conductivity between the opacified and unopacified RSI.

Insulation Thermal Response

The differences in thermal conductivity between the opacified
RSI and the unopacified material are further shown by measuring the
temperature response of each material at equivalent heating rates.
Figure 7 shows the results obtained by convectively heating the
black-coated front surface of each material.   The lower thermal
conductance of the opacified material is demonstrated by the slower
thermocouple response and lower temperatures measured as a function
of depth in the materials.

CONCLUSION

A new reusable surface insulation material has been described.
Its lower thermal conductivity due to opacification with silicon

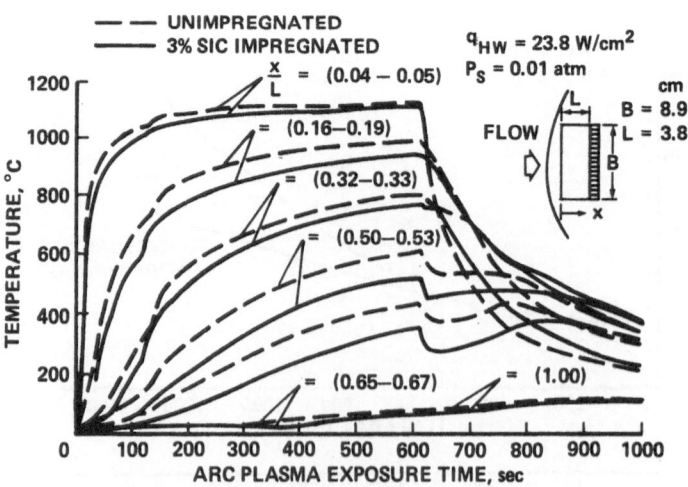

Figure 7.   Temperature history during ARC plasma tests of
0.144 gm/cm^3 silica RSI.

carbide has been shown by measured radiant transmission and thermal conductivities and illustrated with the results of thermal response tests.

REFERENCES

1. Goldstein, H. E., Buckley, J. D., King, H. M., Probst, H. B., and Spiker, I. K., "Reusable Surface Insulation (RSI) Materials Research and Development," NASA Space Shuttle Technology Conference, April 12-14, 1972, NASA TM X-2570.

2. Beasley, R. M., Izu, Y. D., Nakano, H., Ozolin, A. A., and Pechman, A., "Fabrication and Improvement of LMSC's All Silica RSI," NASA TM X-2719.

3. Pigg, O. E., "Results of RSI Thermal-Structure Analysis," NASA TM X-2721, September 1973, pp. 765-792.

4. Cunnington, G., Spectroscopic, Optical and Thermophysical Property Measurements, NASA Contract NAS2-9508, February 1977.

EMITTANCE OF TD-NiCr AFTER

SIMULATED REENTRY HEATING

Ronald K. Clark, Dennis L. Dicus, and W. Barry Lisagor

NASA, Langley Research Center

Hampton, Virginia

ABSTRACT

The effects of simulated reentry heating on the emittance of TD-NiCr were investigated. Groups of specimens were subjected to three different preconditioning treatments. Specimens from each group were exposed to 6, 24, and 30 half-hour simulated reentry exposure cycles in a supersonic arc tunnel at each of three conditions intended to produce surface temperatures of 1255, 1365, and 1475 K. Spectral and total normal emittance were determined at 1300 K on specimens which had been preconditioned only and specimens after completion of reentry simulation exposure. Oxide morphology and chemistry were studied by scanning electron microscopy and X-ray diffraction analysis.

A consistent relationship was established between oxide morphology and total normal emittance. Specimens with coarser textured oxides tended to have lower emittances than specimens with finer textured oxides. However, no systematic relationship between emittance and either exposure time, exposure temperature, or type of preconditioning was evident. No correlation was found between emittance and surface chemistry.

Specimens which were not exposed to simulated reentry heating had emittances on the low end (0.62-0.70) of the range reported in the literature for statically oxidized TD-NiCr (0.6 to 0.9). Specimens exposed to larger numbers of reentry simulation cycles at 1365 and 1475K had emittances as low as 0.52. Results presented in this paper suggest that emittance values previously assumed in design studies of advanced space transportation systems may be too high.

INTRODUCTION

Advanced space transportation systems and hypersonic aircraft which experience severe aerodynamic heating must possess some means of thermal shielding. One shielding scheme employs a reusable high emittance metallic external surface (heat shield) which reradiates most of the incident energy. Nickel base superalloys have been the focus of considerable interest for heat shield applications because of their high temperature strength and oxidation resistance.

One superalloy which has been the subject of extensive laboratory study over the past ten years is TD-NiCr, a dispersion strengthened nickel-chromium alloy. Technical evaluations of this alloy have indicated that it may be usable for temperatures up to 1500 K for heat shield applications (refs. 1 and 2). This is 100 K or more above the maximum use temperature of conventional superalloys.

The successful use of TD-NiCr in radiative TPS applications requires that it have a high, stable emittance in addition to being resistant to temperatures of the order of 1500 K. Design calculations of the temperature distribution of TD-NiCr heat shields for advanced space transportation applications have been based on an assumed emittance of 0.8 to 0.9 (refs. 3 and 4). Statically oxidized TD-NiCr is reported to have a total emittance ranging from 0.6 to 0.9 (ref. 3). However, as shown in reference 5, the oxidation behavior of TD-NiCr and other metallic systems is vastly different under dynamic (high velocity air) as opposed to static conditions (quiescent air). Reference 6 shows that the Cr_2O_3 layer produced during static oxidation is not stable during subsequent arc tunnel exposure and is replaced by a porous NiO layer. It further shows that the oxide scale is repeatedly vaporized and reformed with a period of 30 minutes to 2 hours. The presence of different oxide species may affect the emittance and the catalytic behavior of the surface, thus resulting in different equilibrium surface temperatures under the same stream conditions in flowing air. This paper presents results from a test program aimed at defining the effects of surface preconditioning, exposure temperature, and exposure time on the emittance of TD-NiCr subjected to simulated reentry conditions in an arc tunnel.

PROCEDURE

Three groups of ten specimens were fabricated from 0.64 mm thick TD-NiCr sheet. The specimens were 60-degree circular segments with an area of approximately 14 cm^2. Each group of

specimens was subjected to mechanical/chemical surface treatment
followed by ultrasonic cleaning and static oxidation for 12 hours
in air at 1365 K.   The mechanical and chemical surface treatments
are outlined in Table I.

Reentry Simulation Cycles

Specimens from each group were exposed to 6, 24, and 30
half-hour simulated reentry heating cycles in a supersonic arc
tunnel at each of three conditions intended to produce target
surface temperatures of 1255, 1365, and 1475 K.   During arc tunnel
testing, the specimens were oriented normal to the flow and were
exposed to heating rates of 112 to 216 kw/m^2, mass flow rates of
4.8 to 6.1 g/s, enthalpies of 7.4 to 13.0 MJ/kg and stagnation
pressure of 0.006 atm.   Between successive heating cycles,
specimens were allowed to cool in air for thirty minutes.   These
tests were conducted on contract by the Aerotherm Division of
Acurex Corporation.   Measurements were made of specimen surface
temperature, surface recession and mass loss at six cycle
intervals.   References 7 and 8 present the test conditions and
results of these tests.

Emittance Measurements

Spectral normal emittance measurements were made on specimens
after preconditioning and on specimens after preconditioning and
reentry simulation heating.   Measurements were taken at discrete

Table I. Surface Treatment Used in Preconditioning
TD-NiCr Specimens

Precondition-ing Type	Mechanical Treatment	Chemical Treatment
A	80 Mesh SiC Grit Blast	None
B	120 Mesh SiC Grit Blast	Etched in solution of 1 part* HNO_3, 5 parts HCl and 6 parts Glycerine for 40 min. at 294 K.
C	120 Mesh SiC Grit Blast	Etched in solution of 8 parts HNO_3, 1 part HF and 16 parts H_2O for 40 min. at 344 K.

* By volume

wavelengths over the range from 1 μm to 15 μm at a temperature of
1300 K. Data were taken at 0.5 μm intervals between 1 and 3 μm and
at 1 μm intervals between 3 and 15 μm. These data were obtained by
the radiometric technique described in reference 9. Total normal
emittance (ε_{TN}) was calculated by numerical integration of the
spectral emittance data.

Laboratory Analysis

The morphology of the oxide scale of each specimen was studied
by conventional light microscopy and scanning electron microscopy
(SEM) at magnifications up to 1500X. Elemental analysis of the
oxide scale was carried out during SEM examination by using an
energy dispersive X-ray analysis system (EDAX). The compounds or
phases present in the oxide scale were identified by conventional
X-ray diffraction analysis employing nickel-filtered copper K_α
radiation.

RESULTS

A plot of emittance versus wavelength at 1300 K is shown in
figure 1. The shape of the spectral emittance distribution is
representative of the specimens tested in this program.
Different exposure conditions resulted in varying spectral
emittance values at specific wavelengths especially in the 1 to 6μm

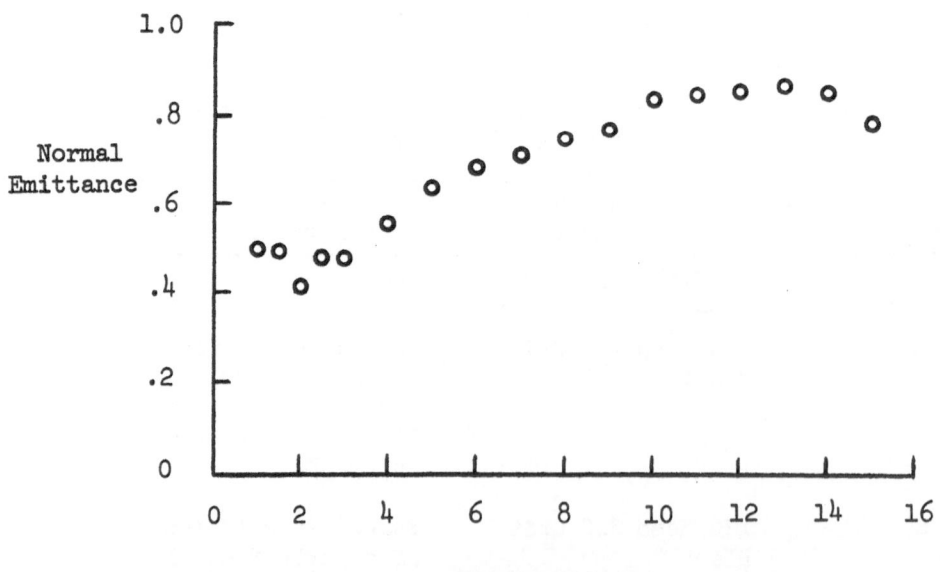

Figure 1. Spectral Emittance Distribution for TD-NiCr at 1300 K
After Type B Preconditioning and 30 Simulated Reentry
Cycles at a Target Surface Temperature of 1365 K.

region, but the general shape of spectral distribution remained the same. Maximum emittance occurred near 13 μm and minimum emittance occurred near 2 μm.

Table II. Emittance and X-Ray Diffraction Results of TD-NiCr Subjected to Various Preconditioning Treatments and Reentry Simulation Conditions

Target Surface Temp., K	No. of Simulated Reentry Cycles	Type of Preconditioning	ε_{TN}	Phases Identified by X-ray Diffraction Analysis				
				Ni	ThO$_2$	Cr$_2$O$_3$	NiO	NiCr$_2$O$_4$
---	---	A	.62	s	vw	vw		
---	---	B	.66	vs	vw	w		
---	---	C	.70	vs	vw	vw		
1255	6	A	.75	vs	vw		m	w
1255	6	B	.76	vs	vw		s	vw
1255	6	C	.76	vs	vw		s	vw
1255	24	A	.62	vs	vw		m	vw
1255	24	B	.61	vs			vs	vw
1255	24	C	.61	s			vs	
1255	30	A	.65	w			vs	vw
1255	30	B	.72	vs	vw		s	
1255	30	C	.65	vs	vw		vs	
1365	6	A	.63	s	vw		vs	vw
1365	6	B	.58	vs	vw		s	w
1365	6	C	.63	s			vs	vw
1365	24	A	.55	w			vs	
1365	24	B	.52	m	vw		vs	
1365	24	C	.57	w	vw		vs	
1365	30	A	.55	m			m	
1365	30	B	.54	vs	vw		vs	
1365	30	C	.57	s	vw		vs	
1475	6	A	.68	vs	vw		m	w
1475	6	B	.67	vs	vw		w	w
1475	6	C	.63	vs	vw		m	w
1475	24	A	.55	w	vw		m	vw
1475	24	B	.63	m	w	m	vw	
1475	24	C	.62	m	vw		s	vw
1475	30	A	.52	w	vw		s	vw
1475	30	B	.52	m	w	s	vw	
1475	30	C	.57	m	vw		s	vw

vs = very strong; s = strong; m = medium; w = weak;
vw = very weak; Blank = not detected.

Table II presents the total normal emittance and the results of X-ray diffraction analysis for each specimen. These data were arranged according to the target surface temperature, number of simulated reentry cycles and type of preconditioning. Specimens with no simulated reentry exposure exhibited total emittances ranging from 0.62 for type A to 0.70 for type C, which is on the low end of the published values of 0.60 to 0.90 for statically oxidized TD-NiCr. Exposure to six simulated reentry cycles at the lower target surface temperature resulted in emittances which were greater than those of the specimens with no simulated reentry exposure. In general, specimens exposed to a greater number of simulated reentry cycles at the lowest target surface temperature exhibited lower emittances than did the specimens exposed to six cycles. The emittances of specimens exposed to the two higher target surface temperatures for large numbers of simulated reentry cycles were noticeably lower (0.50 to 0.60) than those of specimens exposed for six cycles or those of specimens with no simulated reentry exposure and significantly lower than the value of 0.80 to 0.90 used in design studies.

An examination of the X-ray diffraction data in Table II shows that the specimens with no simulated reentry exposure had a surface consisting of Cr_2O_3. On the other hand, both NiO and $NiCr_2O_4$ spinel, but not Cr_2O_3, were detected on the surfaces of specimens exposed to simulated reentry. $NiCr_2O_4$ was not detected in every case. On specimens where NiO and $NiCr_2O_4$ were both detected, their relative intensities were highly variable, but NiO was always the predominant species. The presence of Ni and ThO_2 was also indicated. However, these signals are believed to be the result of X-ray penetration of the oxide scale into the underlying alloy. Overall, little correlation between surface chemistry and emittance was found.

Attempts at correlating emittance data with target surface temperature, exposure time, and specimen preconditioning were not successful. However, a detailed examination of oxide morphology yielded a consistent relationship between the texture of the oxide scale and the total normal emittance. Figures 2 through 5 present SEM photographs with emittance data showing the relation between emittance and morphology as a function of preconditioning type, number of simulated reentry cycles and target surface temperature.

The morphology and total normal emittance resulting from the different preconditioning treatments is shown in figure 2. All three preconditioning treatments resulted in the formation of a continuous oxide scale. EDAX showed the oxide scale to be rich in Cr which is consistent with X-ray diffraction results. However, the specimen with type-A preconditioning had a somewhat

coarser texture and lower emittance than the specimens with
either type-B or type-C preconditioning.

Figure 3 shows SEM photographs and total normal emittance
of type-A preconditioned specimens after exposure to 6, 24, and
30 simulated reentry cycles at 1475 K. The 6-cycle specimen had
an extensive population of small mushroom-shaped oxide clusters

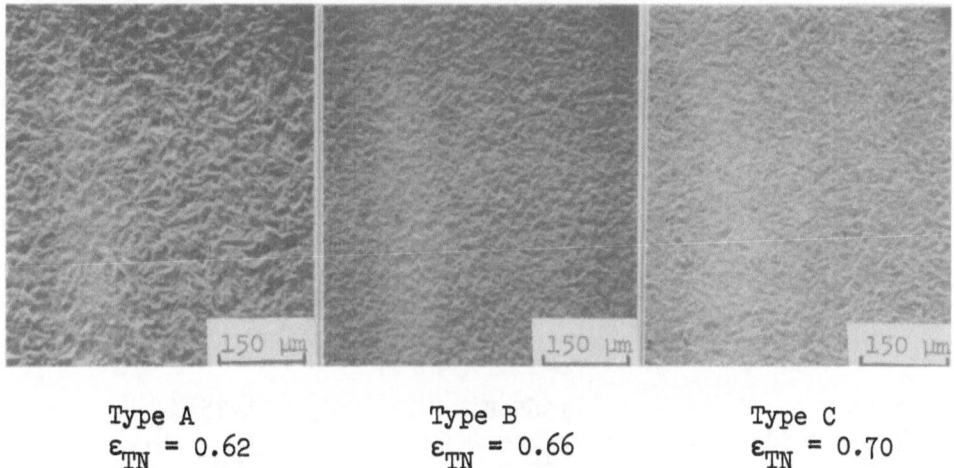

Type A   Type B   Type C
ε_{TN} = 0.62 ε_{TN} = 0.66 ε_{TN} = 0.70

Figure 2. Morphology and Emittance of TD-NiCr With no Simulated
           Reentry Cycles.

6 Cycles   24 Cycles   30 Cycles
ε_{TN} = 0.68 ε_{TN} = 0.55 ε_{TN} = 0.52

Figure 3. Morphology and Emittance of TD-NiCr Subjected to Precon-
           ditioning Type A and Exposed to 6, 24, and 30 Simulated
           Reentry Cycles at a Target Surface Temperature of 1475 K.

embedded in a dark green oxide matrix. EDAX scans indicated that
the "mushrooms" were rich in Ni and that the matrix contained
more Cr than Ni. The 24-cycle specimen had a population of
somewhat larger mushrooms, and the 30-cycle specimen had a heavy
growth of very large mushrooms which had coalesced to yield a
very coarse texture. Stereo microscope examination of these

1255 K   1365 K   1475 K
ε_{TN} = 0.61 ε_{TN} = 0.52 ε_{TN} = 0.63

Figure 4. Morphology and Emittance of TD-NiCr Subjected to Preconditioning Type B and Exposed to 24 Reentry Simulation Cycles
at Target Surface Temperatures of 1255, 1365, and 1475 K.

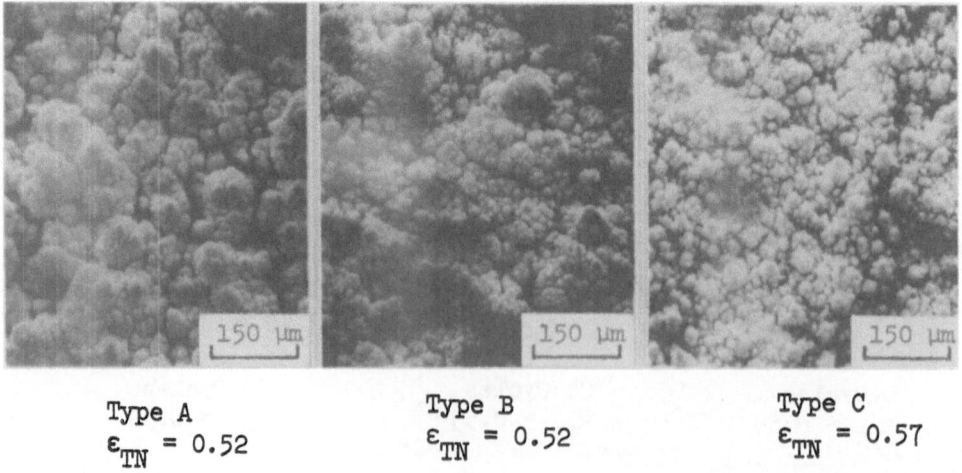

Type A   Type B   Type C
ε_{TN} = 0.52 ε_{TN} = 0.52 ε_{TN} = 0.57

Figure 5. Morphology and Emittance of TD-NiCr Subjected to Various
Types of Preconditioning and Exposed to 30 Reentry Simulation Cycles at a Target Surface Temperature of 1475 K.

specimens showed evidence of local spallation of the mushroom
layer followed by regrowth of smaller mushrooms on the 24-cycle
specimen and fully healed spallation areas on the 30-cycle
specimen. The progressive coarsening of the oxide scale with
increasing exposure time was accompanied by a decreasing total
normal emittance. Although the trend shown in figure 3 is
representative of all specimens exposed at the highest target
temperature, the longest exposure times at the two lower
temperatures did not produce the coarsest textures or lowest
emittances among these specimen groups.

    The morphology and total normal emittance of specimens
subjected to type-B preconditioning and exposed to 24 simulated
reentry cycles at the three target surface temperatures are shown
in figure 4. The mushrooms on the specimens exposed at 1255 and
1475 K were relatively small and did not completely cover the
underlying oxide. The mushrooms on the specimen exposed to the
1365 K had coalesced and were considerably larger than those on
the other two specimens. Likewise, the emittance of the 1365 K
specimen is substantially lower than that of the 1475 and 1255 K
specimens. The emittances of specimens subjected to type-C
preconditioning had a similar relationship to oxide morphology as
shown here, in that the 1365 K specimen which had the coarsest
texture also had the lowest emittance. However, the 1365 and
1475 K specimens subjected to type-A preconditioning had equal
emittances even though the 1365 K specimen had a coarser texture.

    Figure 5 presents the morphology and total normal emittance
of specimens from each preconditioning group that were exposed to
30 simulated reentry cycles at the highest target surface
temperature. All three of these specimens had a relatively
coarse oxide texture composed of fairly large mushrooms. Unlike
the specimens subjected to type-A and type-B preconditioning, the
specimen subjected to type-C preconditioning showed no evidence
of mushroom coalescence. Similarly the emittance of the type-C
specimen was not so low as that of the other two.

    An examination of the emittance results shows no systematic
relationship between oxide morphology and exposure temperature,
exposure time, or type of preconditioning. However, a reasonably
consistent relationship exists between oxide morphology and total
normal emittance; that is, coarser textured surfaces generally
produced lower emittances than did finer textured surfaces. The
repeated vaporization and reformation of the oxide scale
described in reference 6, suggests that the development of
morphology and, therefore, the emittance may occur in a periodic
fashion. Hence, correlation of morphology with exposure
temperature and time may not be possible. The absence of a
correlation between morphology/emittance and preconditioning is
not surprising in view of the finding in reference 10 that the

effects of preconditioning on morphology are limited to
relatively short exposure times.

CONCLUDING REMARKS

An investigation was carried out to determine the effect of
simulated reentry on the emittance of TD-NiCr. Parameters
included in the investigation were: surface preconditioning,
exposure temperature, and time of exposure to simulated reentry.
The emittance of each specimen was measured and the morphology
and composition of the oxide scale on each specimen was studied.

A reasonably consistent relationship between oxide
morphology and total normal emittance was established. Specimens
with coarser textured oxides tended to have lower emittances than
specimens with finer textured oxides.

Wide variations of morphology and emittance were
displayed. No systematic relationship between these variations
and either number of reentry cycles, target exposure temperature,
or type of preconditioning was evident. However, specimens
exposed to 24 and 30 simulated reentry cycles at the two higher
target surface temperatures tended to have significantly coarser
morphology and lower emittance than specimens exposed for only
six cycles or specimens which had been preconditioned only. No
correlation between emittance and surface chemistry was found.

The values of emittance on specimens with no simulated
reentry exposure were on the low end of the range reported in the
literature for statically oxidized TD-NiCr. The emittance of
specimens subjected to simulated reentry heating for large
numbers of cycles were generally 15 to 20% lower than the
emittance of specimens with no simulated reentry exposure. These
results suggest that emittance values previously assumed in
design studies of advanced space transportation systems may be
too high.

REFERENCES

1. Johnson, R., Jr., and Killpatrick, D.H.: Evaluation of
      Dispersion Strengthened Nickel-Base Alloy Heat Shields
      for Space Shuttle Application. NASA CR-2614, 1976.

2. Eidinoff, H.L., and Rose, L.: Thermal-Structural
      Evaluation of TD Ni-20Cr Thermal Protection System
      Panels. NASA CR-132487, 1974.

3. Johnson, R., Jr., and Killpatrick, D. H.: Evaluation
      of Dispersion Strengthened Nickel-Base Alloy Heat
      Shields for Space Shuttle Application. Phase I

Summary Report. NASA CR-132360, 1973.

4. Black, W.E., et al: Evaluation of Coated Columbium
    Alloy Heat Shields for Space Shuttle Thermal Pro-
    tection System Application. Vol. I, Phase I-Environmental
    Criteria and Material Characterization. NASA CR-112119,
    1972.

5. Johnston, James R., and Ashbook, Richard L.: Oxida-
    tion and Thermal Fatigue Cracking of Nickel- and Cobalt
    Base Alloys in a High Velocity Gas Stream. NASA
    TN D-5376, 1969.

6. Tenney, D.R.; Young, C.T., and Herring, H.W.: Oxida-
    tion Behavior of TD-NiCr in a Dynamic High Temperature
    Environment. Metallurgical Transactions, Vol. 5,
    May 1974, pp. 1001-1012.

7. Schaefer, John W.: Thermal Screening of Shuttle
    Orbiter Vehicle TPS Materials Under Convective Heating
    Conditions. Vol I, Final Report, NASA CR-114521,
    1973.

8. Schaefer, John W.: Thermal Screening of Shuttle Orbiter
    Vehicle TPS Materials Under Convective Heating
    Conditions. Vol. II, Tabulation of Test Results.
    NASA CR-114522, 1973.

9. Edwards, S. Franklin, Kantsios, Andronicos G., Voros,
    John P., and Stewart, W.F.: Apparatus Description and
    Data Analysis of a Radiometric Technique for Measure-
    ments of Spectral and Total Normal Emittance. NASA
    TN D-7798, 1975.

10. Young, C.T., Tenney, D.R., and Herring, H.W.: Dynamic
    Oxidation Behavior of TD-NiCr Alloy with Different Surface
    Pretreatments. Metallurgical Transactions A, Vol.
    6A, Dec. 1975, pp. 2253-2265.

SLABS SUBDIVISION EFFECTS ON WALLS THERMAL BEHAVIOUR

V. Ferro, P. Gregorio, and A. Sacchi

Istituto di Fisica Tecnica ed Impianti Nucleari

Politecnico di Torino - Italy

ABSTRACT

A wall has been considered which is composed by a slab having high thermal resistance and low heat capacity, and by a slab having high heat capacity and low thermal resistance. Each slab has than been subdivided into thinner slabs and the multilayer wall, resulting from adding alternatively such thinner slabs, has been conside red.

The parameters which describe the thermal behaviour of the wall have been calculated, following the two methods: the first one, valid for stabilized periodic state, and the later one proposed by N.B.S.

The results indicate that the multilayer walls, often adopted to increase the acoustical insulation of the walls themselves, produce better thermal insulation for variable temperature components.

INTRODUCTION

Thermal studies of multilayer walls, undergoing sinusoidal perturbations, (1), (2) have pointed out the different behaviour of slabs of equal thermal transmittance $K|W/m^2 °C|$ but of different composition because of materials and layers location as well. Similar differences are evidenced by using the calculation method developed by N.B.S, and accepted by ASTM, (3), (4).

The analytical relationships, which describe the heat transfer
phenomena, are similar to the ones describing the propagation of
electrical perturbations in the electical lines and to the ones describing the propagation of acoustical perturbations within the walls
themselves. In these cases the attenuation of the materials is somewhat reduced and the total attenuation results from waves reflections at the surfaces, which separate the different slabs.

As far as thermal phenomena are concerned, the attenuation of
the materials, alone, is 27 dB for each path one wavelength long ,
and therefore the effect of reflections appears less important.

In acoustics it is well known that, for a given value of mass
per unit area,multilayer walls show a higher value of insulation
than the single layer ones (5). Because of this, walls have been
built much lighter but very similar and sometimes even better than
the traditional ones from an acoustical point of view.

In this work the benefit is investigated due to the use of multilayer slabs and to the presence of alternating layers of thermal
insulating materials and of high density materials.

THEORETICAL APPROACH

The basic wall which has been considered consists of a layer
of concrete (thickness = 0.1 m, density = 2100 kg/m^3 , thermal conduc-
tivity=1.28 W/m°C, specific heat=1 kJ/kg°C)and one of expanded poly-
styrene (thickness=0.1 m, density=25 kg/m^3 ,thermal conductivity=0.033
W/m°C, specific heat=1 kJ/kg°C). The wall separates the inner ambient
from the outer one (boundary layer internal heat transfer resistance
= 0.045 m^2°C/W, boundary layer external heat transfer resistance =
$0.125m^2$°C/W). The wall has been placed one time with the insulating
layer facing inside and another time with the insulating layer facing
outside (walls n. 1 and n. 2 in table 1).

Each layer has then been subdivided in 2, 4, 8, 10 sublayers,
assembling one layer of a type behind the layer of the other type.

In this way, for each subdivision, two situations have been obtained where the insulating layer or the compact layer faces outside
(walls n. 3 ÷ 10).

TABLE 1

Wall	Wall composition (from outside to inside) (thickness in cm; C concrete; R expanded polystyrene)	
1	10 C ; 10 R	
2	10 R ; 10 C	
3	(5 C ; 5 R)	two times
4	(5 R ; 5 C)	two times
5	(2.5 C; 2.5 R)	four times
6	(2.5 R; 2.5 C)	four times
7	(1.25 C; 1.25 R)	eight times
8	(1.25 R; 1.25 C)	eight times
9	(1 C; 1 R)	ten times
10	(1 R; 1 C)	ten times
Eq	thickness 20 cm; density 1062 kg/m^3 ; conductivity 0,063 W/m°C ; specific heat 1 kJ/kg°C	

TABLE 2

Wall	Wall composition (from outside to inside) (thickness in cm; C concrete; R expanded polystyrene)	
1	5 C ; 10 R ; 5 C ; 2.5 C	
2	2.5 C ; 5 R ; 5 C ; 5 R	
3	1.25 C; (2.5 R ; 2.5 C) three times; 1.25 C	
4	0.5 C; (1 R; 1 C) nine times ; 1 R ; 0.5 C	

TABLE 3

- Wall n.2 produces E larger than the one of the other walls of Table 1 for periods longer than the following:
(the symbol ∞ means that the above condition is never encountered):

wall	1	3	4	5	6	7	8	9	10	Eq
period T/τ	∞	0,1	10	1,3	3,6	1,7	3,6	1,7	2,3	1,9

- Wall n. 4 produces E larger than the one of the other walls of Table 1 for periods longer than the following:

wall	1	2	3	5	6	7	8	9	10	Eq
period T/τ	∞	7,5	∞	0,3	1,7	0,7	1,7	0,7	1	1

- Wall n. 6 produces E larger than the one of the other walls of Table 1 for periods longer than the following:

wall	1	2	3	4	5	7	8	9	10	Eq
period T/τ	∞	4	∞	1,8	∞	0,2	∞	0,3	0,4	0,3

- Wall n. 1 produces H larger than the one of the other walls of Table 1 for periods longer than the following:

wall	2	3	4	5	6	7	8	9	10	Eq
period T/τ	∞	6	0,3	4	2	2	3	3	3	3

- Wall n. 3 produces H larger than the one of the other walls of Table 1 for periods longer than the following:

wall	1	2	4	5	6	7	8	9	10	Eq
period T/τ	6	∞	∞	2	0,6	1	1	1	1	1

- Wall n. 5 produces H larger than the one of the other walls of Table 1 for periods longer than the following:

wall	1	2	3	4	6	7	8	9	10	Eq
period T/τ	6	∞	2	∞	∞	0,3	0,3	0,3	0,3	0,3

A second series of walls, of symmetrical composition, has been obtained, dividing the compact coating layer in two layers of thickness one half, and placing them as inside and outside coating (see table 2). In this way a symmetrical layout has been obtained, which is more realistic since the insulating layers are protected.

The last solution has been the one of a subdivision in an infinite number of infinitely thin layers, alternatively following each other (Eq. wall of table 1).

The relationships between the k-th sinusoidal components of temperature Θ and thermal fluxes ϕ, internal (index i) and external (index e), is:

$$\begin{vmatrix} \Theta_{ek} \\ \\ \phi_{ek} \end{vmatrix} = \begin{vmatrix} E & F \\ \\ G & H \end{vmatrix} \begin{vmatrix} \Theta_{ik} \\ \\ \phi_{ik} \end{vmatrix} \tag{1}$$

where E, F, G and H are complex parameters characteristic of the type of wall at the k-th frequency under consideration ($f = 2\Pi/T$ where T is the period). For each of such slabs the four parameters E, F, G and H have been calculated and the results are indicated in figs. 1 ÷ 16.

On the abscissa the dimensionless ratios T/τ are plotted as well; τ is the time constant given by:

$$\tau = R \cdot C$$

where $R = 1/K$ and C represents the thermal capacity of the wall. For the case which is considered $\tau = 18$ h.

For the same walls the covolution equation of heat fluxes is (3):

$$\begin{vmatrix} F_{1,\tau} \\ \\ F_{n,\tau} \end{vmatrix} = \sum_{i=o}^{\infty} \begin{vmatrix} X_i & -\Gamma Y_i \\ \\ Y_i & -Z_i \end{vmatrix} \begin{vmatrix} V_{1,\tau-i} \\ \\ V_{n,\tau-i} \end{vmatrix} \tag{2}$$

where:

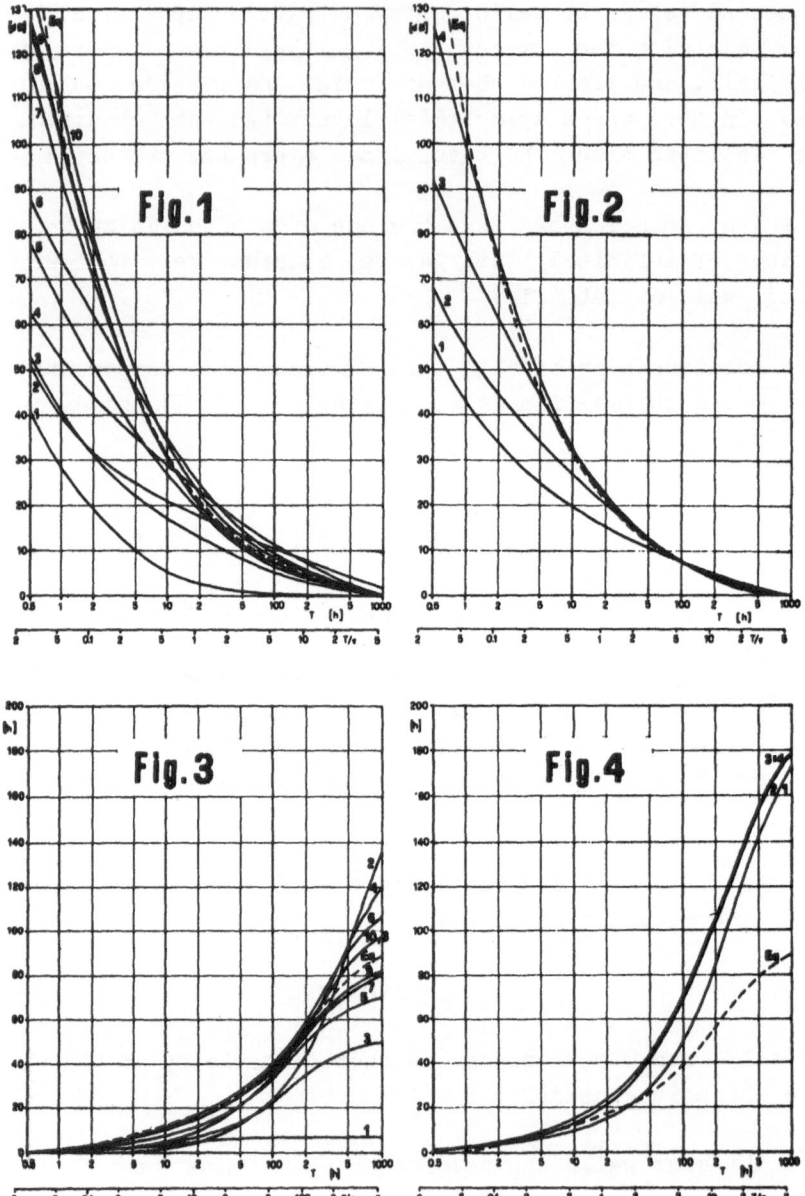

Fig. 1 and 2 - Plot of $\alpha_E = 10 \lg_{10} |E|$ as a function of T and of T/τ
          for non symmetrical and symmetrical walls rispective-
          ly (walls of Table 1 and 2).

Fig. 3 and 4 - Plot of arg E as in fig. 1 and 2.

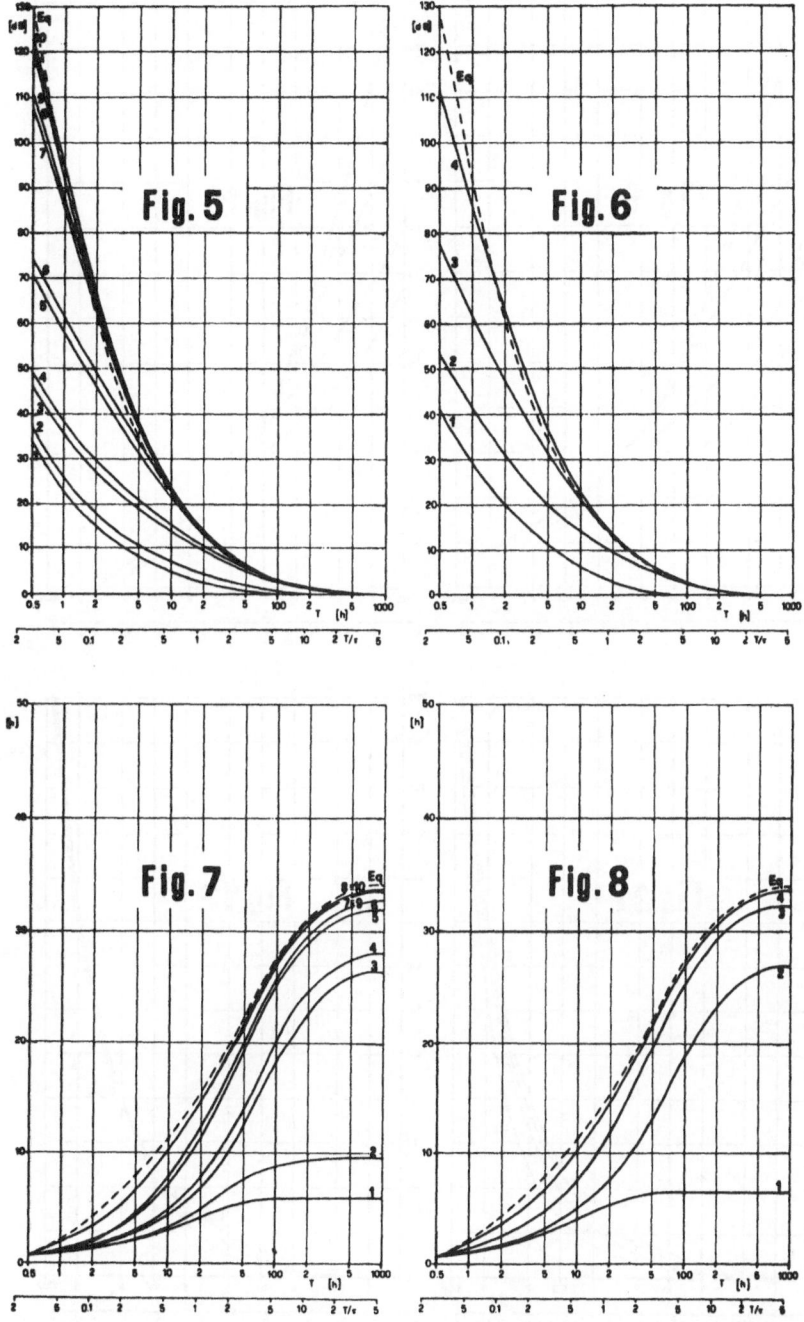

Fig. 5 and 6 - Plot of $\alpha_F = 10 \lg_{10} |KF|$ as in fig. 1 and 2.

Fig. 7 and 8 - Plot of arg F as in fig. 1 and 2

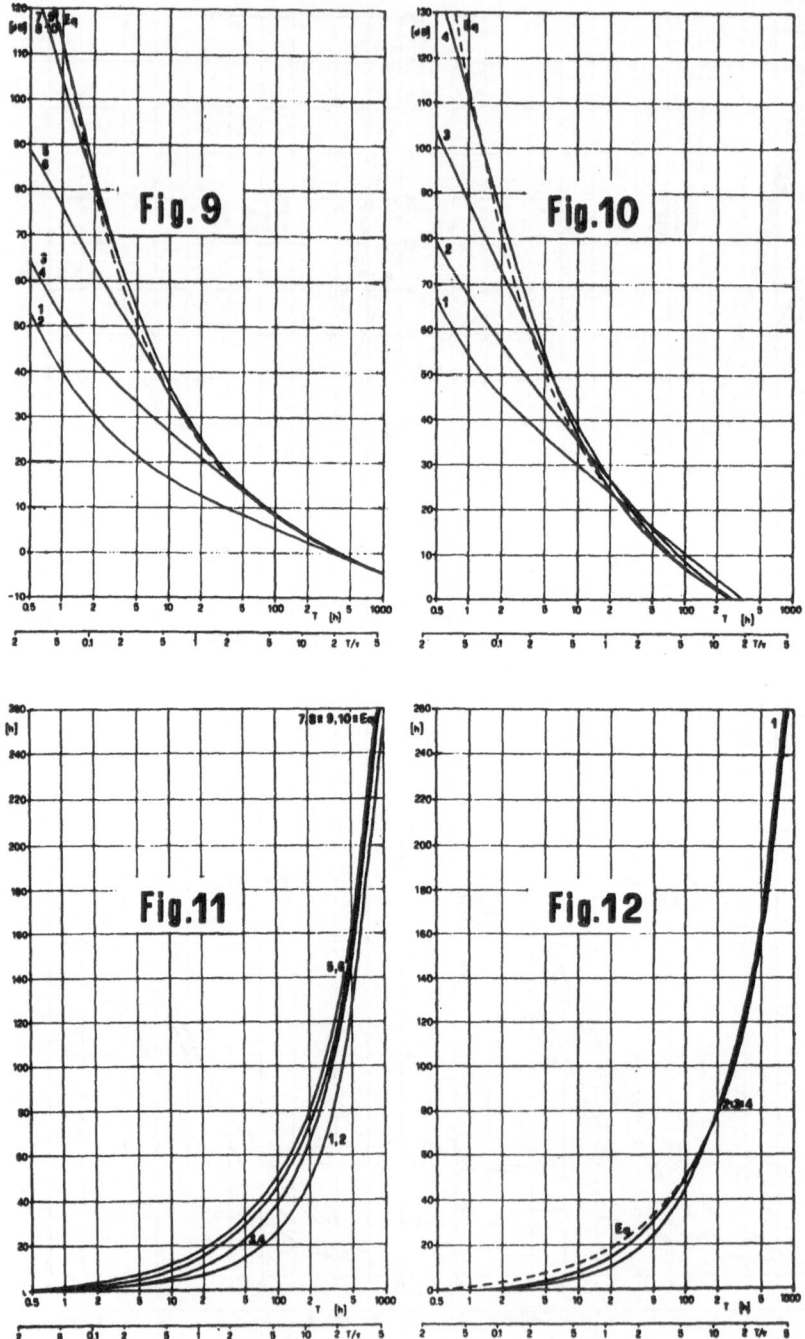

Fig. 9 and 10 – Plot of $\alpha_G = 10 \lg_{10} |G|$ as in fig. 1 and 2

Fig. 11 and 12 – Plot of arg G as in fig. 1 and 2

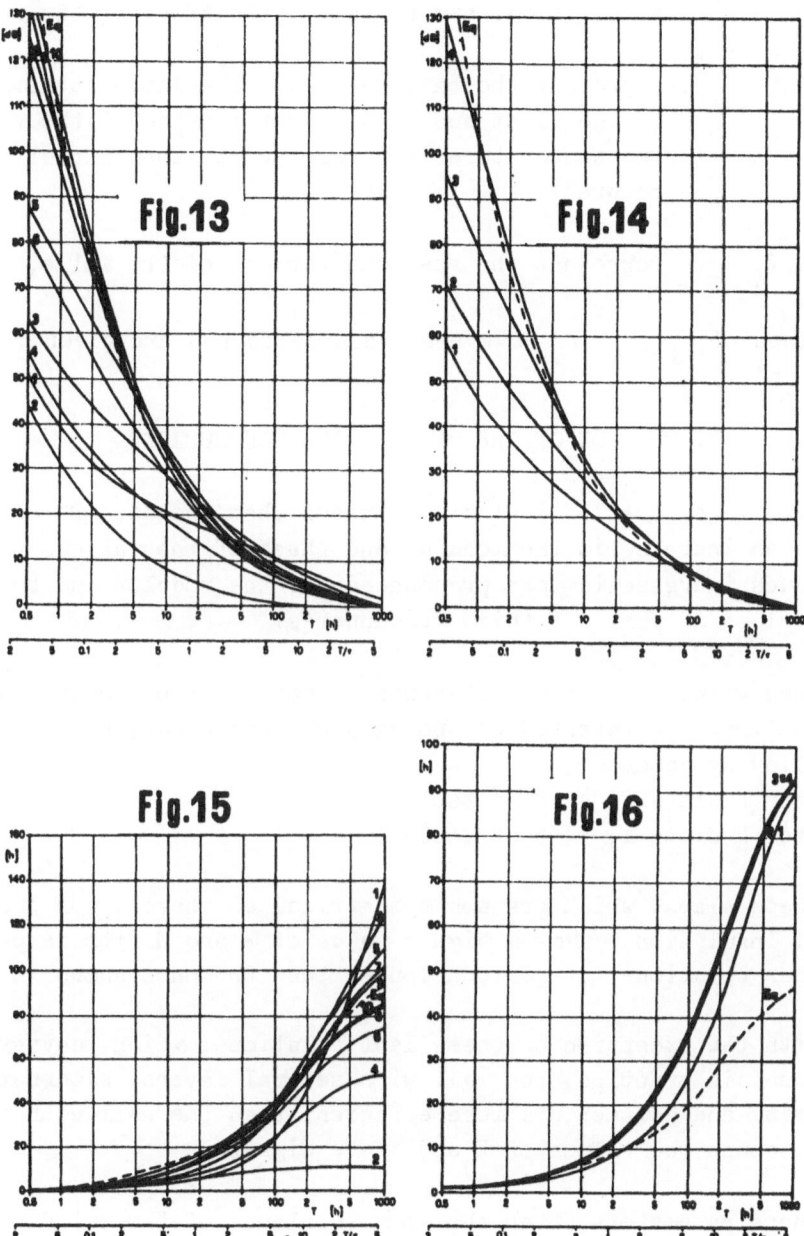

Fig. 13 and 14 - Plot of $\alpha_H = 10 \lg_{10} |H|$ as in fig. 1 and 2.

Fig. 15 and 16 - Plot of arg H as in fig. 1 and 2.

$F_{1,\tau}$, $F_{n,\tau}$ represent the thermal fluxes, calculated at time τ, at the first and at the last surface of the walls;

$V_{1,\tau-i}$, $V_{n,\tau-i}$ represent the temperatures, calculated at time $\tau-i\delta$, of the first and of the last surface of the walls ; the temperatures are referred to the initial temperature T_o;

X_i, Y_i, Z_i represent the responce factors of the walls,

The values of X_i, Y_i, Z_i have been calculated and are given in figs. 17, 18 and 19.

DISCUSSION OF THE RESULTS AND CONCLUSIONS

From figs. 1 ÷ 16 it clearly appears that the use of many layers induces an increase in the modulus and phase of the various parameters. Such increase is more pronounced for the modulus and for the lower values of period T (high frequencies).

When T increases the differences become less pronounced and some situations are inverted but no main changes occur, except than in the following cases:
Parameter E slabs 2, 4, 6 of table 1
Parameter H slabs 1, 3, 5 of table 1

These slabs, which are non symmetrical and have a non protected thermal insulation, show a higher value of E and H with respect to the other solutions for periods longer than the ones shown in table 3.

With the exception of these last solutions, which, anyway, are never used in practice, the wall with several layers, alternatively insulating and compact, is more efficient than the wall with few layers, since the values of E and F are higher.

These parameters enter the thermal balance of the building (5), (6) affecting the thermal flux which enters through the peripheral walls, and has the following expression:

$$\phi_i = \phi_{io} + \sum_{k=1}^{\infty} \left[\left(\frac{1}{F_k} \Theta_{ek} - \frac{E_k}{F_k} \Theta_{ik} \right) \right] \exp\left(j \frac{2\pi}{T_k} t \right) \qquad (3)$$

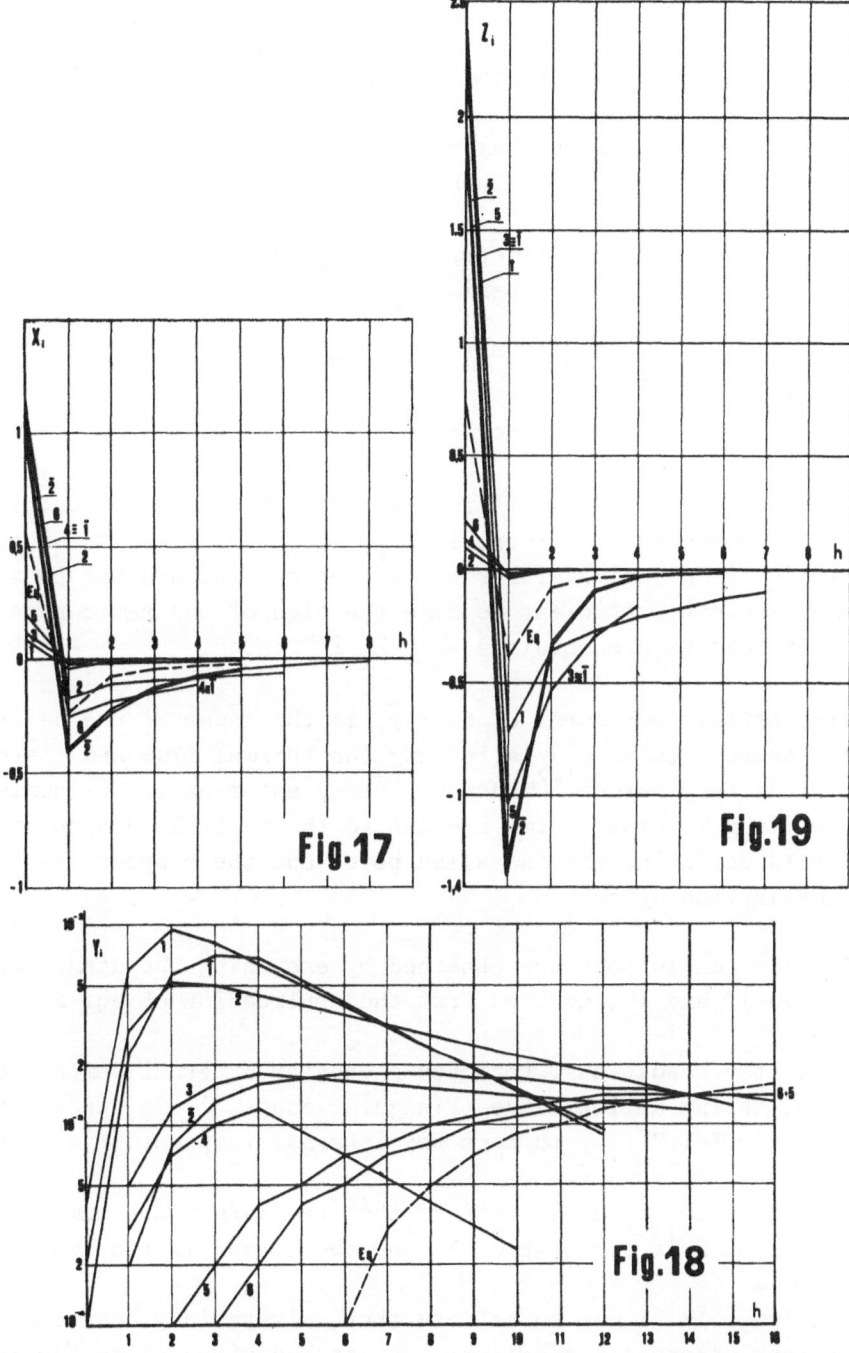

Fig. 17, 18, 19 - Response Factors X_i, Y_i, Z_i VS. time (hours) for
                 the walls described in table 1 and in table 2
                 (These last ones are marked by a line above the
                 identification number).

where:

ϕ_i =   flux entering through the peripheral walls $[W/m^2]$

ϕ_{io} =   average flux component, calculated using
                                 the relationships valid for steady state $[W/m^2]$

F_k and E_k =   wall parameters corresponding to the har-
                                 monic of period T_k

$\Theta_{ek} \exp(j\frac{2\pi}{T_k} t)$ and $\Theta_{ik} \exp(j\frac{2\pi}{T_k} t)$ = harmonics of period T_k

                                 of the inner and outer temperature $[\,^{\circ}C\,]$

From this relationship it follows that the variable contributions of ϕ_i decrease with incresing F_k; moreover, since the second
contribution is negative with respect to the first and the phases of
F_k and E_k are not sufficient to make the sign of the ratio negative,
equivalent results are obtained when E_k increases.

The variable component of flux ϕ_i is the cause of the deviations
from the average value $\phi_{i,o}$ affecting the thermal flux and therefore
the internal temperature. The power peaks entering in the summer time and exiting in winter time are due to the variable component ϕ_i.
These peaks determine the installed power and the response time of
the climatization systems.

Similar conclusions are obtained by examining the diagrams, plotted in figs 17 and 18, derived from the application of eq. 2.

In winter conditions, for each elementar external temperature
variation, having an impulsive triangular shape, it is necessary a
flux $F_{1,\tau} = -\Sigma_i Y_i V_{n,\tau-i}$ to keep the internal temperature constant.

From fig. 18 it may be noticed that the larger the numbers of
layers the lower is the peak of Y_i and the larger is its time lag.

From fig. 17 it may be noticed that, always in winter conditions,
for constant external temperature, for each desired elementar internal temperature variation, having an impulsive triangular shape, it
is necessary to give a thermal flux $F_{1,\tau} = \Sigma_i X_i V_{1,\tau-i}$ which decreases with the peak value of X_i.

Multilayer walls show peak values of X_i sufficiently small and not much larger than the ones of two layers walls having a large thickness of insulation facing inside.

For these reasons the light composite walls in buildings in winter but expecially in summer and in transition seasons, because of the non existance of air conditioning systems, turn out to be too permeable to the external temperature variations, up to rendering, sometime, the rooms impossible to live in.

The use of multilayer walls may reduce these inconveniences if a suitable choices is made.

These multilayer walls behave as they consist of a homogeneous material alone with high mass (and then high thermalcapacity) and with low conductivity; this condition cannot be realized with real homogeneous material.

- This work has been supported by C.N.R. (Consiglio Nazionale delle Ricerche) under the research contracts n° 77.00851.92/B and CT.74. 01060.07. -

REFERENCE

(1) Ferro V., Sacchi A. "Oscillazioni termiche in pareti composte". La Termotecnica - Fascicolo Ricerche n.16 - Vol.XX-1966, p.52-62.

   - Ferro V., Sacchi A., Codegone C. "Thermal attenuation through homogeneous and multilayer slabs in steady periodic conditions - Theory and experiments". Proceedings of the 7th Conference on Thermal Conductivity - NBS Special pubblication N. 302 p.177-196.

(2) Calì M., Sacchi A. "Caratteristiche delle pareti degli edifici in regime oscillatorio". Atti e Rass.Tecnica della Soc.Ingg. e Arch. in Torino, Vol. XXXI, 1977, p.175-181.

(3) Kusuda T. A "Algoritms for Calculating the Transient Heat Conduction by Thermal Response Factors for Multi-layer Structures of Various Heat Conduction Systems". NBS Report 108, August 28,1969.

(4) ASHRAE Handbook of Fundamentals - 1972, p. 425.

(5) Sacchi A. "L'attenuazione delle oscillazioni termiche in pareti semplici e composte". Appendix n.7 to the Volume P.E.Brunelli,C. Codegone - Corso di Fisica Tecnica Vol.II. P. II-ed. V.Giorgio -

Torino 1967 - p.535-568.

(6) Beranek L.L. "Noise Reduction". Mc Graw Hill Co.- New York -
     Toronto - London 1960.

(7) Sacchi A. "Il calcolo numerico delle perturbazioni termiche in-
     dotte in un edificio per effetto di variazioni periodiche delle
     condizioni climatiche esterne" PT IFT 373, giugno 1969.

PLENARY SESSION II

Session Chairman: D.C. Larsen
IIT Research Institute
Chicago, Illinois

PROGRESS IN HIGH TEMPERATURE MEASUREMENT IN LABORATORY

G. Ruffino

Leeds & Northrup

Cso M. D'Azeglio, 60 - 10126 Turin (Italy)

1-INTRODUCTION

The present survey concerns laboratory measurements of temperature. Therefore, industrial measurements and their commercial implementations fall out of the scope of this review.

The borders of "high temperatures" have not yet been the subject of any accepted general convention. A tentative classification of temperature ranges could be the following:
Cryogenic temperatures: below 100K
Low temperatures: between 100K and 273.15K
High temperatures: between 273.15K and 10,000K
Ultra-High temperatures: above 10,000K

The definition of the first range is widely accepted. The last one concerns the plasma thermometry. If we refuse a very fuzzy strip of intermediate temperatures, then we may find two possible borderlines between low and high temperatures.
a) the low limit of "the regime where surface heat losses from a speciman can no longer be ignored." (Laubitz and McElroy, 1971); it is assumed to be 100K;
b) a fixed point close to room temperature, namely the melting point of ice.

The first definition is based on radiation heat exchange and admits no intermediate range between cryogenic and high temperature. The second one depends upon the conditions under which the experiments are run, namely the supply or extraction of heat. We stick to the latter.

The time span to which the present investigation is confined goes from the present day back to 1971, date of the 5th Symposium on Temperature held in Washington, D.C.. Its proceedings were published under the title: "Temperature, its Measurement and Control in Science and Industry," Vol. 4 (Ed. H.H. Plumb, 1972) a book of 2383 pages in three parts, which encompassed the state of the art of temperature measurement at the time of its publication.

Another important conference on temperature measurement took place in Teddington, G.B. in 1975; its proceedings were published under the title: "Temperature Measurement, 1975" (Ed. B.F. Billings and T.J. Quinn, 1975). As the Washington Conference presented the scientific effort which gave birth to the International Practical Temperature Scale of 1968 (IPTS-68) or has been immediately stimulated by its implementation, so an important part of the contributions to the Teddington Symposium was the result of research in perfecting the scale and in improving the measurement techniques. The latter symposium is connected with the amended edition of the IPTS-68 ratified by the General Conference on Weights and Measures in 1975.

The present paper will cover the following topics:
a) the amendment of the IPTS-68;
b) relationship between thermodynamic temperature and the IPTS-
-68;
c) new determination of fixed points;
d) platinum resistance thermometry;
e) thermocouple thermometry;
f) radiation pyrometry.

2-AMENDMENT OF THE IPTS-68

The IPTS-68 (-1969), shortly after its promulgation, showed several imperfections. It was more a matter of imprecisions and inconveniences than of actual errors. The Consulting Committee for Temperature, in its meeting of 1974 (CCT-1974), approved an amended edition of the IPTS-68 which, after the approval of the International Committee of Weights and Measures, was ratified by the General Conference in 1975.

The main features of the amended edition (-1976) are as follows:
a) The relationship between thermodynamic temperatures and temperatures of the IPTS,Kelvin temperatures and Celsius temperatures and their units are clarified.
b) The amended edition does not change the values of the temperatures of the original scale. As a matter of fact, the value of the fixed points remains the same, only with the addition of the triple point of argon as a suggested alternative to the condensation point of oxygen. Among the interpolation formulae, only the reference func-

tion of the platinum resistance temperature below the ice point has been formally reshaped, according to a suggestion of the NCR, Canada (Kirby and Bedford, 1972), so as to give practically the same numerical results with less expenditure of digits in the computation.

c) An addition specifies that the wavelengths to be introduced in Planck's Law to calculate temperatures above the gold point, must be referred to vacuum and not to standard air as suggested by Blevin of the NMR (Australia)(1972) who remarked that all previous pyrometric measurements were affected by errors caused by neglecting the refractive index of air (about 1.0028).

d) The relation which the Pt-10% Rh/Pt thermocouple must satisfy at 630.74^{0}C, the silver and the gold point have been modified to agree with the new international tables.

e) The supplementary information has been pruned and updated.

The table listing the defining fixed points together with their estimated uncertainties has been deleted as the experimental evidence given by the gas thermometer of the NBS showed an excessive optimism in those uncertainty assessments. Similarly, the recommendation of the T_{58} and T_{62} helium vapour scales has been dropped out of the original edition.

Some of the secondary points (as the freezing points of copper and platinum) have been corrected.

History of temperature scales and detailed comments on origin and content of the IPTS-68 are presented by H. Preston-Thomas (1972). The amended edition, along with present trends of primary pyrometry, have been described by Quinn (1975).

3-THERMODYNAMIC TEMPERATURE

Gas thermometry is the classical method of measuring thermodynamic temperature. It may be supplemented by measurements of thermal noise and thermal radiation, which are phenomena governed by laws derived from the principles of thermodynamics.

a-Gas Thermometry

In the NBS, Guildner and co-workers constructed a gas thermometer starting in the early sixties. The high degree of accuracy of this instrument stems mainly from the careful elimination of sorption. The manometer itself has an outstanding precision as it utilizes all expertise in locating the mercury meniscus and in carrying out length measurements. Moreover, the thermal expansion of the bulb, the dead-space volume and its temperature gradient, the thermomolecular pressure in the capillary and gas imperfection were thoroughly

estimated. The development of what may be classified among the most
outstanding metrological achievements of present times, has been published in a series of several papers, the last of which presents
the deviation of the IPTS from thermodynamic temperatures between
273.16K and 730K (L.A. Guildner and R.E. Edsinger, 1976). Results
are synthetized in Fig. 1. All values of the defining points of the
IPTS-68 lay above the thermodynamic temperature and the absolute va-

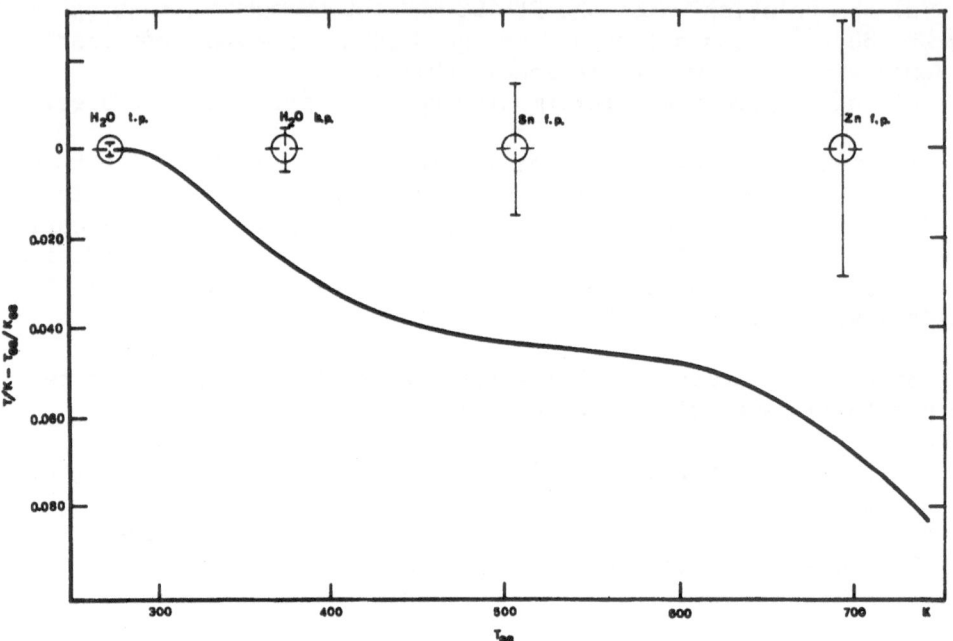

Fig. 1- Difference between thermodynamic temperatures and IPTS-68
between 273.16K and 730K (Guildner and Edsinger, 1976)

lue of the deviation increases monotonically with temperature, no-
where to be found smaller than the estimated uncertainties given by
the original edition of the IPTS-68.

b-Noise Thermometry

    In the Istituto di Metrologia G. Colonnetti (IMGC), L. Crovini
et al. have been working on a noise thermometer in the last thirteen
years and are publishing now their results (1977). Noise thermometry has been successfully applied to cryogenics by C.P. Pickup (1975).

The technique is based on Niquist's equation which relates the r.m.s. voltage of thermal noise generated in a resistance R, within bandwidth Δf, to the Kelvin temperature T:

$$\overline{V} = \sqrt{4kRT\ \Delta f} \tag{1}$$

where k is Boltzmann's constant.

The practical use of this meets enormous difficulties in making resistance noise voltages measurements independent from the noise generated by the elements of the measuring chain, mainly by the amplifiers.

Crovini's final results are summarized in Fig. 2b. It includes part of Guildner's plot, suggesting a reasonable match in the unexplored range.

c-Radiation Thermometry

Both narrow band and total radiometry are suited to measure thermodynamic temperature since Planck's and Stefan-Boltzmann's laws are linked to the principles of thermodynamics through quantum mechanics (the latter law may also be derived directly from the second principle of thermodynamics).

The application of Stefan-Boltzmann's Law to measure temperature may be extended down to the ice point (Ginning and Reilly, 1972). Conversely total radiometry has been used by W.R. Blevin and W.J. Brown (1971) to determine the Stefan-Boltzmann constant on the basis of known temperature (gold point on the IPTS-68):
$\sigma = (5.6644 \pm 0.0075)\ 10^{-8}\ W\ m^{-2} K^{-4}$.

Narrow band radiation pyrometry, which makes use of the Planck Law, avoids geometrical measurements by comparing radiations emitted by two sources; one at a reference temperature and the other at the unknown temperature. It requires the knowledge of the second Planck constant $C_2 = hc/k$ and the thermodynamic temperature of the reference source. Therefore, monochromatic radiation pyrometry only gives reasonably accurate results in determining the thermodynamic temperature interval.

Quinn et al. (1973) measured the departure of the IPTS-68 from thermodynamic temperatures between 725^0C and 1064.43^0C. Their estimated error was 0.1^0C over the range from 790^0C to the gold point and the reference was set at the latter. Bonhoure (1975), in the laboratory of the BIPM, explored the interval between the antimony point (903.89K), taken as a reference, and the gold point. He found $T_{Au} = (1337.53 \pm 0.16)K$ and $T_{Ag} = (1235.16 \pm 0.13)K$, in good agreement with

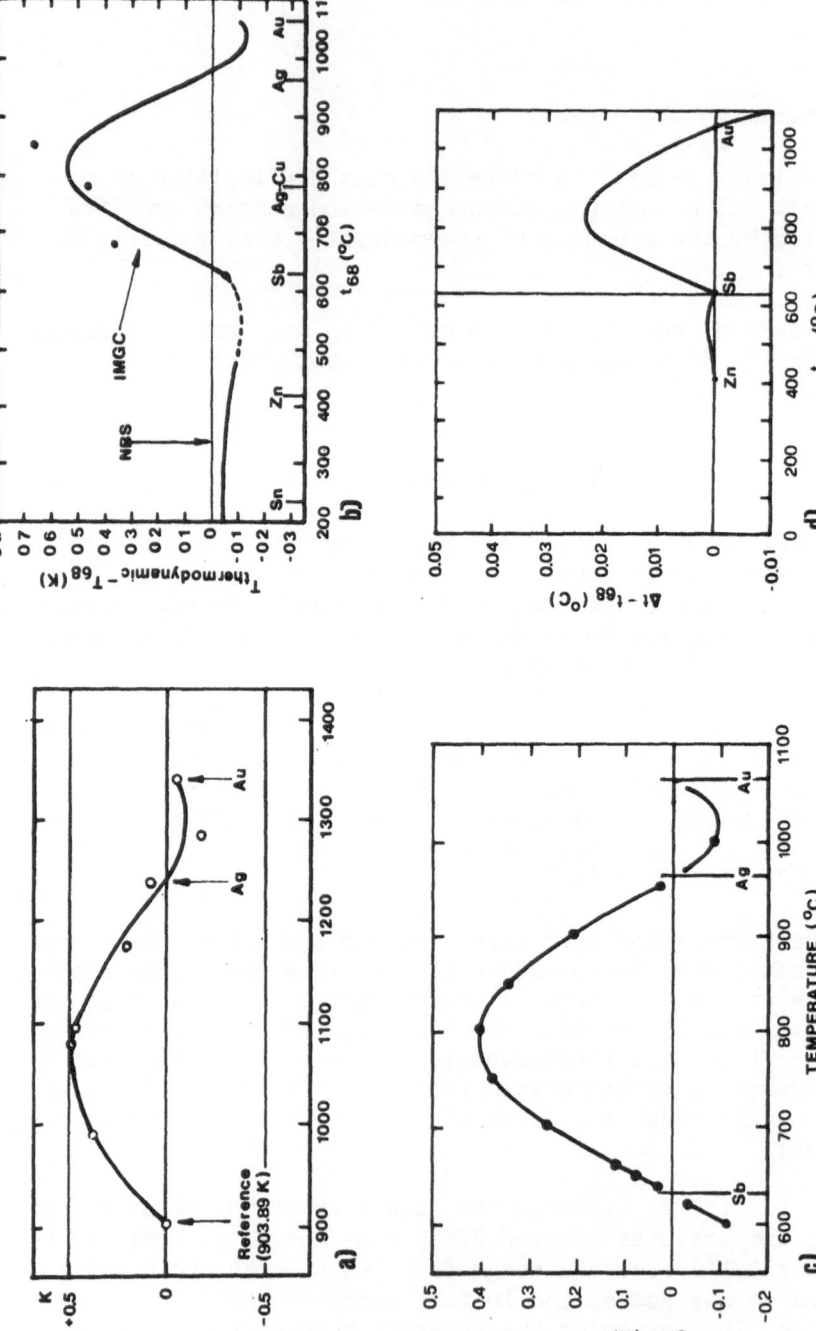

Fig. 2 - Departure from IPTS-68 of: a) BIPM Radiation Scale (Bonhoure, 1975); b) IMGC Noise Thermometer Scale (Crovini and Actis, 1977); c) NBS Pt Resistance Thermometer Scale Above 630°C (Evans and Woods, 1971); d) IMGC Pt Resistance Thermometer Scale Above 630°C (Marcarino and Crovini, 1975).

the IPTS-68 but he got evidence of strong deviations between the antimony and the silver points (Fig. 2a). The thermodynamic temperature interval between silver and gold points has also been measured by Jung (1975), Coslovi et al. (1975) and Ricolfi (1977). All results are listed in Table I.

TABLE I

DIFFERENCE $T(Au) - T(Ag) - T_{68}(Au) - T_{68}(Ag)$

Author	Year	Difference
T.J. Quinn	1972	(-0.13 ± 0.1)K
H.T. Jung	1973	(-0.133 ± 0.04)K
L. Coslovi et al.	1974	(-0.17 ± 0.06)K
J. Bonhoure	1973	(-0.13 ± 0.20)K
T. Ricolfi, F. Lanza	1977	(-0.12 ± 0.04)K

If we give to the results of radiation measurements a weight inversely proportional to their standard deviation, the weighted mean of the interval difference is $\Delta T=(0.14 \pm 0.02)$K.

4-FIXED POINTS

The influence of dissolved oxygen in the melting point of silver and the best conditions in realizing the fixed point have been investigated by Bongiovanni et al. (1975). They conclude that the induced freezing technique allows a precision of about ± 2 mK.

The solid-liquid phase transition of the eutectic Ag-Cu alloy ($\sim 779.5^{0}$C) is the only fixed point available between the aluminum point ($\sim 660^{0}$C) and the silver point ($\sim 961^{0}$C). Bongiovanni et al. (1972) investigated the time dependence of the freezing temperature of this eutectic, which can be reproduced within 10 mK.

The platinum freezing point, besides being a very convenient secondary thermometric point for thermocouple and pyrometer calibration, is the defining point of the candela, a fundamental unit of the SI system. Its value is instrumental in the determination of the spectral luminous efficacy of the radiation (Blevin and Steiner, 1975).

In the seventies, this point has been measured independently in four laboratories. Results are listed in Table II. The average of recent values of the platinum point is $(1768.8\pm0.7)^0$C.

TABLE II

Recent Values of the Freezing Point of Platinum

Laboratory	Temperature/^{0}C
NPL (Quinn and Chandler, 1972)	1767.9 ± 0.3
NML (Jones and Tapping, 1976)	1769.5 ± 0.6
IMGC (Lanza and Ricolfi, 1976)	1769.0 ± 0.4
PTB (Kunz and Lohrengel, 1976)	1768.96 ± 0.5

The freezing point of copper is a convenient alternative calibration point for the gold point, as both temperatures are close to
each other, and the former uses a much less expensive material. Experiments in the seventies demonstrated its easy feasibility and excellent reproducibility. Besides, they corrected the original value
of the IPTS-68. Results are listed in Table III. It may be seen that
they agree within a few millikelvins.

TABLE III

Recent Values of the Freezing Point of Copper

Laboratory	Temperature/^{0}C
IMGC (Righini et al., 1972)	1084.88 ± 0.1[*]
IMGC (Ricolfi and Lanza, 1977)	1084.87 ± 0.02
NPL (Coates and Andrews, 1977)	1084.87 ± 0.04

[*] Corrected value introduced in the amended edition of the
IPTS-68.

Fixed points above 2000^0C are of paramount importance in cali-
brating pyrometers in the IPTS. For this reason the Committee for
High Temperatures and Refractories of the IUPAC set up a Task Group
to determine the fixed points of alumina, yttria and beryllia. Other
refractory materials have been studied, some with new techniques,
as the high speed automatic data acquisition systems introduced by
Cezairliyan. A summary of results, with profuse literature has been

published by Kenisarin, Chekhovskoi et al. (1976). Their proposed melting points (except for yttria) are listed in Table IV. As to yttria, the results of the task group were summarized by Foëx (1977).

TABLE IV

Proposed Values of Freezing Points Above 2000K

Material	Temperature/K
Alumina	2326±3
Yttria	2712±12
Niobium	2748±5
Beryllia	2840±10
Molybdenum	2896±3
Tungsten	3695±8

Methods required by measurements of melting points at high temperature are described by Cabannes et al. (1972, 1976). The effects of environment for alumina have been stressed by Schneider and Mc-Daniel (1967). As another example of careful statistical analysis of the errors may be quoted the work of Kenisarin et al. (1972) on the melting point of molybdenum.

Cezairliyan proposed the use of radiance temperature at known wavelength of melting points for checking pyrometers. We quote his work on molybdenum as another example (Cezairliyan et al., 1975).

5-PLATINUM RESISTANCE THERMOMETRY

Basic knowledge of platinum resistance thermometry, as well as essential literature, can be found in the NBS Monograph 126 (Riddle, 1973).

The IPTS-68 extended down to 13.81K the temperature range of the platinum resistance thermometer as an interpolating instrument. The uniqueness of this scale may be impaired by the oxygen activated cycling effects (Berry, 1974, 1975).

The extension of the range of the platinum resistance thermometer to the gold point has been a long effort culminating in the 1971 Temperature Symposium (Washington, D.C.) where a session was devoted to reports of the NBS, NPL and NRLM (Japan). They investigated the repeatibility of this sensor in the 630-1064°C range. At that time

it appeared to be of the order of 10 mK, against 0.2K repeatibility
of the standard thermocouple.

Subsequent problems are the interpolating formula of the resistance thermometer and the comparison of its scale with the IPTS-68.
Evans and Wood (1971) intercompared nine high temperature resistance thermometers and eight thermocouples to determine the difference
between the resistance thermometer scale and the IPTS-68. The temperature difference data were fitted by the equation:

$$F(\theta)=(\theta-1064.43)(\theta-961.93)(\theta-630.74)(A_0+A_1\theta) \qquad (2)$$

with $A_0=-8.492168.10^{-8}$, $A_1= 1.768905.10^{-10} {}^0C^{-1}$. Experimental points
and fitting curve of temperature difference are represented in Fig.
2c.

Similar results were obtained by Quinn et al (1973). Marcarino
and Crovini (1975) repeated Evans' comparisons. Their preferred interpolation formula is:

$$t= t'+0.045 \frac{t'}{100^0C} (\frac{t'}{100^0C}-1)(\frac{t'}{419.58^0C}-1)(\frac{t'}{630.74^0C}-1) \times$$

$$\times 1-\varepsilon(\frac{t'}{1064.43^0C}) \qquad (3)$$

Constants ε and n are adjusted to make the resistance thermometer
scale to match the IPTS-68 at the gold point and to achieve the
smallest deviation at the silver point. Their deviation curve is represented in Fig. 2d.

6-THERMOCOUPLES

The life of the Pt-10% Rh/Pt thermocouple as a defining instrument of the IPTS-68 in the 630-1064^{0}C range is seriously threatened
by the platinum resistance thermometer and by the radiation pyrometer. The reasons why it is still kept alive are because metrologists
are waiting for the final solution of three problems:
    a) determination of the difference between thermodynamic tempe-
rature as defined by gas thermometry and the IPTS-68;
    b) assessment of the reproducibility of the Pt resistance ther-
mometer between the silver and the gold point;
    c) comparison of resolution, reproducibility and practicality
of the said sensor and the radiation pyrometer below the gold point.

The amended text of the IPTS-68 includes the following statement about the standard type thermocouple: "However, in general use,

precautions may not ensure an accuracy of better than ±0.2°C because of continually changing chemical and physical inhomogeneities in the wires in the regions of temperature gradients." These facts were established by a careful experimental study of McLaren and Murdock (1972).

A concise and complete review of the state of the art of the thermocouple thermometry up to 1975 has been written by Burns and Hurst (1975), with exhaustive list of references.

Standardized thermocouples are object of documents of the American National Standards Institute (ANSI 1973 C96.2) and the ASTM (E230-72). The reference tables in the NBS Monograph 125 (Powell et al. 1974), provide the basis of the present ASTM and ANSI standards. The monograph includes the new international tables for types R and S (Bedford et al. 1972) which were established to bring into accord UK and US standards for these thermocouples.

Three types of W and W-Re alloys thermocouples are commercially available and used for high temperatures up to 2300°C. Their positive elements, W, W-3%Re and W-5%Re respectively, are combined with the negative element, a W-Re alloy with 25% to 26% Re. Reference tables of these thermocouples have been published by ASTM, simply as proposed tables without any status as standards (1974). EMF--temperature relationships of some non-standardized thermocouples are represented in Fig. 3.

In order to minimize environmental and structural instabilities that are inherent to type K (chromel-alumel) thermocouples, Burley developed new nickel-base alloy thermocouples named Nicrosil/Nisil and reported on them to the Fifth Symposium on Temperature (Burley 1972). He has shown that the thermoelectric stability of nickel-base thermocouples can be significantly enhanced by increasing alloy solute levels above those required to cause a transition from internal to external modes of oxidation, and by selecting solutes which preferentially oxidize to form impervious diffusion barrier films. Furthermore, the short term EMF change can be virtually eliminated by the choice of higher solute levels at which this structure dependent effect is not evident. Based upon these considerations, the Materials Research Laboratories of the Australian Government Department of Defence, developed two alloys called Nicrosil (Ni-14.2Cr--1.4Si) and Nisil (Ni-4.4Si-0.1Mg).

These alloys have been fabricated with close composition tolerances by five major manufacturers (three in the USA, one in the UK and one in Sweden). Wires from these sources were supplied to the NBS (USA) and to the NML (Australia) for further investigation. Both laboratories presented reports on their results to the temperature symposium of Teddington (Burley et al. 1975, Burley and Jones,

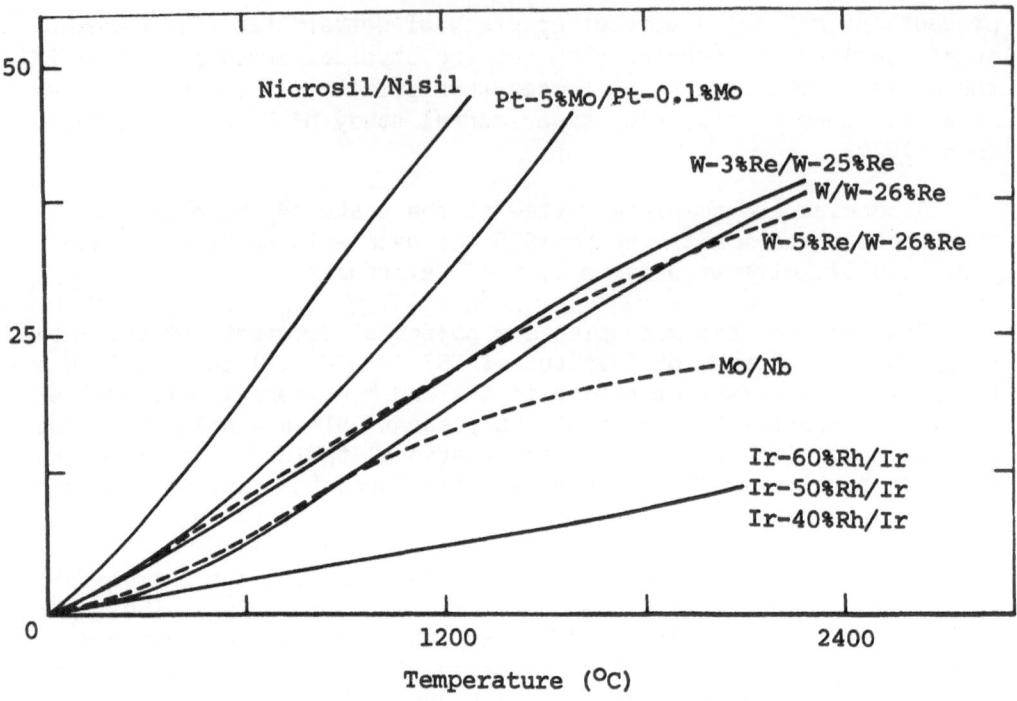

Fig. 3- Thermal EMFs of some thermocouples.

1975). They agreed in confirming the superiority of this thermocouple over the type K thermocouple as far as immunity from oxidation and stability are concerned (see also NBS Monograph by Burley et al. 1975).

7-RADIATION PYROMETRY

Improvements or renewals of photoelectric pyrometers which keep the IPTS-68 in National Laboratories have taken place. The NPL developed a photon counting pyrometer by modifying the existing one (Coates 1975) by use of a new photomultiplier, with modified range S-20 cathode. This instrument is intended to compare the radiation scale with the IPTS-68 and as a candidate substitute for the Pt-10% Rh/Pt thermocouple. Similar recent work has been described by Jones and Tapping (1972) in the NML, by Kunz and Kaufman (1975) in the PTB, by Lanza and Ricolfi (1975) in the IMGC, and by Takata and Hattori (1974) in the NRLM.

Intercomparison of temperature scales above the gold point was carried out in 1970 by circulating a set of tungsten strip lamps among NBS, NPL and PTB. They agreed within 0.4°C up to 1550°C. All but the NML agreed within $\pm 2.0^{\circ}$C at 2200°C. Results were reported at the 5th Symposium on Temperature (Lee et al. 1972). Some disagreement, due to the reflection on this strip of the internal pyrometer lamp, has been detected and explained in the NML (Jones and Tapping, 1976).

Spectral emissivity of tungsten, the material used to fabricate the strip lamps, has been measured lately by Latyev et al. (1969, 1970). Recently (1976) they made a critical evaluation of existing data and proposed a reference table of tungsten normal monochromatic emissivity in the temperature range of 1200-2600K and wavelength range of 0.3-5 μm.

Data reduction in primary pyrometry has been further investigated in the seventies by four authors who developed algorithms to calculate temperatures on the basis of pyrometer signal ratio through Planck Law.

Bezemer (1974) introduces a constant reference wavelength as a pyrometer parameter to give a value of temperature which is subsequently adjusted by means of a radiance correcting factor. Similarly Coates (1971) calculates the correct signal ratio by multiplying the ratio of radiances at the central wavelength λ_0 through a correcting factor. This method is similar to the one of Bezemer, but it presents a clearer insight into physical parameters and into the limits of applicability of the effective wavelength.

Jung (1974) and Ruffino (1977) propose algorithms, valid within pre-defined temperature limits, to calculate, the former, the logarithms of spectral radiance integrals or, the latter, the ratio of those integrals.

Two-colour pyrometry, although very attractive in principle, has been rather unsuccessful in the past. However, increased precision may be achieved (Ruffino, 1975a, 1975b) by introducing effective wavelengths that have the same properties in two-colour as well as in monochromatic pyrometry. They allow interpolation formulas to be written thus leading to a calibration process capable of good precision. On the basis of these principles, an instrument has been designed and made (Pasta et al. 1973) in which two signals corresponding to each wavelength are generated; an original analog-to-digital converter yields directly the temperature in digital form. A general calibration procedure for this kind of instrument has been set up (Pasta et al. 1975).

High speed radiation pyrometry has been developed in connection with the pulse heating technique for the study of thermophysical properties. New impetus and higher precision were given to this method after Cezairliyan in the NBS associated to it digital electronic techniques. The NBS equipment embodies a fast pyrometer designed by Foley (1970). A similar system has been installed in the IMGC (Righini et al. 1972) which includes a pyrometer designed by Ruffino. Increase in speed, and reduction in target size of the latter, at the expense of bandwidth have been achieved later on (Righini et al. 1975).

Multiwavelength high speed pyrometry, with microsecond time resolution has been implemented in Livermore Laboratory (Gathers et al. 1976, Shaner et al. 1976) and in the Euratom of Karlsruhe (Ohse and Kinsman, 1977). Another two-wavelength pyrometer with microsecond resolution is in progress in the NBS. An analysis of the problems connected with high speed pyrometer has been carried out by Ruffino (1976).

REFERENCES

ASTM, 1974, Annual Book of ASTM Standards, part 44, 730.

Bedford, R.E., Ma, C.K., Barber, C.R., Chandler, T.R., Quinn, T.J., Burns, G.W., Scroger, M.G., 1972 in Plumb (1972), part 3, 1585.

Berry, R.J., 1974, Metrol., 10, 145.

Berry, R.J., 1975 in Billing and Quinn (1975), 99.

Bezemer, J., 1974, Metrol., 10, 47.

Billing, B.F. and Quinn, T.J. (ed.), 1975, Temperature Measurement 1975 (Inst. of Phys., London and Bristol, 1975).

Blevin, W.R. and Brown, W.J., 1971, Metrol., 7, 15.

Blevin, W.R., 1972, Metrol., 8, 146.

Blevin, W.R. and Steiner, B., 1975, Metrol., 11, 97.

Bongiovanni, G., Crovini, L., Marcarino, P., 1972, High Temp.-High Press., 4, 573.

Bongiovanni, G., Crovini, L., Marcarino, P., 1975, Metrol., 11, 125.

Bonhoure, J., 1975, Metrol., 11, 141.

Burley, N.A., 1972 in H.H. Plumb (1972), part 3, 1677.

Burley, N.A., Burns, G.W., Powell, R.L., 1975a in Billing and Quinn (1975), 162.

Burley, N.A. and Jones, T.P., 1975 in Billing and Quinn (1975), 172.

Burley, N.A., Powell, R.L., Burns, G.W., Scroger, M.G., 1975, NBS. Monograph (under publication).

Burns, G.W. and Hurst, W.S., 1975 in Billing and Quinn (1975), 144.

Cabannes, F., Simonato, J., Foëx, M., Coutoures, J.P., 1972, High Temp.-High Press., 4, 589.

Cabannes, F., Vu Thien Loc, Coutoures, J.P., Foëx, M., 1976, High Temp.-High Press., 8, 359.

Cezairliyan, A., Coslovi, L., Righini, F., Rosso, A., 1975 in Billing and Quinn (1975), 287.

Coates, P.B., 1975 in Billing and Quinn (1975) 238.
Coates, P.B., 1977, Metrol., 13, 1.
Coates, P.B. and Andrews, 1977 (under publication).
Coslovi, L., Rosso, A., Ruffino, G., 1975a, Metrol., 11 , 85.
Coslovi, L., Righini, F., Rosso, A., 1975b, Alta Freq., 44, 592.
Crovini, L. and Actis, A., 1977 (under publication).
Evans, J.P. and Woods, S.D., 1971, Metrol., 7, 108.
Foëx, M., 1977 (to be published in High Temp.-High Press.).
Foley, G.M., 1970, Rev. Sci. Instr., 41; 827.
Gathers, G.R., Shaner, J.W., Brier, R.L., 1976, Rev. Sci. Instr.,
        47, 471.
Ginnings, D.C., Reilly, M.L., 1972 in Plumb (1972), part 1, 339.
Guildner, L.A., Edsinger, R.E., 1976, Journ. Res. NBS, 80A, 703.
- International Practical Temperature Scale of 1968 (The), 1969,
        Metrol., 5, 35.
- International Practical Temperature Scale of 1968 (The). Amended
        edition of 1975, 1976, Metrol., 12, 7.
Jones, T.P., Tapping, J., 1972, Metrol., 8, 4.
Jones, T.P., Tapping, J., 1976a, Metrol., 12, 19.
Jones, T.P., Tapping, J., 1976b, Metrol., 12, 41.
Jung, H.J., Verch, J., 1973, Optik, 38, 95.
Jung, H.J., 1975 in Billing and Quinn (1975), 107.
Kenisarin, M.M., Berezin, B.Ya., Chekhovskoi, V.Ya., 1972, High
        Temp.-High Press., 4, 707.
Kenisarin, M.M., Chekhovskoi, V.Ya., Berezin, B.Ya., 1976, High
        Temp.-High Press., 8, 367.
Kirby, C.G., Bedford, R.E., 1972, Metrol., 8, 82.
Kunz, H., Kuafman, H.J., 1975 in Billing and Quinn (1975), 244.
Kunz, H., Lohrengel, J., 1975, Jahresbericht der PTB, 190.
Kunz, H., Lohrengel, J., 1976, CCT-76, 11th Sess.
Lanza, F., Ricolfi, T., 1975, Alta Freq., 44 589.
Lanza, F., Ricolfi, T., 1976, High Temp.-High Press., 8, 217.
Latyev, L.N., Chekhovskoi, V.Ya., Shestakov, E.N., 1969, Teplofiz.
        Visokich. Temp., 7, 666.
Latyev, L.N., Chekhovskoi, V.Ya., Shestakov, E.N., 1970, High Temp.-
        -High Press., 2, 175.
Latyev, L.N., Chekhovskoi, V.Ya., Shestakov, E.N., 1974, High Temp.-
        -High Press., 4, 679.
Laubitz, M.J., McElroy, D.L., 1971, Metrol., 7, 1.
Lee, R.D., Kostkovski, H.J., Quinn, T.J., Chandler, P.R., Jones, T.
        P., Tapping, J., Kunz, H., 1972 in Plumb (1972), part 1, 377.
Marcarino, P., Crovini, L., 1975 in Billing and Quinn (1975), 107.
McLaren, E.H., Murdock, E.G., 1972 in Plumb (1972), part 3, 1543.
Ohse, R.W., Kinsman, P.R., 1976, High Temp.-High Press., 8, 209.
Pasta, M., Ruffino, G., Soardo, P., Toselli, G., 1973, High Temp.-
        -High Press., 5, 99.
Pasta, M., Ruffino, G., Soardo, P., 1975, High Temp.-High Press.,
        7, 595.
Pickup, C.P., 1975, Metrol., 11, 151.

Plumb, H.H. (ed.), Temperature, its Measurement and Control in Scien-
    ce and Industry, vol. 4, (ISA, Pittsburgh, 1972).
Powell, R.L., Hall, W.J., Hyink, C.H., Jr. Sparks, L.L., Burns, G.
    W., Scroger, M.G. and Plumb, H.H., 1974, NBS Monograph 125.
Preston-Thomas, H., 1972 in Plumb (1972), part 1, 3.
Quinn, T.J., Chandler, T.R.D., 1971, Metrol., 7, 132.
Quinn, T.J., Chandler, T.R.D., 1972 in Plumb (1972), part 1, 295.
Quinn, T.J., Chandler, T.R.D., Chattle, M.V., 1973, Metrol., 9, 44.
Quinn, T.J., 1974, Metrol., 10, 115.
Quinn, T.J., 1975 in Billing and Quinn (1975), 1.
Ricolfi, T., Lanza, F., 1977 (to be published in High Temp.-High
    Press.).
Riddle, J.L., Furukawa, G.T., Plumb, H.H., Platinum Resistance Ther-
    mometry, NBS Monograph 126.
Righini, F., Rosso, A., Ruffino, G., 1972a, High Temp.-High Press.,
    4, 471.
Righini, F., Rosso, A., Ruffino, G., 1972b, High Temp.-High Press.,
    4, 597.
Ruffino, G., 1975a, Rév. Int. Htes Temp. et Réfract., 12, 172.
Ruffino, G., 1975b, in Billing and Quinn (1975), 264.
Ruffino, G., 1976, High Temp.-High Press., 8, 143.
Schneider, S.J., McDaniell, C.L., 1967, Journ. Res. NBS, 71A, 317.
Shaner, J.W., Gathers, G.R., Minichino, High Temp.-High Press., 8,
    425.
Takata, S., Hattori, S., 1974, CCT-74, Annexes 15 and 16.

SESSION I   NUCLEAR WASTE DISPOSAL, COAL

Session Chairmen: R.U. Acton
                      Sandia Lab.
                      Albuquerque, New Mexico

                      P. Wagner
                      Los Alamos Scientific Lab.
                      Los Alamos, New Mexico

THERMAL TRANSPORT STUDIES IN DEEPSEA SEDIMENT*

W. P. Schimmel, Jr. and C. E. Hickox

Fluid and Thermal Sciences Department
Sandia Laboratories
P. O. Box 5800
Albuquerque, NM   87115

INTRODUCTION

Thermal problems associated with the emplacement of radioactive wastes in the deepsea sedimentary layer have been studied. In particular, the nature of the temperature field surrounding and the interstitial water velocity arising from a buried cask have been examined. For moderately low levels of heat generation by the decaying radionuclides, conservative estimates indicate that the velocity field will be extremely weak.[1] Because of the low interstitial water velocity, thermal conduction models can be used to predict the temperature field in the surrounding sediments as well as the cask surface temperature. This is equivalent to "decoupling" the energy and momentum conservation relationships thus simplifying the solution of the temperature field.

The present work considers in some detail the temperature field surrounding a vertical, isothermal, circular "cylinder" located a distance below a horizontal, isothermal, plane surface. Actually, the isotherm corresponding to the cylindrical cask surface is approximated by an ellipsoid of revolution but the error will be small for large values of the length to diameter ratio. The resulting expression can be used to estimate temperature of the cask surface for material degradation studies and the effect of temperature upon the ion transport process in the sedimentary layer.

_____

*This work was supported by the Department of Energy.

ANALYSIS

General Considerations

In this section, a simplified mathematical model is proposed for the description of free convection about a cask buried in a fluid saturated porous medium. The medium is assumed to be homogeneous and isotropic and the fluid incompressible, with density changes occurring only as a result of changes in the temperature according to,

$$\rho = \rho_{\infty}\left[1 - \beta(T - T_{\infty})\right] \quad , \tag{1}$$

where ρ is the density, T is the temperature, β is the coefficient of thermal expansion, and the subscript refers to reference conditions. In accordance with the usual Boussinesq approximation, density changes are accounted for only in the buoyancy term in the equations of motion. It is furthermore assumed that all physical properties are constant and that thermal and hydrodynamic dispersion effects are negligible. In non-dimensional form, the steady state equations of continuity, motion (Darcy's law), and thermal transport are,

$$\nabla \cdot \underset{\sim}{v} = 0 \quad , \tag{2}$$

$$\underset{\sim}{v} = -\nabla p - \Theta\nabla h \quad , \tag{3}$$

$$N_R\underset{\sim}{v} \cdot \nabla\Theta = \nabla^2\Theta \tag{4}$$

where $\underset{\sim}{v}$, p, and Θ are, respectively, the non-dimensional velocity, pressure, and temperature, and h is the upward vertical direction. The parameter, N_R, is the Rayleigh number defined by,

$$N_R = \frac{K\rho_{\infty}\beta g\ell\Delta T}{\mu\alpha} \quad , \tag{5}$$

where K is the intrinsic permeability, μ is the fluid viscosity, α is the effective thermal diffusivity, ℓ is a reference length, and ΔT is a reference temperature difference.

Physically, the Rayleigh number represents a measure of the relative effectiveness of thermal transport by free convection to that by conduction. Thus, for small values of the Rayleigh number, the effects of free convection are relatively unimportant and Equation (4) reduces to the classical Laplace equation for heat conduction. For typical seabed disposal schemes, the Rayleigh number is $O(10^{-3})$. The temperature field can thus be determined from a conduction analysis.

Conduction Model

In this section, an expression will be developed for the temperature field surrounding a finite length line heat source located vertically below an isothermal surface. This approximates the burial of a heated vertical cylinder in a semi-infinite conducting medium which is essentially the situation that prevails in the placement of a waste canister in the deepsea sediments. The line of reasoning taken in the analysis follows that proposed in Reference 2 for zero burial depth. Unfortunately, the results presented in this reference are incorrect due to a fundamental error in the analysis, thereby precluding direct application to the problem at hand. It is thus necessary to develop an appropriate solution from first principles.

Consider a line of steady point heat sources located along the negative y axis from $y = -H$ to $y = -H - N$ as shown in Figure 1. In order to ensure that the surface $y = 0$ remains at T_O, the initial temperature of the entire system, an identical row of negative steady point heat sources (heat sinks) is located from $y = H$ to $y = H + N$. The mathematical solution of the temperature field surrounding the source can be changed to that of the sink by reflecting it about the $y = 0$ plane and changing the sign. Because the resulting "negative" temperature field is with respect to the initial uniform temperature, no philosophical problems should arise. The analysis requires the specification of only a single medium property, thermal conductivity and it will be assumed to be constant.

The temperature rise in a medium initially at zero temperature due to the presence of a point heat source located at the origin is,

$$T = \frac{Q}{4\pi k r} , \qquad\qquad (6)$$

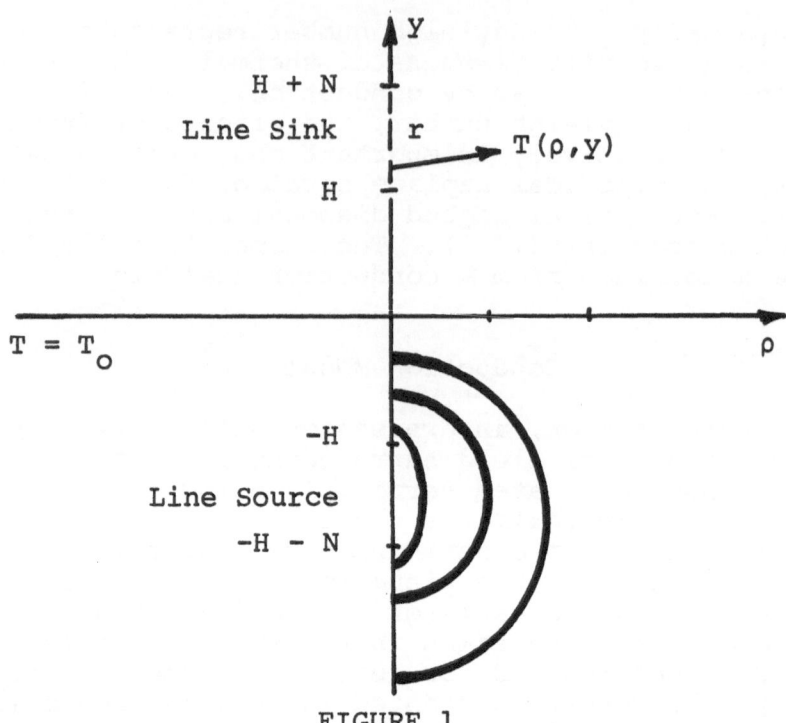

FIGURE 1

Analytical Model and Calculated Isotherms for H/N = 2

where $r = \sqrt{\rho^2 + y^2}$ in units of, say m, Q is the strength of the source in watts (W) and k is the medium thermal conductivity in W/m-K.

This expression is now summed for a series of point sources located along the vertical axis. In the limit, the temperature increase due to the line source can be expressed as an integral. In terms of the coordinate system of Figure 1, this becomes,

$$T_1 = \int_{-H-N}^{-H} \frac{Q'}{4\pi kr} \, ds \quad , \tag{7}$$

where Q' is the strength per unit length and ds is an infinitesimal length of the line source. Note that the contribution of the finite line source at $r \to \infty$ is zero and thus $T_1 \to 0$ as desired. In Reference 2, a similar argument for the case of H = 0 was made except that the sign of the integral was taken as negative. If the sign of the integral is negative and one evaluates the integral along an r = constant contour, it can

readily be shown that the temperature rise will be negative for all positive source strengths. This is an
obvious contradiction: the sign, therefore, must be
positive.

It is now stipulated that a line of heat sinks be
located along the vertical axis from y = H to y = H + N.
Arguing as before, this results in a temperature "rise"
at point r given by,

$$T_2 = -\int_H^{H+N} \frac{Q'}{4\pi k r} \, ds \quad . \tag{8}$$

This temperature rise is, of course, negative because
Q' corresponds to a line sink.

Now the net temperature rise at some arbitrary
position, r, will be given by the sum of the two integrals. From Figure 1,

$$r^2 = \rho^2 + (y - s)^2 \quad , \tag{9}$$

and summation of Equations (7) and (8) results in,

$$T = \frac{Q'}{4\pi k} \left\{ \int_{-H-N}^{-H} \frac{ds}{\sqrt{\rho^2 + (y - s)^2}} - \int_H^{H+N} \frac{ds}{\sqrt{\rho^2 + (y - s)^2}} \right\} \quad .$$

$$\tag{10}$$

Carrying out the integration and combining the natural
log functions yields,

$$T = \frac{Q'}{4\pi k} \ln \left\{ \frac{[y + H + N + \sqrt{\rho^2 + (y + H + N)^2}]}{[y + H + \sqrt{\rho^2 + (y + H)^2}]} \right.$$

$$\left. + \frac{\ln[y - H - N + \sqrt{\rho^2 + (y - H - N)^2}]}{[y - H + \sqrt{\rho^2 + (y - H)^2}]} \right. \tag{11}$$

Letting the burial depth, H, become zero results in,

$$T = \frac{Q'}{4\pi k} \ln \qquad\qquad\qquad\qquad\qquad (12)$$

$$\left\{ \frac{\left[y + N + \sqrt{\rho^2 + (y + N)^2} + y - N + \sqrt{\rho^2 + (y - H - N)^2} \right]}{\left[y + \sqrt{\rho^2 + y^2} \right]^2} \right\}.$$

In Reference 2, an incorrect expression for this case, is given which, when evaluated for the surface $y = 0$, yields a non-constant value for the temperature. Therefore, the assumption of the $y = 0$ surface remaining isothermal is clearly violated and the result of Reference 2 must be wrong. Equation (12) evaluated for $y = 0$ yields,

$$T = \frac{Q'}{4\pi k} \ln(1) = 0 \quad , \qquad\qquad\qquad (13)$$

which is in agreement with the assumption of zero initial temperature and zero temperature along the $y = 0$ plane for all values of X and Z.

Returning to Equation (11) for the general case, the shape of the isotherms can be determined. Multiplying both sides by $4\pi k/Q'$ and exponentiating results in,

$$\exp \frac{4\pi kT}{Q'} = \left[\frac{(y + H + N) + \sqrt{\rho^2 + (y + H + N)^2}}{(y + H) + \sqrt{\rho^2 + (y + H)^2}} \right]$$

$$\left[\frac{(y - H - N) + \sqrt{\rho^2 + (y - H - N)^2}}{(y - H) + \sqrt{\rho^2 + (y - H)^2}} \right] \quad . \qquad (14)$$

In the region near the line source, the following approximation can be made,

$$\left. \begin{array}{l} y \approx -H - N/2 \\[2mm] \rho \approx 0 \quad . \end{array} \right\}$$

The contribution due to the line sink thus becomes,

$$\lim_{r \to 0} \left[\frac{(-2H - 3N/2) + \sqrt{\rho^2 + (2H + 3N/2)^2}}{(-2H - N/2) + \sqrt{\rho^2 + (2H + N/2)^2}} \right] = 1 \text{ as } H \to \infty \quad . (15)$$

Hence, near the source (for large values of H),

$$T \approx \frac{Q'}{4\pi k} \ln \left[\frac{(y + H + N) + \sqrt{\rho^2 + (y + H + N)^2}}{y + H + \sqrt{\rho^2 + (y + H)^2}} \right] \quad . \quad (16)$$

Along an isotherm, the argument of the logarithm will be equal to a constant,

$$\frac{y + H + N + \sqrt{\rho^2 + (y + H + N)^2}}{y + H + \sqrt{\rho^2 + (y + H)^2}} = c \quad . \quad (17)$$

If the substitution is made for y such that,

$$\xi = y + H + m \quad , \quad (18)$$

where,

$$m = \frac{N}{2} \quad ,$$

then,

$$\frac{\xi + m + r_2}{\xi - m + r_1} = c \quad , \quad (19)$$

where $r_2^2 = \rho^2 + (\xi + m)^2$ and $r_1^2 = \rho^2 + (\xi - m)^2$. The constant, c, is evaluated at $\xi = 0$, where $\rho = b$, a constant. For r_1 and r_2, this yields,

$$r_1 = r_2 = \sqrt{m^2 + b^2} \quad . \quad (20)$$

The constant, m, is arbitrarily set equal to $a^2 - b^2$, where a is another constant, so

$$r_1 = r_2 = a \quad . \tag{21}$$

Solving for c results in,

$$c = \frac{a + m}{a - m} \quad . \tag{22}$$

For the general case, Equation (19) specifies,

$$(\xi + m + r_2)(a - m) = (\xi - m + r_1)(a + m) \quad , \tag{23}$$

which simplifies to,

$$\frac{\rho^2}{a^2 - m^2} + \frac{\xi^2}{a^2} = 1 \quad , \tag{24}$$

but $a^2 - m^2 = b^2$, so finally,

$$\frac{\rho^2}{b^2} + \frac{\xi^2}{a^2} = 1 \quad . \tag{25}$$

This is the equation of an ellipse of major axis 2a and minor axis 2b centered at the origin. In terms of physical coordinates,

$$\frac{\rho^2}{b^2} + \frac{(y + H + N/2)^2}{a^2} = 1 \quad , \tag{26}$$

which is the same size ellipse centered about the $\rho = 0$, $y = -H - N/2$ point. The same procedure applied to the second part of Equation (14) indicates that the negative isotherms about the line sink are ellipses centered about $y = H + N/2$. The resulting isotherms due to the presence of both the source and sink can be obtained using either form of the equation. Isotherms computed from Equation (14) are plotted in Figure 1 for the case H/N = 2.

Since the temperature of the ellipsoid surface is constant, it can be evaluated at any value of (ρ, y) consistent with the ellipse equation. Equation (12) is thus evaluated at $\rho = b$, $y = -H - m$. As before, m = N/2.

$$T = \frac{Q'}{4\pi k} \left\{ \ln\left[\frac{m + \sqrt{b^2 + m^2}}{-m + \sqrt{b^2 - m^2}}\right] \right.$$

$$\left. + \ln\left[\frac{-2H - 3m + \sqrt{b^2 + (2H + 3m)^2}}{\sqrt{b^2 + (2H + m)^2}}\right]\right\} \quad . \quad (27)$$

The final temperature expression is,

$$\frac{T}{Q'/2\pi k} = \ln\left(\ell + \sqrt{\ell^2 - 1}\right) + \frac{1}{2} \ln\left[\frac{2h + 3n - r_3}{2h + n - r_4}\right] \quad , \quad (28)$$

where,

$$h = H/b$$

$$n = m/b = \sqrt{\ell^2 - 1}$$

$$r_3 = \sqrt{1 + (2h + 3n)^2}$$

$$r_4 = \sqrt{1 + (2h + n)^2} \quad .$$

All the depth of burial dependence is in the second term so that as h tends to infinity, only the first term remains. This is in essential agreement with Reference 3 which considers a finite length cylinder in an infinite medium.

CONCLUSION

Equation (14) could be evaluated for various dimensionless lengths as a function of burial depth and these results could be used to estimate the surface temperatures of candidate waste containing casks. Because of space limitations, this will not be done in the present work but is presented in a recent technical memorandum.[4]

REFERENCES

1.   Hickox, C. E., and Watts, H. A., "Steady Thermal
      Convection from a Concentrated Source in a Porous
      Medium," SAND76-0562, Sandia Laboratories,
      Albuquerque, NM, November 1976.

2.   Eckert, E. R. G., and Drake, R. M., "Analysis of
      Heat and Mass Transfer," Fourth Edition, McGraw-
      Hill Book Co., NY, pp. 103-106, 1972.

3.   Easton, C. R., "Conduction from a Finite-Size
      Moving Heat Source Applied to Radioisotope Capsule
      Self Burial," Thermal Conductivity, U.S. Department
      of Commerce, Spec. Publ. 302, pp. 209-218, 1968.

4.   Schimmel, W. P., Jr., and Hickox, C. E., "Application
      of Thermal Conduction Models to Deepsea Disposal
      of Radioactive Wastes," SAND77-0752, Sandia Labor-
      atories, Albuquerque, NM, August 1977.

THERMAL CONDUCTIVITY OF ROCKS ASSOCIATED WITH ENERGY EXTRACTION FROM HOT DRY ROCK GEOTHERMAL SYSTEMS*

W. L. Sibbitt, J. G. Dodson, and J. W. Tester

Los Alamos Scientific Laboratory, Univ. of California

Los Alamos, New Mexico 87545

Abstract

Because the lifetime of and heat extraction rate from a hot dry rock (HDR) geothermal reservoir can be substantially controlled by the in situ rock thermal conductivity, information concerning the dependence of thermal conductivity on moisture content and temperature is important for proper design and management of the reservoir. Results of thermal conductivity measurements are given for 14 drill core rock samples taken from two exploratory HDR geothermal wellbores (maximum depth of 2929 m (9608 ft) drilled into Precambrian granitic rock in the Jemez Mountains of northern New Mexico. These samples have been petrographically characterized and in general represent fresh competent Precambrian material of deep origin. Thermal conductivities, modal analyses and densities are given for all core samples studied under dry and water-saturated conditions. Additional measurements are reported for several sedimentary rocks encountered in the upper 760 m (2500 ft) of that same region. A cut-bar thermal conductivity comparator and a transient needle probe were used for the determinations with fused quartz and Pyroceram 9606 as the standards. The maximum temperature range of the measurements was from the ice point to 250°C. The measurements on wet, water-saturated rock were limited to the temperature range below room temperature. Conductivity values of the dense core rock samples were generally within the range from 2 to 2.9 W/mK at 200°C. Excellent agreement was achieved between these laboratory measurements of thermal conductivity and those obtained by in situ measurements used in the HDR wellbores. By using samples

---

*Work done under the auspices of the U.S. Department of Energy.

of sufficient thickness to provide a statistically representative
heat flow path, no difference between conductivity values and their
temperature coefficients for orthogonal directions (heat flow parallel or perpendicular to core axis) was observed. This isotropic
behavior was even found for highly foliated gneissic specimens.
Estimates of thermal conductivity based on a composite dispersion
analysis utilizing pure minerallic phase conductivities and detailed
modal analyses usually agreed to within 9% of the experimental
values.

INTRODUCTION AND SIGNIFICANCE

The objective of this work was measurement of the thermal conductivity of deep crustal crystalline rock obtained from cores taken
at depths to 3 km. This material is unique in the sense that it is
free of surface weathering, reasonably competent with well-sealed
microfractures (permeability <1 μdarcy) and is part of a hot dry
rock (HDR) geothermal reservoir system in Precambrian granite at
200°C. Detailed information on the dependence of thermal conductivity on temperature and moisture content as well as on mineralogy
was required for predictive performance modeling of an HDR reservoir
for the following reasons [Smith, 1975].

The operational feasibility of hot dry rock (HDR) geothermal
systems requires that heat be transferred efficiently from hot
reservoir rock to a circulating fluid (water). In most of the concepts being pursued that involve fractured rock reservoirs, the
thermal conductivity of the formation critically affects the lifetime and thermal capacity of the reservoir. [Tester and Smith,
1977]. Of equal importance are the amount of surface area accessible to fluid circulating across the fracture faces and the rate of
fluid flow. As given by Murphy [Tester and Smith, 1977], the recoverable power, P(t), in watts, for a single ideal disc shaped
fracture system in granite can be estimated for uniform water flow
conditions as:

$$P(t) = \dot{m}_w C_w (T_i - T_{min}) \mathrm{erf}\left(\sqrt{\frac{(\lambda \rho C)_r}{t}}\; \frac{\pi R^2}{\dot{m}_w C_w}\right) \tag{1}$$

where
   R = fracture radius, m
   C_r = heat capacity of granite, ~1000 J/kgK [Clark, 1966]
   C_w = heat capacity of water, ~4200 J/kgK
   $\dot{m}_w$ = water flow rate through fracture, kg/sec
   t = time, sec
   T_i = mean initial rock temperature, °C

T_{min} = fluid temperature as it enters the reservoir, °C
λ_r = thermal conductivity of granite, W/m K
ρ_r = rock density, kg/m³

The error function term in eq (1) describes the finite thermal resistance for heat transport by transient conduction through the
rock to the circulating fluid. Even in situations with non-uniform
flows and non-ideal fracture geometries, recoverable power levels
will depend on the parametric form of eq (1). [Murphy and McFarland,
1976; Harlow and Pracht, 1972.] Thus in order to predict reservoir
performance, one must have reasonable values for the thermal conductivity of the rock formation.

In addition to influencing reservoir performance, and therefore reservoir-related costs, thermal conductivity also affects
the economics of HDR systems in another way. For a given regional
heat flow, the thermal conductivity of the overlying rocks will
control the mean geothermal temperature gradient observed. Thus
the drilling depth required to reach rock of a given temperature
will vary inversely with the mean gradient. Because drilling costs
increase exponentially with depth [Tester and Smith, 1977] thermal
conductivity-heat flow characteristics are important in determining
costs associated with developing the reservoir.

The conventional definition of thermal conductivity λ using the
Biot/Fourier formulation of unidirectional conductive heat flow has
been applied to mineral rock systems [Birch and Clark, 1940].
Because rocks are all diathermanous materials, the thermal conductivity λ of laboratory-sized samples can depend upon specimen
thickness as well as upon specimen composition and its intrinsic
physical condition. All rocks transmit infrared radiation; therefore,
there are always two parallel paths for heat. Thus λ has been called
an "apparent conductivity" since it is not a true physical property
of the material. The apparent conductivity is the sum of the true
conductivity (a physical property) and the "radiation conductivity."
The contribution due to radiation however is usually negligible for
temperatures below 800 K and does not influence the data presented
in this paper.

Heat flow in rock composites is rather complex: it is always
three-dimensional on a macroscopic scale although in many cases it
can be treated as unidirectional flow. In a typical granitic rock
of interest to our geothermal project, microcapillaries occur along
mineral contacts and along cleavage planes of feldspars and iron-
magnesium silicates (biotites). Quartz tends to form microcracks
around and across quartz grains. The volume fraction of these voids
is usually small (less than 0.001) but the interfacial surface is
very large and substantially reduces the apparent radiation conductivity but only moderately reduces the apparent lattice

conductivity. For situations where confining pressures are sufficiently high, microcracks closure can be complete and result in
almost no change in phonon transport. In almost any case the inter-
crystalline resistance to heat flow in relatively fresh, wet igneous
rocks is low at moderate temperatures. [Walsh and Decker, 1966].

If apparent conductivity values are to be used to estimate the
geothermal heat flux, then experimental rock specimens should be of
such a size that average conductivity values will be characteristic
of large masses of rock eliminating the effects of microscopic
heterogeneities. The thickness of the specimen should be of at least
an order of magnitude greater than the grain size.

Practical problems, however, influence the choice of a method
for measuring the thermal conductivity of rocks. Larger sized
specimens usually result in longer measurement times. In addition,
igneous rocks are difficult to machine: complicated shapes and
especially long holes of small diameter present problems. Flat
circular discs can be obtained by coring, sawing, grinding, and then
lapping the surfaces to some degree of flatness. However, when the
temperature gradient is imposed on an aggregate of non-cubic crystals,
surface warping may limit sample flatness [Birch and Clark, 1940].

In measuring the surface temperature of a flat rock, good
thermal contact between the temperature sensor and the non-metallic
surface of the rock must be maintained. Errors also may be introduced by local cooling in the area of temperature sensor contact
because of its generally higher conductivity than the rock. Furthermore, temperature averaging effects result when a finite sized sensor
is placed in a temperature gradient.

APPARATUS DESCRIPTION

Most of the problem areas cited above were eliminated completely
or reduced in magnitude so that their effects were negligible by
employing a properly designed cut-bar thermal conductivity comparator
for steady state and a needle conductivity probe for transient
measurements with carefully prepared samples and fully characterized
standard materials.

Steady-State Measurements

Measurements of conductivities parallel to the axis of the cores
were made on a guarded, steady-state, divided-bar apparatus (cut-bar
thermal-conductivity comparator) [Birch and Clark, 1940]. A schematic assembly is shown in Figure 1. This comparator provides a
secondary method for determining λ: a disc rock specimen of unknown
conductivity is placed in series between two standard discs of known
conductivity and then the temperature drops are measured with a

constant, steady-state heat flux imposed. Two known standards were used to verify that the heat flow is unidirectional. The comparator consists of a sandwich structure with a rock sample fixed in place between two heat meter integral units composed of a comparison standard glued between two copper temperature sensor discs. For dry rock measurements, a high conductivity acryloid-silver composite cement was used to provide minimum contact resistance at all inter-faces, as shown in Figure 1. For wet rock measurements, a loaded water-glycerol-silver paste was used. All gluing operations and conductivity measurements were carried out with an axial load main-tained on the [heat source-heat meter-rock sample-heat sink] stack assembly. A pressure controlled, total reflux condenser employing different pure organic fluids (ethylene glycol, acetone, ethanol, methanol, methylene chloride) or water was used as the heat source operating at fixed temperatures over a range from 0° to 250°C. An ethylene glycol, controlled temperature bath was used as the heat sink. The space between the guard heaters and the stack of discs is filled with a silica Aerogel insulating powder. The practical working range for this comparator is from 0.1 to 10 W/m K.

There is a finite heat flux through the insulation from the top (adjacent to the heat source) to the bottom (adjacent to the sink) because of the imposed temperature gradient. This energy might be supplied, in part, by either the stack of discs or optimally by the guard heaters. Consequently, the guard heaters are maintained at a somewhat higher temperature than the disc stack at the same axial position. The approximate magnitude of the required temperature mismatch can be established by a numerical analysis for the appro-priate useful ranges of thermal and geometrical parameters and/or by a series of experiments with different degrees of mismatch.

Flat disc samples were selected because they were convenient to produce in uniform sizes. The faces of all rock discs were made as nearly parallel and flat as possible. However, because of varia-tions in hardness of the minerals comprising the composite, slight differences in surface perfection resulted. A suitable sample thick-ness of approximately ∿1 to 3 cm was used.

A constant linear temperature gradient in both the stack of discs and in the guard heaters is the ideal situation; thus, known conductivity standards should be selected to match the sample. This is a major problem since the choice of standard materials is very limited. Probably the only materials with any justifiable claim to being standards in the range of conductivities from 1 to 4 W/m K (at temperatures of interest) are Pyroceram 9606 and fused silica. The four thin discs containing the temperature sensors are made of oxygen-free, high-conductivity copper; consequently, they present a negli-gible resistance to heat flow and because of their thinness do not cause an important deviation from the desired linear temperature

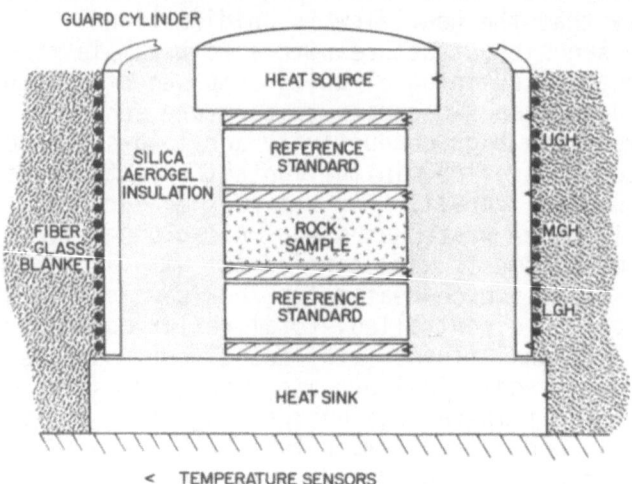

Figure 1.   Schematic of cut-bar thermal conductivity comparator.

gradient.   Therefore, by varying the cross-sectional areas of the
standards, the temperature gradient through them can be matched to
the gradient through a given rock sample.

In this modified divided-bar apparatus, two adjacent temperature sensors measure the temperature drop across one sample disc
plus two interfaces (copper and rock) plus two copper discs of one-
half thickness (see Figure 1).   The total temperature difference
across the sample and reference standard composite was usually
between 10 to 20°C.   In each case the measured temperature drop was
corrected to obtain the actual temperature drop across the rock
disc.

The measured film thickness between two small ($\sim$40-mm-diam)
fused quartz disc standards was about 2×10^{-3} mm but increased to
about 6×10^{-3} mm for the large discs ($\sim$60 mm diam).   The conductance
across such an interface in an air environment, while reasonably high
is not reproducible.   Consequently, a heat-transfer media was used
to decrease the contact resistance and thus obtain a high, reproducible conductance at all interfaces in the stack of discs.   A
number of semi-liquid and liquid contact materials were tried.
Excellent results were obtained with a mixture of Acryloid A-10
(Rohn and Hass) and silver pigment (Silflake 131, Handy and Harmon).
This particular cement composite provides a strong bond with uniform
high conductance, which can be used in ultrahigh vacuum, is bakeable
to 450°C, and yet can easily be removed from the sample when necessary.   Cement film thicknesses ($\leq 5 \times 10^{-2}$ mm) were measured for each

experimental set-up and then the corresponding temperature drop
correction was applied to the data. The corrections were of the
order of 0.2 of one percent.

Transient Measurements

A transient needle conductivity probe was used to measure conductivities perpendicular to the core axis. The needle probe is
about 0.91 mm in diam by 36.5 mm in length and required that 1-mm-
diam holes be drilled into samples using an expensive sonic technique.

If a line source of heat is placed in an infinite homogeneous
isotropic medium initially at a uniform temperature and the heat
is generated by this source at a constant rate q per unit source
length then the temperature rise (ΔT) (above the initial temperature) at a distance r from the line source of heat is given as a
function of time t by the following equation: [van der Held, 1949]

$$\Delta T = \frac{-q}{4\pi\lambda}\ Ei\left(\frac{-r^2}{4\alpha t}\right) \tag{3}$$

where
    Ei = exponential integral
    α = thermal diffusivity
For large values of time, this may be approximated by:

$$\Delta T = \frac{q}{4\pi\lambda}\left[\ell n t + \ell n\ \frac{4\alpha}{r^2} - \ell n \gamma\right] \tag{4}$$

where
    $\ell n \gamma$ = ℓn (Euler's constant) = 0.5772

Therefore, for fixed values of r and α, the temperature increases logarithmically with time. A plot of temperature rise
versus the logarithm of time should give a straight line whose slope
is equal to $q/4\pi\lambda$. This technique is customarily used with the line
heat source method to evaluate the thermal conductivity. For probes
where the heater and temperature sensor are located within the same
sheath, the value of r is indeterminate; however if the probe approximates a line heat source then the value of r is immaterial as
long as it is constant for a given series of measurements. This
method of calculation gives correct values of conductivity under
most conditions achieved in practice. However the probe diameter,
length and construction materials must be selected so as to minimize
a common deviation from ideal line source behavior - namely, the
"initial lag effect" at short experimental times which is caused by
the thermal properties of the probe (both intensive and extensive).

To a first approximation, if t >> r_p^2/α (r_p = hole radius) then the logarithmic dependence of eq (4) should be followed. The contact resistance between the probe and sample contribute to this lag effect. An axial heat flow error occurring at long experimental times caused by non-radial heat flow from the probe to the sample also introduces errors [Wechsler, 1966]. These probes were calibrated by a series of measurements on fused quartz and Pyroceram 9606 standard samples using the λ values recommended by the Thermophysical Properties Research Center (T.P.R.C.) [1964].

EXPERIMENTAL RESULTS

The experimental effort cited below concentrates on thermal conductivity measurements for fresh, competent Precambrian crystalline rock core material obtained from two different deep wellbores, GT-1 and GT-2, which are part of the LASL hot dry rock geothermal field demonstration project [Tester and Smith, 1977]. The important feature we emphasize in this paper is that accurate thermal conductivity measurements are reported on samples that have been petrographically characterized in detail with corresponding modal analyses presented for all crystalline rock measurements. [See also Sibbitt, 1976].

In 1972, an exploratory hole located in the Jemez Mountains of north central New Mexico was completed to a depth of 785 m (2572 ft) (Geothermal Test Hole No. 1, GT-1). The bottom 50 m (165 ft) of the hole penetrated Precambrian basement rock and was cored continuously. The Precambrian section was characterized by a wide range of crystalline rock compositions with quartz (SiO_2) microcline or potassium feldspar ($KAlSi_3O_8$), plagioclase [albite ($NaAlSi_3O_8$): anorthite ($CaAl_2Si_2O_3$)] and biotite [$\sim K_2(Fe,Mg)_2(OH)_2AlSi_3O_{10}$ (with numerous substitutions)] as major constituent minerals [Perkins, 1973]. The lower 40% of this core section was essentially biotite-amphibolite veined by tonalite-aplite. The upper 60% was essentially granite and granodiorite, partly gneissic in texture. A number of thermal conductivity measurements from 0° to 250°C were made on samples from this core section (see Table 1 and Figure 3). These are averaged values based on approximately 20 determinations on each sample. The samples are designated by the indicated depth below the surface as measured from the drilling platform. The petrography of the rock types in this drill core was studied in some detail [Perkins, 1973] and modal analyses are given in Table 2.

A second hole, GT-2 located at Fenton Hill about 2.5 km (1.5 miles) south of GT-1, was drilled to a depth of 2930 m ($\sim$9600 ft) into Precambrian-age basement granitic rocks. Figure 2 is a lithologic log depicting the formations penetrated and temperature profile in GT-2. The lithology of GT-2 is approximately as follows (starting from a surface ground elevation of 2652 m (8702 ft): First a

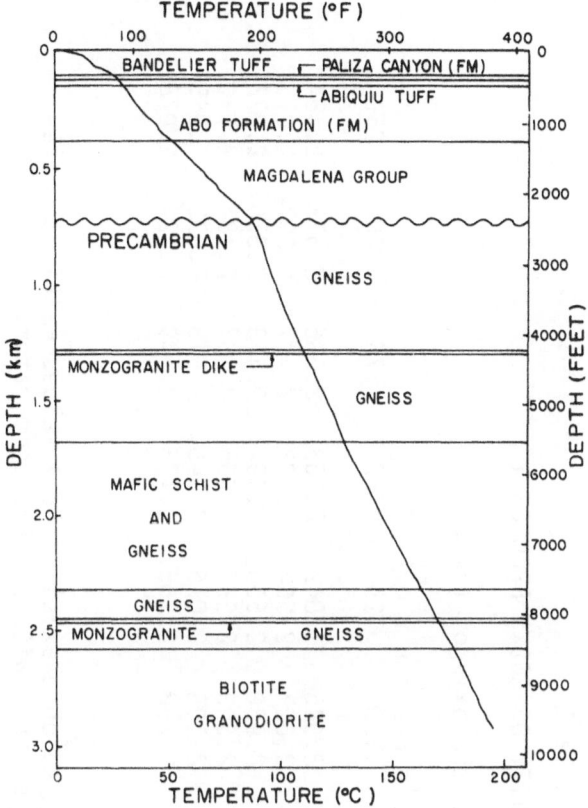

Figure 2. Lithologic log of GT-2 showing temperature versus depth.

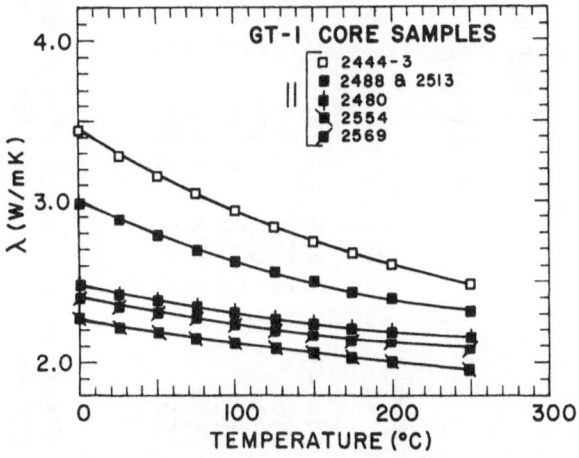

Figure 3. GT-1 core sample thermal conductivity. See Table 2 for modal compositions.

Table 1

THERMAL CONDUCTIVITY OF ROCKS FROM GT-1

(λ in $\frac{W}{mK}$ parallel to the axis of the core)*

Sample No.	Temperature °C									
	0	25	50	75	100	125	150	175	200	250
2444-3	3.435	3.287	3.165	3.052	2.945	2.842	2.763	2.686	2.615	2.484
2488	2.978	2.865	2.777	2.697	2.627	2.564	2.503	2.447	2.397	2.313
2513	2.985	2.872	2.784	2.702	2.630	2.559	2.494	2.433	2.376	2.277
2524	2.480	2.423	2.379	2.338	2.303	2.267	2.235	2.207	2.181	2.140
2554	2.272	2.213	2.178	2.140	2.108	2.076	2.046	2.018	1.993	1.945
2569	2.407	2.354	2.310	2.272	2.238	2.205	2.176	2.149	2.124	2.078

*Measurements taken with needle probe ($\perp$ to core axis) yield λ values within the pre-
cision of data shown for $\parallel$ measurements taken with cut-bar comparator.

Table 2

PROXIMATE ANALYSIS OF GT-1 ROCKS [PERKINS, 1973]

(volume percentages of minerals)

Sample No.	K-feldspar (microcline)	Plagioclase	An %	Plagioclase altered	Quartz	Biotite	Myrmekite	Chlorite	Epidote	Sphene	Hornblende	Opaques	Rock Type	Density Ratio
2444-3	35	21	25	13	20	6						1	Granite, adamellite	2.69
2488	25	28	30	5	32	8						1	(Gneissic granite, adamellite)	2.70
2513	11	36	29	10	25	14	1	1	1				(Gneissic biotite-granodiorite)	2.70
2524		17	29	19	16	15					32		Biotite-amphibolite	2.96
2554		23	36	11	34	20					12		Biotite-amphibolite	2.97
2569		23	36	13	13	19		1		1	29		Biotite-amphibolite	2.97

Pleistocene layer of Bandelier tuff followed by thin layers of
Paliza Canyon and Abiquiu tuffs which terminate at a depth of about
140 m (460 ft); next a Permian layer, the Abo Formation (red beds;
shale, sandstone with limestone stringers) which terminates at a
depth of about 380 m (1250 ft); next the Magdalena group comprised
of a Pennsylvanian layer of Madera Limestone (limestone with clay
and shale layers) which terminates at a depth of 660 m (2165 ft)
and a Mississippian layer of Sandia Formation (limestone with shale
and sandstone layers) which terminates at a depth of 734 m (2404 ft);
finally, the Precambrian granitic rocks. The Precambrian section
although showing a wide range of lithologic composition can generally
be characterized as a competent crystalline section of low permeability (0.01 to 1.0 μ darcy) containing natural fractures with a
frequency 1-5 per cm of core which have been completely sealed by
calcite and/or silica [see Laughlin and Eddy, 1977 for details].

     Table 3 and Figure 4 summarize the conductivity values from $0°$
to $250°C$ as measured on selected GT-2 samples which are described
in Table 4. The sample designation corresponds to the indicated
depth (in feet) along the wellbore below the surface as measured
from the Kelly Bushing on the drilling rig platform.

     Sample 9608 was described as a foliated gneiss; therefore it
was selected for conductivity determination in two directions;
parallel to the axis of the core and perpendicular to the axis of
core - the foliation was parallel to the axis of the core. The conductivity values and their temperature coefficients for orthogonal
directions were nearly equal for a sample thickness of 3.75 cm.
The conductivity parallel to the foliations increased as the sample
thickness was decreased: for example, for a sample thickness of 0.8 cm
the conductivity in the parallel direction was more than 20% greater
than the conductivity in the perpendicular direction. Thus as we
indicated earlier, measurements made on foliated rocks can be very
misleading until a sufficiently thick sample is used to provide a
statistically representative path for heat flow. For this core,
3.75 cm was adequate. For all other granitic samples tested, there
was no observable difference between parallel (steady state cut-bar
comparator) and perpendicular (transient needle probe) measurements
of λ on the same sample for thicknesses of 1 cm or more.

     A number of determinations were made on sedimentary rock samples
from the Abo Formation. These rocks have not yet been completely
characterized. One red shale sample of compacted minute particles
which immediately disintegrated in water (dry density ratio of
2.475) had a dry conductivity of 2.056 W/m K at $0°C$ and 2.109 W/m K
at $125°C$. A weakly compacted red sandstone (dry density ratio 2.355)
had a dry conductivity of 2.160 W/m K at $0°C$ and 1.810 W/m K at $125°C$.
A strongly compacted red sandstone (dry density ratio of 2.407) had a
dry conductivity of 3.11 W/m K at $45°C$. Based on the assumption of

Table 3

THERMAL CONDUCTIVITY OF ROCKS FROM GT-2

(λ in $\frac{W}{m \cdot K}$ parallel to the axis of the core)*

Sample No.	Temperature °C									
	0°C	25	50	75	100	125	150	175	200	250
3-2580-43	3.785	3.615	3.475	3.330	3.206	3.098	2.998	2.903	2.820	2.680
12-4918	3.800	3.622	3.475	3.336	3.209	3.091	2.981	2.882	2.797	2.646
5964-2A	2.900	2.796	2.714	2.635	2.565	2.498	2.440	2.387	2.341	2.260
6153-3A	3.475	3.343	3.222	3.103	2.992	2.908	2.836	2.770	2.713	2.608
6153-3B	3.413	3.264	3.143	3.026	2.921	2.882	2.735	2.653	2.584	2.466
17-6156-1	2.908	2.866	2.777	2.693	2.625	2.560	2.503	2.446	2.393	2.292

(λ in $\frac{W}{m \cdot K}$ perpendicular to the axis of the core)*

Sample No.	0°C	25	50	75	100	125	150	175	200	250
8579-1	3.125	3.062	2.990	2.916	2.852		2.750		2.660	2.595
9608-1	3.115	3.005	2.906	2.813	2.728		2.587		2.473	2.376

*$\perp$ (needle probe) and $\parallel$ (cut-bar comparator) measurements yield results within the precision of data shown.

Table 4
PROXIMATE ANALYSIS OF GT-2 ROCKS [LAUGHLIN AND EDDY, 1977]
(volume percentages of minerals)

Sample No.	K-feldspar	Plagioclase	% An	Quartz	Biotite	Chlorite	Muscovite	Myrmekite	Eipdote	Amphibote	Opaques	Sphene	Rock Type/Location	Density Ratio
3-2580-43	36+2	32+2	34	25+2	1	3+1	1				1		Leucocratic Monzogranite gneiss	2.661
12-4918	29+3	29+3	30	36+3	3+1	1		1					Leucocratic Monzogranite gneiss	2.648
5964-2A***	1	40+3	38	6	2					47+3	1		Amphibolite	2.882
6153-3A	22+2	42+3	33	28+3	5+1		1						Leucocratic granodiorite gneiss	2.635
6153-3B	22+2	42+3	33	28+3	5+1		1						Leucocratic granodiorite gneiss	2.630
17-6156-1	9+2	46+3	37	35+3	6+1		1	1			2		Leucocratic granodiorite gneiss	2.727
8579-1	21+3	34+3	31	29+3	11+2				1		1		Granodiorite gneiss	2.723
9608-1	12+1	43+2	36	31+2	10+1				1		2	2	Biotite granodiorite gneiss	2.715
Tonalite*	-	50	45	28	15					7			Val Verde, CA	2.735
Westerly*	33	40	100	19	6								Granite-Westerly, RI	2.643
Rockport*	64**	-	-	28						6			Granite-Rockport, MA	2.610
Quartz Monzonite*	27	33	27	34	4.5					0.2			Porterville, CA	2.637

*From Birch and Clark [1940].
**Microperthite.
***Modal analysis actually from amphibolite 5983-3B.

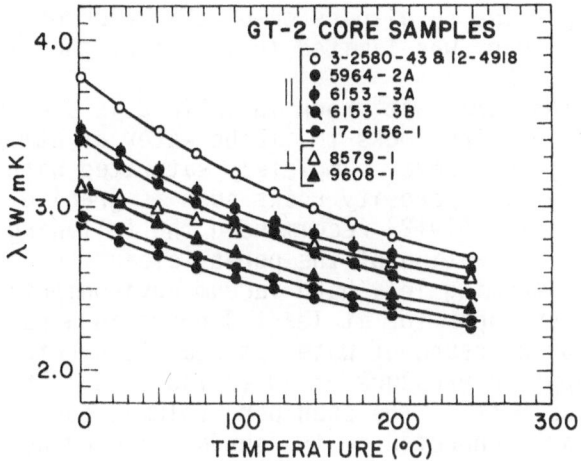

Figure 4. GT-2 core specimen thermal conductivity. See Table 4
          for modal compositions.

uniform heat flux between the Precambrian and sedimentary section of
GT-2, the higher average temperature gradient in GT-2 from 0 to
0.8 km of ∿100°C/km versus ∿50-60°C/km for the basement section from
0.8 to 3 km indicates that the in situ conductivity of the sedimentary section should be lower than the measured experimental
values of 1.8 to 3.1 W/m K. The much higher gradient in the vol-
canic tuff portion is consistent with normally lower (<1.0 W/m K)
conductivities associated with these low density, highly porous
rocks.

                              DISCUSSION

     The conductivity values are given to four places to facilitate
interpolation and numerical analysis; they are not an indication
of the accuracy of the measurements. The cut-bar thermal-conductivity comparator and the transient needle conductivity probe are
secondary instruments. Therefore the accuracy is limited by the
accuracy of the conductivity values assigned to the standard comparison materials. The values recommended by T.P.R.C. for fused
quartz and Pyroceram 9606 [TPRC, Purdue University, 1964] were used
since they were mutually consistent in a series of comparison
experiments in both instruments. The recommended values are given
in Table 5. A direct comparison was made of our measurements with
those made by the U.S. Geological Survey at Menlo Park, CA (see
Table 6). Both LASL and USGS measurements were in agreement and

were internally precise (a dispersion of the order of ± 1%) but
their absolute accuracy was limited to ± 4-5% by the standards used.

The use of the term "wet" rock in Table 6 is a misnomer. Both
USGS and LASL allowed the rocks to imbibe water at atmospheric
pressure thus they were never completely saturated with water. Complete saturation of low-porosity rocks such as granite is very
difficult. Hirschwald [1912] recommended the following procedure
to obtain optimum saturation of the pores: clean the rock sample;
remove the air by warming in a hard vacuum environment; let the
sample imbibe water vapor for at least 3 hours in a partial vacuum
equal to the vapor pressure of water at room temperature; cover with
water and then apply a pressure of 50 to 150 bars. This technique
probably saturates most of the open pore-volume; pores which inter-
communicate and are connected to the surface including the "dead-
ended" pores. However, we presently know of no method of proving
that the rock is ever completely saturated.

The conductivities of these so-called "wet" rocks were of the
order of 1-4% higher than the conductivities of these same dry rocks.
The dry rocks were prepared by warming at 70°C in a hard vacuum for
3 hours. The temperature effects on dry rocks were reversible to
at least 250°C. This is consistent with results cited by Birch
and Clark [1940]. Thermal conductivity values for GT-1 and GT-2
rocks up to 250°C are plotted in Figures 3 and 4. A comparison
between GT-2 λ values in the biotite granodiorite section and data
presented by Birch and Clark [1940] for a number of granitic rocks
is shown in Figure 5. The gradual decrease of conductivity with
increasing temperatures reflects the classical dependence expected
for dense crystalline materials in the anharmonic phonon coupling
region. As Birch and Clark [1940] point out, plots of thermal
resistivity $(1/\lambda)$ versus temperature are linear for materials of
this type. The crosshatched region of Figure 5 indicates the expected variation of λ in the biotite granodiorite section, where
fluid circulation experiments are underway at LASL [Tester and Smith
(1977)]. The temperature effect on thermal conductivity is similar
for the selected Birch and Clark [1940] data for tonalite, Rockport
and Westerly granite, and quartz monzonite in comparison to the GT-2
cores at 8580 and 9608 ft. The agreement between GT-2 9608 and
quartz monzonite is fortuitous in the sense that the modal compositions are different as shown in Table 4. The Birch and Clark
[1940] thermal conductivity data represent the most comprehensive
study available in the high temperature region where complete modal
analyses are provided. Heating above 400°C sometimes resulted in
permanent changes with a slight decrease in the conductivity.

Murphy and Lawton [1977] were able to estimate in situ thermal
conductivities of rock contained around GT-2 and a second deep wellbore, EE-1, drilled nearby. Basically, they extended the transient

Table 5
THERMAL CONDUCTIVITIES OF STANDARD MATERIALS
F u s e d   Q u a r t z

Temperature (°C)	TPRC[b] (W/m K)	Ratcliffe (W/m K)	Birch & Clark, 1940 (W/m K)
-23	1.28	1.275	--
0	1.33	1.323	1.36
27	1.38	1.374	1.40
77	1.45	1.431[a]	1.46
127	1.51	--	1.51
Error	+4%	+2%	∿+1%

	Pyroceram[b] 9606 (W/m K)
0	4.13
27	3.99
77	3.79
127	3.65
227	3.45
327	3.31
Error	+5%

---

[a]Extrapolated.

[b]TPRC, Purdue University, 1964.

[c]Error given as average deviation from mean.

Table 6
COMPARISON OF CONDUCTIVITY VALUES FOR SAMPLE GT-1, 2425 FT
(Density, wet, 2.66 g/cm^3)

USGS, wet at ∿25°C	3.78 W/m K
LASL, wet, 1st run at 25°C	3.797 + 0.040 W/m K
LASL, wet, 2nd run at 25°C	3.789 $\mp$ 0.035 W/m K

NOTE: Between Runs 1 and 2, the stack was disassembled;
the sample was cleaned, dried, and resaturated with water;
and then the stack was reassembled.

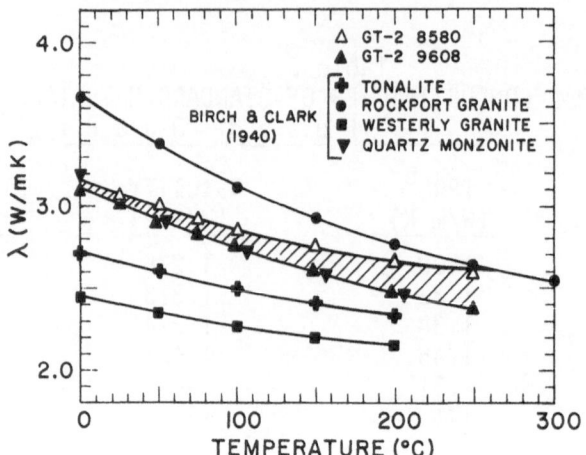

Figure 5.   Comparison of GT-2 biotite granodiorite thermal conductivities with those for selected igneous rocks. See
Table 4 for modal compositions.

line source method described earlier to include effects caused by
flowing fluid in the wellbores. By comparing the conductive heat
flux from the rock to the convective heat transported by the wellbore fluid, Murphy and Lawton [1977] showed that temperature measurements made between ∿0.25 and 100 hours provide meaningful and sufficient data for independently estimating a mean conductivity λ and
diffusivity α of the formation. Numerical solutions in the form of
dimensionless temperature-time type curves were used to analyze
experimental flowing temperature logs of the wellbores. λ was estimated at 2.9 W/m K and α = 1.0 x $10^{-6}m^2$/sec [Murphy and Lawton,
1977]. By using mean values for ρ_r = 2700 kg/m^2 and C_r = 1050 J/kgK,
a λ of 2,8W/m K was calculated from the estimate of α. This internal
consistency for in situ measurements coupled with the excellent
agreement with the laboratory measurements of λ supports the hypothesis that core material in terms of its thermal conduction properties
is representative of the actual conditions that exist in the reservoir.

A number of simplistic schemes can be used to estimate the
thermal conductivity of massive dense igneous rocks which are

macroscopically isotropic and homogeneous and do not contain large volume fractions of either quartz or potassium feldspar. We applied the estimating method developed originally by Maxwell for composite materials and later modified by Birch and Clark [1940], Powers [1961], and Mitoff [1968]. Although other techniques such as the statistical approach of Hashin and Shtrikman [1962] have been applied to igneous rock systems [Horai and Baldridge, 1972] with reasonable success, we felt that variations in modal analysis for thin sections taken from the same core sample (See Tables 2 and 4) were sufficiently large to limit the accuracy of any prediction method and consequently we adopted a dispersion analysis modification of the Maxwell approach using thermal conductivity values for the pure minerallic phases as provided by Birch and Clark [1940] and others [Diment, 1967; Haskin and Shtrikman, 1962; Horai and Simmons, 1969; and Horai and Baldridge, 1972].

The discrepancies between the measured and estimated λ values using a series combination of resistances were usually less than 9%. This type of discrepancy is acceptable since a small degree of anisotropy and inhomogeneity is inherent in solid rock samples and the modal compositions vary appreciably at different cross sections in the specimens although the density may appear to be invariant. However a single value of thermal conductivity cannot be assigned to a modal mineral when it is defined to exist over a range of chemical compositions: for example the conductivity of the plagioclase feldspar series has a minimum value at an intermediate composition of anorthite and albite [Horai and Simmons, 1969]. Variations based on chemical impurities and structural defects may also be necessary to refine estimates of λ.

Since these simplistic schemes all indicate that the conductivity of the rock is essentially determined by the volume fractions of constituent rock-forming minerals; then the crystal boundaries apparently present a low, constant resistance to the flow of heat. This generalization apparently applies to all of the accessory minerals, and the major minerals, biotite, the plagioclase feldspars, and quartz, but not to the potassium feldspar and quartz composite whose interface must present a lower heat flow resistance. The conductivity of the potassium feldspar is about 25 percent greater than that of the plagioclase feldspar series but this difference is not sufficient to explain the very high conductivity of rocks which contain large volume fractions of both quartz and potassium feldspar. Figure 6 shows a rough empirical correlation between λ and the volume fraction of potassium feldspar and the temperature. The four different lines correlate different compositional regions for the GT-2, GT-1 and Birch and Clark [1940] data plotted in Figures 3, 4, and 5. For a given granite of fixed composition, $1/\lambda$, the resistivity, is linearly proportional to temperature; and at a fixed temperature for a particular type of granite, λ increased linearly with the volume fraction of potassium feldspar present (x(K-feldspar)) as determined by modal analysis of thin sections. Consequently, for empirical reasons we selected x(K-feldspar)/T as a correlating parameter for λ. At very low K-feldspar concentrations of <1 volume percent it was difficult to plot all the data because x(K-feldspar)/T varied over a small range 1 to 4 x $10^{-5}K^{-1}$. Therefore only some of the points for the low K-feldspar granites are shown in Figure 6.

If a thin specimen of a rock with a coarse texture is used, then these simplistic schemes are not appropriate: just the modal analysis alone is not sufficient to predict λ for this specimen. The continuous phases must be identified and then the appropriate relationship can be selected to estimate the effective conductivity of the specimen.

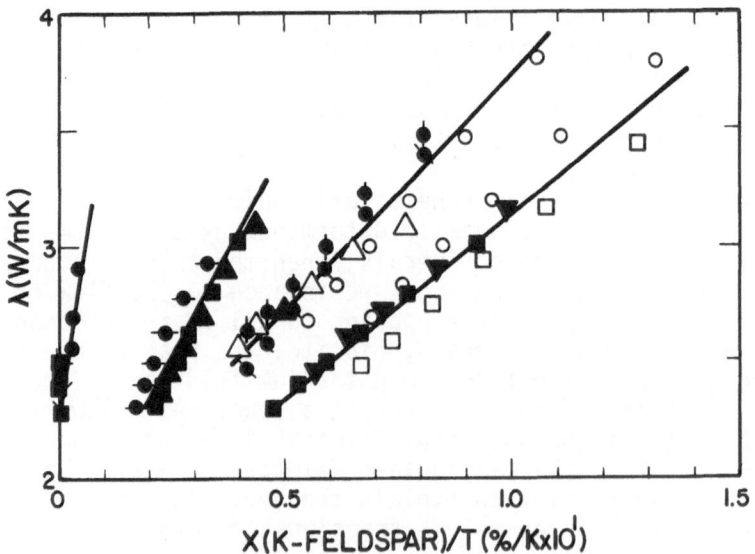

Figure 6.   Empirical representation of the dependence of thermal conductivity of GT-2 and GT-1 rocks on K-feldspar volume fraction and temperature. Symbols correspond to those used in Figures 3 and 4. In addition data for quartz monzonite (▼, Birch and Clark [1940]) are presented.

ACKNOWLEDGEMENTS

The authors would like to thank T. Shankland, W. Laughlin, P. Wagner, R. Potter, R. L. Aamodt, H. Murphy, A. Eddy, and A. Blair for their comments and helpful suggestions in reviewing our work. The financial support of the Division of Geothermal Energy of the U. S. Department of Energy is gratefully acknowledged.

NOMENCLATURE

A = cross-sectional area perpendicular to heat flow direction, m^2

C_r = heat capacity of granite, $\sim$1050 J/kgK

C_w = heat capacity of water, $\sim$4200 J/kgK

$\dot{m}_w$ = water flow rate through fracture, kg/sec

q = radial heat flux per unit length, W/m

q_z = heat flow along z axis, J/sec or W

r = radial distance from line source, m

R = fracture radius, m

t = time, sec

T_i = mean initial rock temperature, °C

T = temperature, K or °C

T_{min} = fluid temperature as it enters the reservoir, °C

Z = heat flow direction, m

α = thermal diffusivity, m^2/sec

λ = thermal conductivity, W/m K

λ_r = thermal conductivity of granite, W/m K

ρ_r = density of granite, kg/m^3

REFERENCES

1. Birch, F. and H. Clark, "The Thermal Conductivity of Rocks and Its Dependence Upon Temperature and Composition," Part I Am. J. Sci. 238, 8, pp. 529-558, August 1940.

Part II Am. J. Sci. 238, 9, pp. 613-635, September 1940.

2. Clark, H., "The Effects of Simple Compression and Wetting on the Thermal Conductivity of Rocks," Trans. Am. Geophysical Union, pp. 543-544, Reports and Paper, Tectonphysics--1941.

3. Clark, Jr., S. P., Editor, Handbook of Physical Constants (The Geological Society of America, In., Memoir 97, 1966).

4. Diment, W. H., "Thermal Conductivity of Some Rock-Forming Minerals," abs IUGG meeting, Zurich, 1967.

5. Harlow, G. H. and W. E. Pracht, "A Theoretical Study of Geo-
    thermal Energy Extraction," J. Geophy. Res. 77 (35), 7038, 1972.

6. Haskin, Z. and S. Shtrikman, "A Variational Approach to the
    Theory of the Effective Magnetic Permeability of Multiphase
    Materials," J. App. Phys. 33, 3125, 1962.

7. Hirschwald, "Handbuch der bautechnischen Gesteinprüfung,"
    Berlin, Borntraeger, 1912.

8. Horai, K. and S. Baldridge, "Thermal Conductivity of Nineteen
    Igneous Rocks, I. Application of Needle Probe Method to the
    Measurement of the Thermal Conductivity of Rock and II. Esti-
    mation of the Thermal Conductivity of Rock from the Mineral
    and Chemical Compositions," Phys. Earth Planet Interiors, 5,
    151-156, 157-166, 1972.

9. Horai, K. and G. Simmons, "Thermal Conductivity of Rock-Forming
    Minerals," Earth and Planetary Science Letters 6, 359-368, 1969.

10. Laughlin, A. William and A. Eddy, "Petrography and Geochemistry
    of Precambrian Rocks from GT-2 and EE-1," Los Alamos Scientific
    Laboratory Report LA-6930-MS, August 1977.

11. McFarland, R. D. and H. D. Murphy, "Extracting Energy from
    Hydraulically Fractured Geothermal Reservoirs," in proceedings
    of 11th Intersociety Energy Conversion Engineering Conf.,
    State Line, NV, 1976.

12. Mitoff, S. P., "Properties Calculations for Heterogeneous
    Systems," in Advances in Materials Research, ed. H. Herman,
    John Wiley, NY, 3, 305, 1968.

13. Murphy, H. D. and R. G. Lawton, "Downhole Measurements of
    Thermal Conductivity in Geothermal Reservoirs," 1977 Energy
    Technology Conference, Houston, TX, Sept. 1977.

14. Perkins, P. C., "Petrography of Some Rock Types of the Pre-
    cambrian Basement Near the Los Alamos Scientific Laboratory
    Geothermal Test Site, Jemez Mountains, New Mexico," Los Alamos
    Scientific Laboratory report LA-5129, June 1973.

15. Powers, A. E., "Conductivity in Aggregates," Knolls Atomic
    Power Laboratory report KAPL-2145, March 1961.

16. Sibbitt, W. L., "Preliminary Measurements of the Thermal Con-
    ductivity of Rocks from LASL Geothermal Test Holes GT-1 and
    GT-2," Los Alamos Scientific Laboratory report LA-6199-MS,
    January 1976.

17. Smith, M. C., "The Los Alamos Scientific Laboratory Dry Hot Rock Geothermal Project," (LASL Group Q-22), Geothermics, 4, 1-4, 27-39, 1975.

18. Tester, J. W. and M. C. Smith, "Energy Extraction Characteristics of Hot Dry Rock Geothermal Systems," in proceedings of 12th Intersociety Energy Conversion Engineering Conf., p. 816, Washington, D.C., Aug. 28-Sept. 2, 1977.

19. TPRC, Thermophysical Properties Research Center, Data Books, Purdue University, 1964.

20. van der Held, E. F. M., and F. G. van Drunen, Physica, 15, 865, 1949.

21. Walsh, J. B. and E. R. Decker, "Effect of Pressure and Saturating Fluid on the Thermal Conductivity of Compact Rocks," J. Geophys. Res. 71, 12, pp. 3053-3061, June 1966.

22. Wechsler, A. E., "Development of Thermal Conductivity Probes for Soils and Insulations," Tech. Rpt. 182, Arthur D. Little, Inc., October 1966.

ENERGY LOSS BY THERMAL CONDUCTION AND
NATURAL CONVECTION IN ANNULAR SOLAR RECEIVERS*

A. C. Ratzel, C. E. Hickox, and D. K. Gartling

Fluid and Thermal Sciences Department
Sandia Laboratories
Albuquerque, New Mexico 87115   USA

INTRODUCTION

An effective device for the collection of solar energy is the so-called parabolic-cylindrical solar collector. In this device, a circular receiver tube is enclosed by a concentric glass envelope and situated along the focal line of a parabolic trough reflector. The heat transfer processes which occur in the annular space between the receiver tube and the glass envelope are important in determining the overall heat loss from the receiver tube. In typical high temperature receiver tube designs the rate of energy loss by combined thermal conduction and natural convection is of the same order of magnitude as that due to thermal radiation, and can amount to approximately 6% of the total rate at which energy is absorbed by the solar collector. The elimination of conduction and natural convection losses can significantly improve the performance of a large collector field.

In this paper, several techniques useful for the reduction of energy loss by thermal conduction and natural convection are considered. The receiver configuration chosen for study is typical of those used in the Solar Total Energy System at Sandia Laboratories. The receiver tube has a "black chrome" selective coating and is 2.54 cm in outside diameter. The inside diameter of the glass envelope is approximately 4.4 cm. Typical operating temperatures of the receiver tube and glass envelope are approximately 573 K and 373 K, respectively.

---

*This work was supported by the United States Department of Energy.

CONDUCTION HEAT LOSS

Of the three modes of heat transfer, the most significant heat loss savings for an annular receiver can be accomplished by limiting conduction losses. Convection losses are negligible so long as the annular space is properly sized. Radiation losses, being primarily fixed by the receiver tube selective surface properties, are more difficult to reduce. Variations in electroplating parameters to reduce the receiver thermal emittance properties may result in lower solar absorptivity and perhaps poor durability properties.

Attempts to limit heat transfer through the annular space will be discussed in the following sections. Techniques studied include (1) evacuation of the annulus gas, (2) oversizing the annular space, and (3) using gases other than air for the heat transfer medium.

Effect of Vacuum

A review of the literature on vacuum technology indicates that the thermal conductivity of a gas is a function of the mean free path of the gas molecules [1,2]. An expression relating the mean free path of a gas to the enclosure pressure and gas temperature is

$$\lambda = 2.331(10^{-20})T/P\delta^2 \, , \tag{1}$$

where T, P, and δ are given in deg K, mm Hg, and cm, respectively. The governing equation for the effective heat transfer coefficient for the annular space is [1]

$$k_{ef} = \frac{k}{r_i Ln(r_o/r_i) + b\lambda(r_i/r_o + 1)} \, , \tag{2}$$

where

$$b = \frac{2 - a}{a} \, \frac{18\gamma - 10}{8\gamma + 8} \, .$$

The reduction of receiver tube heat loss by evacuation has been analytically modeled and experimentally verified. Energy balances were made on the receiver tube and glass surfaces and were incorporated in a computer analysis. The analysis assumed steady state conditions and utilized standard correlations for the effects of pressure, wind, geometry, and temperature on the conduction and convection terms. Receiver heat losses were measured using a tubular resistance heater to provide heat to the Sandia Laboratories Phase IV-B annular receiver design [3]. Receiver tube temperatures and annulus pressures were carefully monitored throughout the testing program.

Early experimental work involved maintaining a fixed receiver assembly heat loss while varying the annulus pressure. Variations in receiver tube coating properties necessitated bracketing the experimental data with analytical results calculated for receiver tube emissivities of 0.2 and 0.3 at 589 K, with the emissivities decreasing linearly with temperature to 0.15 to 373 K. (For receiver tube temperatures above 589 K, the emissivity data were extrapolated.) The receiver tube temperatures are seen to be independent of annulus vacuum for pressures above 1 mm Hg in Fig. 1. This is also implied in Fig. 2, which provides the heat loss data as a function of annulus pressure. Significant heat loss reductions of nearly 50% are seen to result if vacuums below 0.01 mm Hg can be maintained. Similar work has been reported by Ortabasi [4] for a tubular flat plate collector. Differences in geometry between the two designs result in considerably different vacuum requirements for significant conduction heat loss reduction.

Effect of Annulus Gap Sizing

Optimum sizing of the annular space for operation at atmospheric pressure requires that the energy transferred across the gap be by thermal conduction and radiation heat transfer. Incorrect sizing could result in enhanced convective energy transport which would increase the net heat loss. Work to be discussed concerning natural convection has indicated that the effects of natural convection will be suppressed as long as the Rayleigh number is maintained below a value of 1000.

Figure 2 indicates that the Phase IV-B receiver sizing has not been optimized since a heat loss reduction occurs when the annulus pressure is reduced below atmospheric pressure. By decreasing the pressure, the Rayleigh number is reduced below 1000 through lowering the annulus gas density. Based on these trends, the computer model was utilized to vary the gap spacing for a fixed receiver tube radius and temperature. From data presented in Reference [5], the effective conduction coefficient used for a particular spacing was generated using the following correlation:

$$k_{ef} = k \text{ for } N_{Ra} < 1000 \quad , \tag{3}$$

$$k_{ef} = 0.1558k \, N_{Ra}^{0.2667} \text{ for } N_{Ra} > 1000 \quad ,$$

where N_{Ra} is the Rayleigh number

$$N_{Ra} = \rho g \beta \ell^3 \Delta T / \mu \alpha \quad . \tag{4}$$

The analysis indicated that the heat loss is minimized for an annular space of 0.81 cm. For the same test conditions, the Phase IV-B gap of 0.92 cm results in a Rayleigh number of 1550. Despite the discrepancy in sizing, it was found that oversizing the gap results in minimal increased heat loss compared to that obtained for reducing the gap size to maintain the Rayleigh number below 1000. This results from the fact that the insulating effect of the gas, due to both the low gas thermal conductivity and the gap space, is not optimized for undersized annular spacings.

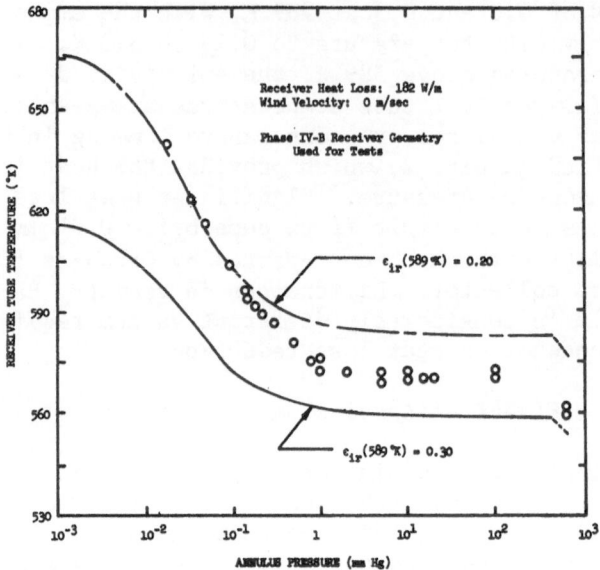

Figure 1.   Receiver tube temperature as a function of annulus
            pressure.

Further investigation into varying the annular space to reduce heat loss resulted in the data summarized in Figure 3.   By
maintaining annulus pressures below 200–300 mm Hg and oversizing the
gap, heat loss savings of between 15 and 30 W/m may be obtained over
that lost by a receiver design sized to eliminate convection heat
transfer at atmospheric pressure.   Since the reduction of heat loss
by 30 W/m for the "correctly sized" annular space necessitates using

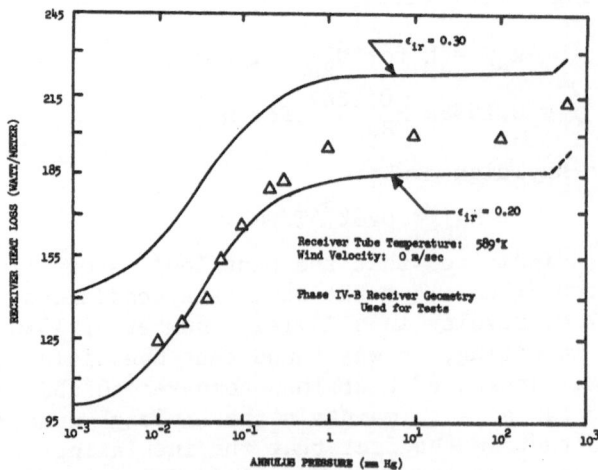

Figure 2.   Effects of pressure on receiver tube heat loss.

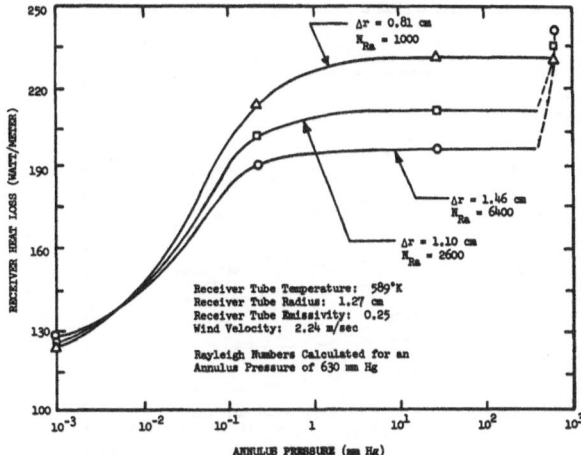

Figure 3.   Receiver annulus oversizing for reducing heat loss.

vacuums below 0.1 mm Hg, it can be noted that oversizing may allow for energy loss savings without requiring hard vacuum systems.

Effect of Gases Other Than Air

Utilization of gases other than air in the receiver annulus should reduce the conduction heat loss so long as (1) the gas thermal conductivity is less than that of air, and (2) the effective Rayleigh number is similar to that of air for a given geometry. Calculations were performed for argon and carbon dioxide. Although carbon dioxide has a lower thermal conductivity than air (0.031 W/m-K compared to 0.048 W/m-K at 477 K), its other physical properties necessitate small gap spacings to minimize natural convection heat loss. The insulating effect of the lower thermal conductivity is thus lost.

Heat loss savings of 25 to 30 W/m may be realized by replacing with argon. Such savings are comparable to oversizing the annulus and maintaining a partial vacuum of 100-300 mm Hg. The advantage gained using argon gas to reduce the heat loss is that due to the similarity in sizing the gap for air or argon, a complete loss of argon and replacement with air will still minimize the heat loss. As can be seen in Fig. 3, an increase in heat loss will result in the event of loss of vacuum for the oversized geometry.

In summary, Fig. 4 is provided to show the relative heat loss savings for each technique previously discussed. Additional alternatives are also shown. Of the options, it appears that evacuation can best eliminate conduction heat loss, although the relative costs of each heat loss reduction scheme should be considered in selecting the best option.

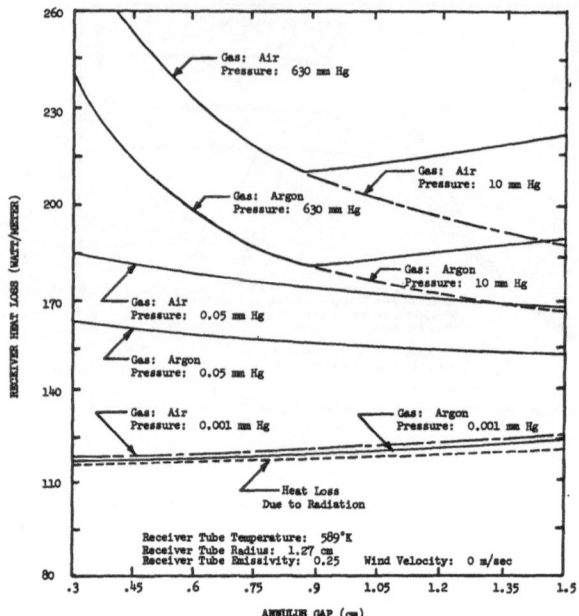

Figure 4.   Summary data on heat loss reduction techniques.

NATURAL CONVECTION IN AN ANNULUS

In the previous section, the classical experimental result for
natural convection heat transfer between horizontal, concentric,
circular cylinders as originally presented by Kraussold [5], was used
in the analysis of the heat transfer process between the receiver tube
and glass envelope. Recently Kuehn and Goldstein [6] have compiled a
comprehensive review of the available experimental results for natural
convection heat transfer between circular cylinders and proposed correlating equations using a conduction boundary layer model. It is
evident from this review that almost all the results are valid only
for horizontal, concentric, circular cylinders with uniform temperatures. Typically, the temperature distribution on a receiver tube
is not uniform, nor is the receiver tube concentric with the glass
envelope. In order to assess the effects of these nonuniformities,
a numerical study of the natural convection process was performed
using the finite-element computer program, NACHOS [7].

Heat Transfer Between Concentric Cylinders

In order to demonstrate that the results of the numerical analysis are compatible with existing experimental results, an initial
series of calculations was performed for horizontal, concentric, circular cylinders with uniform temperatures. The temperatures of the
inner and outer cylinders were held constant at 583 K and 333 K, respectively. These values were selected to correspond to the average

operating conditions expected for the existing Sandia Laboratories
Solar Total Energy collector field. The radius of the inner cylinder was held constant at 1.27 cm and the outer radius allowed to
vary from 2.24 cm to 4.32 cm, producing radius ratios in the range
1.4 to 3.4. Rayleigh numbers ranging from approximately 300 to
97,000 were thus obtained with the Rayleigh number defined in the
usual way, as given by Eq. (4).

Following the accepted standard, results of the analysis are
presented in Fig. 5 as a plot of the heat loss ratio versus the Rayleigh number, where the heat loss ratio is defined to be the ratio
of the energy loss per unit length due to natural convection to that
due to thermal conduction acting alone. In Fig. 5, the cross-hatched
area indicates the region occupied by the experimental data as compiled by Kuehn and Goldstein [6]. Typical streamlines and isotherms
as computed for a Rayleigh number of approximately 12,000 are shown
in Fig. 6.

It is evident that, in general, there is rather good agreement
between the computed and experimentally determined values of natural
convection heat transfer rates. As expected, the heat loss is seen
to initially deviate from that due to thermal conduction at a Rayleigh number of approximately 1000. The numerical method employed
is apparently of sufficient accuracy to warrant its application to
the study of the non-ideal situations described previously which
have not, as yet, been extensively investigated.

Nonuniform Temperature Distribution

Two situations involving nonuniform temperature distributions
on concentric, circular cylinders were selected for study. In both

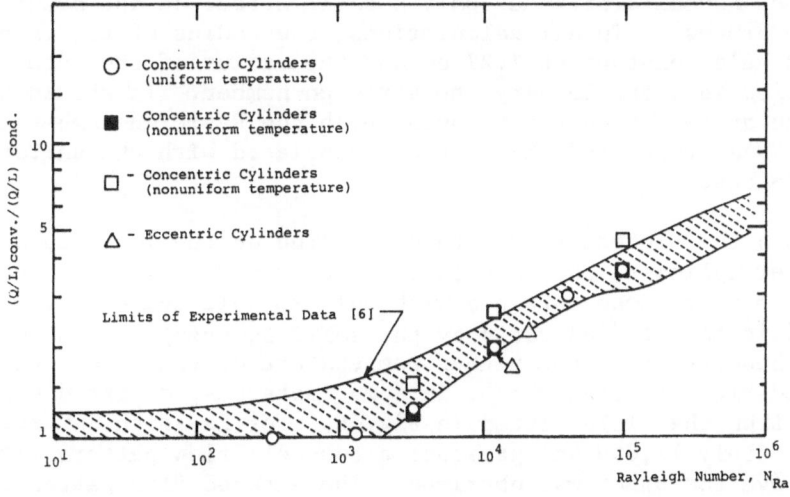

Figure 5.   Natural convection heat transfer in an annulus. Comparison of computed and experimental results.

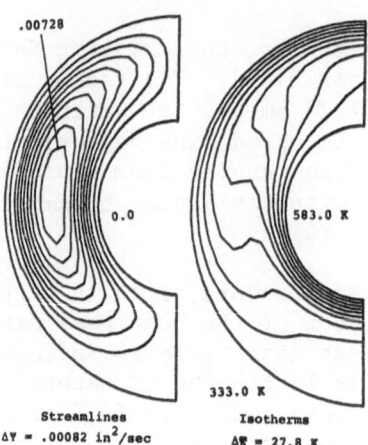

Figure 6.   Steamlines and isotherms for natural convection between concentric, circular cylinders, uniform temperature, N_{Ra} = 12,142.

cases the temperature of the outer cylinder was held constant at 333 K and the temperature of the inner cylinder allowed to vary according to

$$T = T_m \pm T' \cos\theta \quad , \tag{5}$$

where T_m is the mean temperature (583 K), T' the perturbation to the mean, and θ is the angle measured from the bottom of the cylinder. A perturbation of 139 K was used in all calculations, yielding a total variation of 278 K around the inner cylinder. The two cases studied are distinguished by the choice of sign in Eq. (5); the positive sign corresponding to the occurrence of greatest temperature on the lower surface of the inner cylinder and vice versa. A variation of 278 K is far in excess of that encountered in conventional receiver tubes but, as will subsequently be shown, even this amount of nonuniformity has only a small effect on the natural convection process. In all calculations, the radius of the inner cylinder was held constant at 1.27 cm and the radius of the outer cylinder varied in order to vary the Rayleigh number. It should be noted that the average temperature upon which the Rayleigh number is based has the same value (458 K) as that associated with the uniform temperature case.

When the highest temperature occurred on the lower surface of the inner cylinder, the calculated results were virtually indistinguishable from those obtained with uniform wall temperatures as plotted in Fig. 5 (indicated by the solid squares). It should be noted, however, that the higher temperature on the lower surface produced flow patterns which, in some instances, differed significantly from that illustrated in Fig. 6. For Rayleigh numbers of approximately 12,000 and greater, a two-cell flow pattern with one cell above the other was obtained. The altered flow pattern, however, produced no appreciable change in the overall heat transfer characteristics.

When the higher temperature occurred on the upper surface of
the inner cylinder, the natural convection heat transfer rate was
enhanced over that obtained with uniform temperatures, as shown in
Fig. 5. Departure from the thermal conduction curve still occurs
at a Rayleigh number of approximately 1000.

From the results described in this section, it is evident that
highly nonuniform temperature distributions are required in order
to appreciably affect the natural convection process between concentric cylinders. Since the nonuniformity of temperature in typical receiver geometries is substantially less that that used in the
current study, it is anticipated that the effects of nonuniform
temperature can be neglected in most instances.

Eccentric Cylinders

The geometry chosen for the study of eccentric cylinders consisted of inner and outer cylinders of radii 1.27 cm and 2.79 cm,
with the inner cylinder displaced downward a distance equal to one-
half the gap width yielding an eccentricity of 0.76 cm. Each cylinder was held at a uniform temperature. Although other geometries
could be analyzed without difficulty, the selection of the case
chosen for study was influenced by certain practical considerations.
First, a downward displacement of the inner cylinder enhances the
convection process more than an upward displacement. Secondly, an
eccentricity of one-half the gap width is greater than that encountered in practical receiver designs. Finally, it was convenient to
maintain symmetry about a vertical plane in order to simplify the
numerical computations.

For proper interpretation of subsequent results, it should be
recalled that the heat loss by thermal conduction between eccentric,
circular cylinders with uniform wall temperatures is given by

$$Q/L = 2\pi k \Delta T / \left(\ln x + \sqrt{x^2 - 1} \right) , \qquad (6)$$

where

$$x = (r_o + r_i^2 - \epsilon^2)/2r_o r_i . \qquad (7)$$

The overall heat transfer results obtained for two different
Rayleigh numbers are plotted in Fig. 5. The calculated streamlines and isotherms are similar to those illustrated in Fig. 6. If
the Rayleigh number is based on the mean gap size of 1.52 cm and
the cylinders held constant at the values previously used in the
analysis of concentric cylinders, a Rayleigh number of approximately
12,000 results. For the remaining situation the temperature of the
inner cylinder was increased and that of the outer cylinder decreased,
by equal amounts, in order to produce a Rayleigh number of approximately 22,000 while simultaneously maintaining the mean temperature

constant at 458 K. This latter calculation was performed in order
to tentatively assess the effect of Rayleigh number on the overall
heat transfer process.

It is apparent from Fig. 5 that the results for eccentric cylinders coincide with those for concentric cylinders when the appropriate conduction solution is used as a reference and the Rayleigh
number is based on the mean gap size. However, it should be noted
that the overall heat transfer is slightly enhanced because of the
increased effect of conduction for eccentric cylinders. For the
variables used, the heat transferred by conduction is 14.75% greater
for eccentric cylinders than for concentric cylinders.

REFERENCES

1. Dushman, S., Scientific Foundations of Vacuum Technique, J. M.
   Lafferty, Editor, Wiley, New York, 1962.
2. Holkeboer, D. et al, Vacuum Engineering, Boston Tech. Pub., Inc.
   Cambridge, 1967.
3. Treadwell, G. W., "Design Considerations for Parabolic-Cylindrical
   Solar Collectors," Sandia Laboratories Report SAND 76-0082.
4. Ortabasi, U., "Indoor Test Methods to Determine the Effect of
   Vacuum on the Performance of a Tubular Flat Plate Collector,"
   ASME Paper 76-WASOL-24.
5. Kraussold, H., "Wärmeabgabe von zylindrischen Flüssigkeits-
   schichten bei natürlicher Konvektion," Forsch. Hft. Ver. Dt. Ing.,
   Vol. 5, No. 4, 1934, pp. 186-191.
6. Keuhn, T. H. and Goldstein, R. J., "Correlating Equations for
   Natural Convection Heat Transfer Between Horizontal Circular
   Cylinders," Int. J. Heat Mass Transfer, Vol. 19, 1976, pp. 1127-
   1134.
7. Gartling, D. K., "Convective Heat Transfer by the Finite Element
   Method," Comp. Meth. Appl. Mech. Eng., (to appear in 1977).

NOMENCLATURE

a - accommodation coefficient	α - thermal diffusivity
g - acceleration of gravity	β - coefficient of volumet-
k - thermal conductivity	ric thermal expansion
k_{ef} - effective thermal conducti- vity	δ - molecular diameter
	ϵ - eccentricity
L - length of receiver tube	ϵ_{ir} - thermal emissivity
$\ell, \Delta r$- gap size	
N_{Ra} - Rayleigh number	γ - ratio of specific heats
	λ - mean free path
P - pressure	μ - dynamic viscosity
Q - heat loss	ρ - density
r_o - inner radius of glass envelope	
r_i - outer radius of receiver tube	
ΔT - temperature difference	

HEAT TRANSFER THROUGH COALS AND OTHER NATURALLY OCCURRING CARBONACEOUS ROCKS[*]

N.E. Vanderborgh, J.P. Bertino, G.E. Cort and P. Wagner

Los Alamos Scientific Laboratory
P. O. Box 1663, MS-734
Los Alamos, New Mexico 87545 USA

ABSTRACT

The understanding of heat transfer through solid fossil fuels is essential to the phenomonology of pyrolysis, gasification and combustion of these fuels. While coals have thermal conductivities of 0.1-0.5 W/mK at 300 K, heat transfer measurements are complicated by the changes found in these fuels caused by the heating processes. Such complications are clearly shown when one looks at thermal conductivity differences between virgin and heat-treated materials.

Coals, upon heating, undergo a variety of chemical and physical modifications. Initially, these materials lose low molecular weight gases; additional heating removes moisture. Such pyrolytic processes result not only in a significant mass decrease (as much as 50% for low-rank coals), but a marked alteration in the internal structure of the material. In virgin coals, mass transfer is dominated by a system of pores. Drying these materials typically alters the flow mechanisms and consequently the permeability. Heat transfer becomes dominated by the convective transport of products generated within the specimen during the heating process.

Studies are described that explore the concurrent and counter-current heat and mass transfer problems through semiporous materials such as coals and other model specimens.

---

[*]Work performed under the auspices of the Department of Energy.

INTRODUCTION

Even though vast quantities of subbituminous coals are known to exist under the semiarid regions of the Southwestern United States, the expanded development of this resource presents unusual problems. These coals show enough seam dip so that surface mining is rather limited; conservative estimates[1] suggest that less than 5% of the known reserve is within surface mining range. Moreover past experience with underground mining suggests that unusual technical problems result from the limited H_2O supply, generally friable roofs, and the lenticular nature of these deposits. Thus the coal supply may permit only a modest expansion of current production rates, unless new extraction technologies are developed.

The Los Alamos Scientific Laboratory is exploring an approach that appears to have promise in expanding coal production in the Southwestern U.S. This approach involves an initial in situ preconditioning of coal. During this step the coal is chemically modified to offer low resistance (flow enhancement), and suitable chemistry underground. This thermal processing simultaneously yields a stream of hydrogen-enriched products and leaves a porous semichar that is used for underground gasification. This is shown schematically in Figure 1.

"Chemically mining" coal for underground conversion of this fossil resource to gaseous or liquid fuels, is not a new idea.[2] The concept of utilizing coal without the coincident societal and environmental costs of conventional mining has intrigued mankind for decades. Early experiments, mainly in Europe,[3-6] have identified the severe technical problems of developing and maintaining flow and reactivity conditions that may be dominated by the frequently unfavorable naturally occurring coal-seam properties. Initial successes resulted from the realization[7] that flow modification (reverse combustion) is necessary not only to change overall flow resistance but also to achieve a single controlled flow path of high permeability. The two-stage process described in this paper attempts to achieve the same type of control by first drying specific regimes in the coal and then effecting pyrolysis in those dried regions to release gaseous and liquid hydrocarbons. Once that process has occurred, then the highly porous seam will be subject to a variety of different processing options. This process concept has been developed for dry, Southwestern coals.

DRYING AND PYROLYSIS OF SUBBITUMINOUS COALS

The structure of naturally occurring coals is complicated by the highly anisotropic nature of these materials. Here we assume we are dealing with "average coal". The flow behavior through

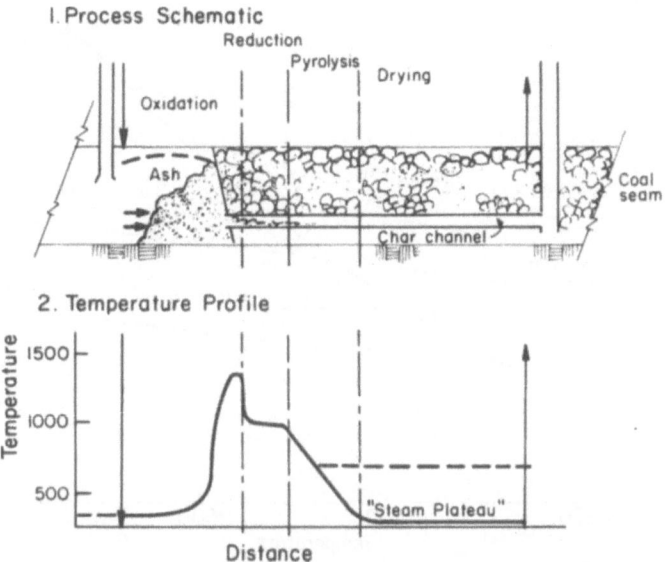

Figure 1.   Underground Gasification of Southwestern U.S. Coal.

coals is explained by assuming that coal is a hardened gel, with
various-sized, interconnecting pores and capillaries. Such connecting paths are typically emphasized in one direction. The majority of these pores, in virgin coal, are less than 50 nm. Due
to these pores, coals exhibit a molecular sieve property excluding
certain molecules while entrapping others. This sort of molecular
discrimination results from chemical properties as well as from
molecular size.

     Moisture plays an essential role in initial flow distinctions.
Water is typically absorbed in coal micropores. Moreover, water is
a pore-filling material due to the fact that tetragonal bonding between a chemisorbed water molecule and an additional water molecule

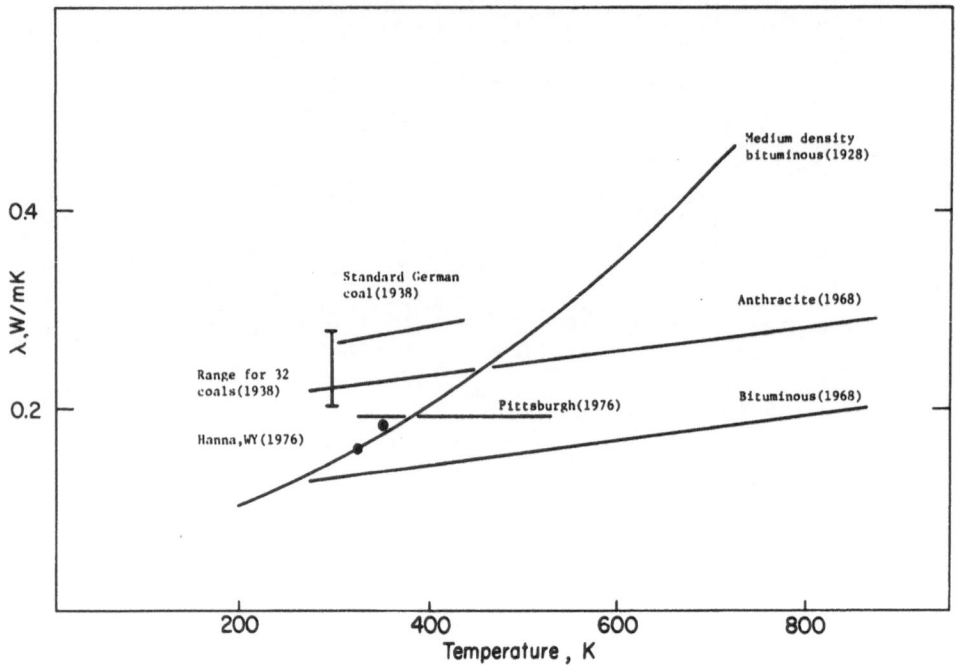

Figure 2. Thermal Conductivities of Coals.

is possible. Thus, the first significant flow enhancement process
is pore-dewatering of these coals. This type of pore drainage is
thought to be the first step in any drying process; pore dewatering
apparently is the first essential mechanism in the promotion of
methane drainage from coals.

Heating the coal is one possible approach to accomplish
pore dewatering. Drying subbituminous coals leads to the formation
of shrinkage cracks. Thus the removal of moisture changes the flow
regime from one dominated by pores to that of one dominated by
cracks. This process is known to lead to an increase in flow permeability, the volume of the gas moving in unit time through a fixed volume of coal caused by a pressure gradient, of approximately
two to three orders of magnitude. It is obviously essential that
the dewatering (drying) and subsequent shrinkage of these candidate
coals be understood if one is to learn how to optimize the underground processes.

HEAT TRANSFER THROUGH COALS

Figure 2 is a compilation of reported values[8-12] of thermal
conductivities (λ) of various coals. It is clear from the figure

that the differences in the reported values for λ of virgin (i.e., unheated) coals are not large and appear to be independent of origin and rank. Not included in the figure are the room temperature data that we obtained on the coal samples from the open pit mine at Farmington, NM; however, these too fall in the 0.2 W m^{-1}K^{-1} vicinity that seems to be characteristic for λ of coal at about 300 K. It is seen from the figure that there is no simple relationship between temperature and λ.

The behavior of the thermal conductivities of coals with temperature gives us a clue to the mechanism of heat transport within coal. The thermal conductivity of a solid (neglecting higher order terms) is

$$\lambda = \lambda_{el} + \lambda_1,$$

where λ_{el} is the electronic transport term and λ_1 is the lattice component. The electronic term is evaluated from the Weidemann-Franz law by

$$\lambda_{el} = L\,\sigma\,T.$$

L is a constant and σ is the electrical conductivity. For coals, σ is very low and λ_{el} is $\ll \lambda_1$ so $\lambda \sim \lambda_1$. Peierls[13] showed, in 1929, that λ_1 goes as 1/T above the Debye temperature, θ_D. Since there are probably a family of Debye temperatures for coal, let us consider coal by analogy with graphite which has an effective Debye temperature just above room temperature (this is stated this way because θ_D is different for the a-b plane and for the c-axis direction). Thus above 300 K the observed thermal conductivity for coal should drop with increasing temperature. But it doesn't, it goes up instead. One is forced to the conclusion that the classical rate determining mechanisms for thermal conductivity are not valid for coal- or perhaps we should start thinking in terms of coal systems. One mechanism that is consistent with the experimental observations is that of convection within the pores and cracks of the coal itself. It is postulated that in addition to the gases normally present in the coal pores, volatilization of the organic components of the coal provides a continuing source of gaseous matter with increasing temperature thus generating a system in which the flow of mass was really caused by the flow of heat and vice versa. Therefore it would be expected that steady state measurements of λ for coal done as a function of temperature would reflect an ever increasing convective heat transfer contribution which would show up as an increase of λ with increasing T -- which is what we see in Figure 2. Modelling of the coal gasification in terms of combined heat and mass transfer is consistent with the described heat transfer mechanism.

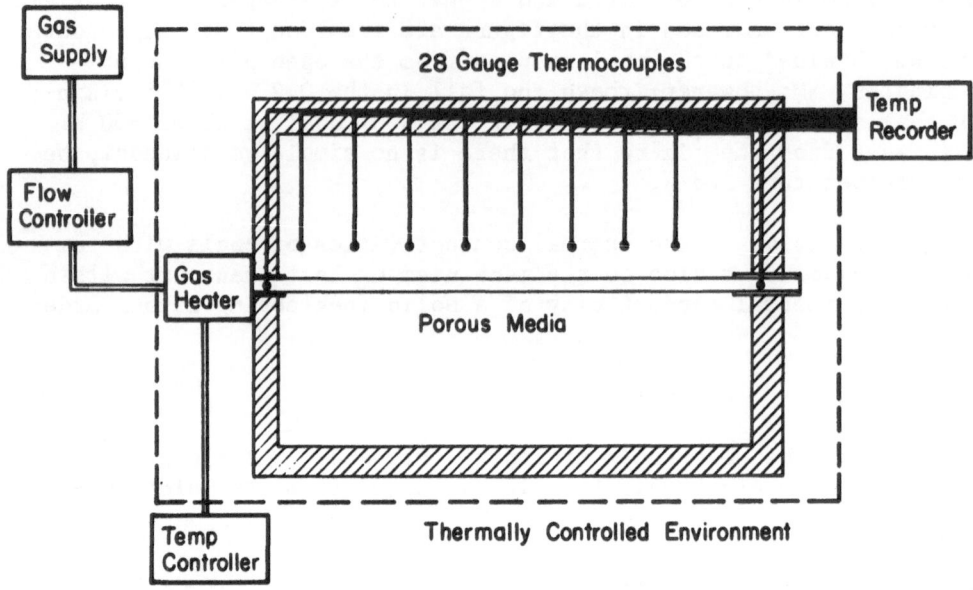

Figure 3. Schematic of Apparatus for Combined Heat and Mass
          Transfer Experiments.

EXPERIMENTAL

The experimental and calculational tasks are to describe the
states of flow that occur through media as the media are altered by
the drying processes. Although several geometries may be possible,
we chose to work with heating through a narrow channel bored within
the center of a candidate sample. Figure 3 is a schematic of the
experimental arrangement. The entire sample is bounded by a high-
flow-resistance material so that the only mass transfer into the
specimen is through the axial channel and the only transfer path
out is also through the exit of that channel. This would be somewhat analogous to a low flow-resistance link, (for example, an ex-
plosively driven penetrator channel) connecting two vertical bores
in a coal seam. It represents the type of geometry that might be
best for the two-stage process.

Two different media were selected for study. The ideal material for such testing of the flow codes would have a highly homogeneous and well characterized permeability and porosity. Moreover,
it should exhibit known moisture content and moisture transfer

rates as well as heat transfer rates. It was decided that these properties would be best approached by using well known ceramic materials which could be cast in place around a central rod and around a series of temperature and pressure transducers. The rod would be removed to form the central channel. The relative dimensions and thermocouple positions of the first ceramic test section are shown in Figure 3. This specimen used highly porous gypsum, $CaSO_4 \cdot 2H_2O$, as the ceramic.

Experiments consisted of passing gases into the block at fixed inlet temperatures and at known flow rates. In this way, a known and controlled heat flux was introduced into the specimen and then, using thermocouples, the time dependent spatial distribution of that thermal energy was determined. Simultaneous heat and mass transfer with convective terms that caused the predominate effects were determined in this manner.

Other blocks of similar dimensions made from Southwestern subbituminous coals were prepared. In these cases, the central channel, with length/diameter ratio of 50, was bored through the specimen. Thermocouples were inserted into holes bored through the outer epoxy insulating sheath and then into the coal for a predetermined distance. After the thermocouples were inserted, holes were filled and sealed with additional epoxy. Both types of experiments were photographed using X-ray techniques to ascertain the exact positions of various transducers and flow channels.

MODELLING THE COMBINED HEAT AND MASS TRANSFER

The calculations to describe heat transfer in these porous media were bounded by these particular flow geometries. These calculations were performed using a generalized finite-element heat conduction program, PANCHO. This code, which has been described by Vanderborgh et al,[14] describes the simultaneous flow and heat transfer in semiporous materials. The analytical approach specifies an estimated axial pressure distribution along a flow channel wall; this distribution is used as a boundry condition for the Darcy porous flow calculation. Conservation of mass is then invoked to adjust the flow in the central channel and the resulting wall pressure. The new wall pressure is used in succeeding iterations until the net flow in and out of the porous wall is zero, and the axial pressure distribution is converged. The flow solution is then included in the convective heat transfer terms, $C_p(V_r dT/dr + V_z dT/dz)$, which are used in the finite element heat conduction solution.

The model is also capable of extension to the following cases:
- inclusion of non-isotropic and spatially varying
  properties.
- inclusion of terms due to heat sources and sinks due
  to condensing and vaporizing fluids, including steam.
- inclusion of terms that compensate for mass sources
  and sinks due to the above.
- inclusion of fixed external pressure and temperature
  boundary conditions instead of sealed, insulated
  surfaces.
- inclusion of temporal variation of properties caused
  by changes in flow resistivities.

RESULTS

Some of the early results from our combined heat and mass
transfer experiments are shown in Figures 4 and 5. Figure 4 shows
the thermal profiles measured in the gypsum block and compares
those values to results predicted using the combined heat and mass
transfer code, PANCHO. The mass transfer boundary, epoxy, and
ceramic interface are shown. As indicated, the temperatures peak
at this interface even though the thermal conductivity of the epoxy
is considerably higher than that for the ceramic test specimen
(0.48 W/m-K). These results show the final, steady state behavior
of this system. Currently the code also depicts the dynamic thermal changes and can be used to study semiporous systems that have
different mass transfer resistances during the course of the
system.

Studies were also done to study the effects of drying ceramic
blocks. Although it would have been feasible to introduce a fixed
amount of water initially into the ceramic system, the difficulties
of accomplishing this without adversely influencing the physical
properties of the ceramic are apparent. Then, too, one would have
to model the effects of a two-phase flow (gas and liquid), greatly
increasing the complexity of the analytical model. Consequently
the inverse experiment, that of wetting, was done. In this case
liquid water was pumped at known, low rates (5 ml/min) through the
inlet heater (400 K) and into the block which was maintained at a
sufficiently high temperature that steam condensation should predominate. Results of this experiment were well predicted by the
PANCHO code. First heating effects were concentrated at the inlet
side of the block, this resulted in not only heat deposition but
also partial blockage of the mass transfer channels. As a direct
consequence, the steady state experiment showed that the exhaust
side of the specimen saw the highest temperature.

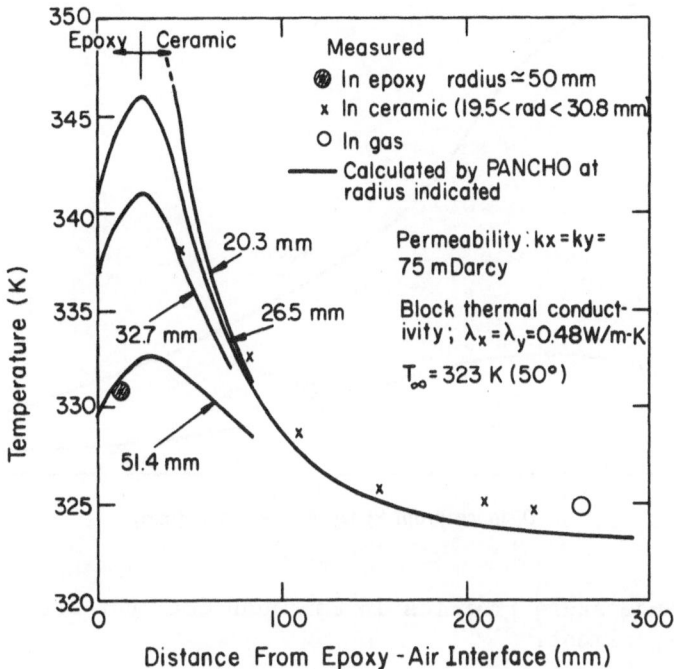

Figure 4.   Temperature Profiles in $CaSO_4 \cdot 2H_2O$ Experiment.

Results on coal specimens are shown in Figure 5. These studies show that initially the heating removes moisture to cause a permeability increase in the inlet region. Indications are that the permeability increases from the initial approximately 0.1 md (milli-darcies) to approximately 10 mD. This permeability variation changes the heat transfer mechanism from one dominated by thermal conductivity variations to one dominated by mass transfer.

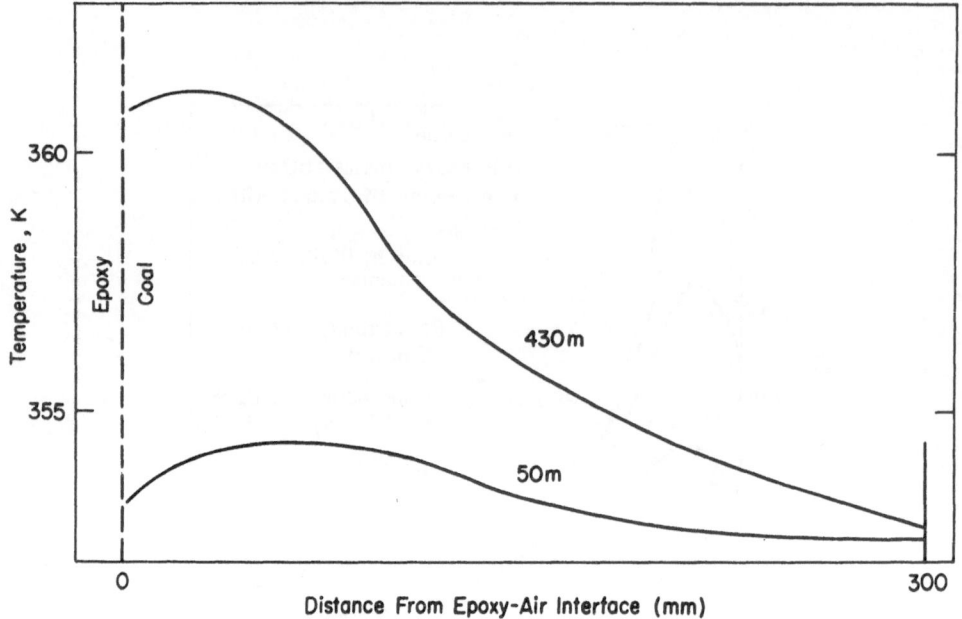

Figure 5.   Temperature Profiles in San Juan Coal Hot-Gas Drying
            Experiment.

SUMMARY

The temperature-time histories show clearly that heat transfer
in these porous materials results almost entirely from convective
heat transfer. The range of permeability studied here, near 3.5 mD,
is what we expect to find in the middle of the drying-pyrolysis processing of these Southwestern coals. These data strongly suggest
that once this pyrolysis process begins, convective heat transfer
can be efficiently used through out the resulting semi-porous media.

REFERENCES

1. J. W. Shomaker et al., "Strippable Low-Sulfur Coal Resources
    of the San Juan Basin in New Mexico and Colorado," Memoir 25,
    NMIMT, State Bureau of Mines and Mineral Resources, Socorro,
    NM, 1971.

2. G. R. B. Elliott, "Chemically Mining Coal," Mining Engineering,
    Sept. 1973, p. 73.

3. D. I. Mendeleev, Severvy Vestnik (St. Petersburg), 1888, No. 8,
    Sect. 2, 27-73; No. 9, Sect. 2, 1-35.

4. R. Loison and J. Venter, "Underground Gasification, Russian,
    Italian, Belgian and French Experiments," Centre d'etudes et
    recherches des charbonnages de France (Cerchar), Paris, France,
    and Instit National de l'Industrie Charbonniere (Inichar),
    Liege, Belgium, Feb. 1952, 37 pp.

5. G. O. Nusinov, "Underground Gasification of Coal," Moscow-
    Lenigrad Gosstoptekhnisdat, 2nd ed., 1941, 140 pp.

6. J. J. Dowd et al., U. S. Bur. Mines, Rept. Invest. No. 4164,
    (1947), 62 pp.

7. G. C. Cambell et al., "Preliminary Investigation of Under-
    ground Coal Gasification at Hanna, Wyo," U. S. Bur. Mines,
    BM-TPR-82, Oct. 1974.

8. Proceedings of the Second Conference on Bituminous Coal, 2
    (1928).

9. W. Fritz and H. Diemke, "The Thermal Conductivity of Natural
    Coals and Coke," Feuerungstechnik (Berlin) 27, 1939.

10. A. A. Agroskin, E. I. Goncharov, and L. V. Lovetskii, "Combined
    Determination of the Thermal Characteristics of Solid Fuel
    During High Temperature Pyrolysis," Khim. Tverd. Topl., No. 4,
    (1968).

11. R. U. Acton, "Thermal Conductivity of Hanna, WY Coal Silt
    Materials," Sandia Laboratories, NM (personal communication).

12. R. P. Tye, A. O. Desjarlais, and J. S. Singer, "Thermal Conduc-
    tivity Behavior of Pittsburgh Seam Coal," Proceedings of the
    15th International Conductivity Conference, Ottawa, Canada
    (1977).

13. R. Peierls, "Zur Kinetischen Theorie der Wärmeleitung in
    Kristallen," Ann Physik, 3 (1929).

14. N. E. Vanderborgh, E. M. Wewerka, J. M. Williams, J. P. Bertino,
    and G. E. Cort, "Heat and Mass Transfer Through Southwestern
    Coals," Third Underground Coal Symposium, Lake Tahoe, CA (1977).

SESSION J   DIFFUSIVITY, CONTACT CONDUCTANCE

Session Chairmen: H.E. Schmidt
Euroatom, W. Germany

J.G. Hust
National Bureau of Standards
Boulder, Colorado

THERMAL DIFFUSIVITY OF POTASSIUM MAGNESIUM TITANATES HAVING A HOLLANDITE STRUCTURE

V.V. Mirkovich, D.H. Quon and T.A. Wheat

Canada Centre for Mineral and Energy Technology

Dept. of Energy, Mines and Resources, Ottawa, Canada

ABSTRACT

As part of a wider program to develop solid-state ionically conducting materials for potential application in energy storage and conversion systems, the thermal diffusivity of a series of K-ion conducting electrolytes having the composition $K_xMg_{x/2}Ti_{8-x/2}O_{16}$, where $1.6 \leqslant x \leqslant 1.8$, was determined between room temperature and 700°C using a transitory heat-pulse technique. It was found that the diffusivity decreased with an increase in x at any given temperature. The rate of decrease of the thermal diffusivity with increasing temperature was found to be dependent on composition within the composition range of x from 1.6 to 1.8.

INTRODUCTION

In recent years, there has been an increasing interest shown in a variety of ionically conducting solid-state electrolytes as they offer considerable promise for exploitation in various applications. At present, over thirty ceramic materials are known to be ionically conductive (1) and some of the more important members of this group of materials include the refractory anionic conductors such as doped ZrO_2 that are exploited in fuel cells and the cationic conductors such as $AgRb_4I_5$ (Ag conductor) and beta-alumina (Na conductor) that are under development for use in a solid state coulometer and a high energy-density battery respectively.

One of the attractive features of this group of materials is that they are generally non-corrosive, unlike the more common fused-salt systems. In addition, they also simplify the design of many devices by acting as a structural component as well as an electrolyte, e.g., ZrO_2 in high-temperature fuel cells and beta-alumina in Na-S cells for the development of high energy-density batteries for automotive and other uses. The properties and other applications of solid-state electrolytes have recently been reviewed by Worrell (2).

Recently, the ionically conducting properties of a number of materials having the hollandite structure (essentially a substituted alpha-MnO_2 (3)) have been recognized. Crystallizing in the tetragonal system, these materials develop an open-tunnel structure parallel to the c-axis through which group I cations diffuse rapidly (4). As part of a program to determine the properties of this potentially important class of materials, a series of potassium magnesium titanates was synthesized and characterized and the thermal diffusivity as a function of composition was determined.

EXPERIMENTAL PROCEDURE

Material and Specimen Preparation

The technique for producing single-phase, sintered discs of $K_xMg_{x/2}Ti_{8-x/2}O_{16}$ has been reported in detail elsewhere (5). In essence, a wet-chemical process was used in which a $TiO(OH)_2$ gel was precipitated from 99.5% $TiCl_4$ by NH_4OH and washed to less than 10 ppm Cl^- with distilled water (Figure 1). By control of the pH of the system, the gel was subsequently dispersed using $CH_3.COOH$ and doped with the appropriate amount of $CH_3.COOK$ and $(CH_3.COO)_2Mg$ solutions prior to spray freezing and freeze drying. The dried amorphous powder was finally calcined to produce a finely divided, highly reactive single-phase raw material that could be readily fabricated into sintered bodies having densities greater than 95% of theoretical using conventional cold pressing and sintering techniques. In order to avoid the possible formation of partially reduced material during the firing stage, all the specimens were fired in a flowing O_2 atmosphere.

Measurement of Thermal Diffusivity

The boundary conditions for determining the thermal diffusivity of materials by a radial heat-flow technique have been discussed elsewhere (6) and the restrictions that apply to the measurement of electrically non-conductive materials in a normal atmosphere have also been reported (7). Briefly, the method requires the measurement of the temperature at the centre and at a point near the surface of a nominally infinite cylinder of material. With a constant rate of heating of the surface, the temperature differen-

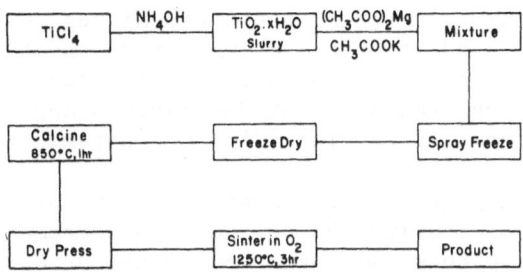

Figure 1. General flow sheet
for the preparation of samples.

tial is plotted until it reaches a maximum. An equation derived
for this model relates the increase in the dimensionless temperature differential to a dimensionless factor which includes thermal
diffusivity, time and the geometry of the cylinder.

A schematic diagram of the apparatus is given in Figure 2.
Experimentally, the sample column is formed by stacking five or
seven discs to give a cylinder at least 60 mm long and 25.4 mm in
diameter (G). To reduce axial losses, a cylindrical ceramic insulator (D) is placed at each end of the column which is held together
by two pairs of stainless steel wires (not shown) binding the whole
assembly together. A 0.51-mm Nichrome heating wire (F) is subsequently non-inductively wound in a helix around the sample column
and partially over the cylindrical insulators and covered with a
thin layer of refractory cement to restrict the movement of the

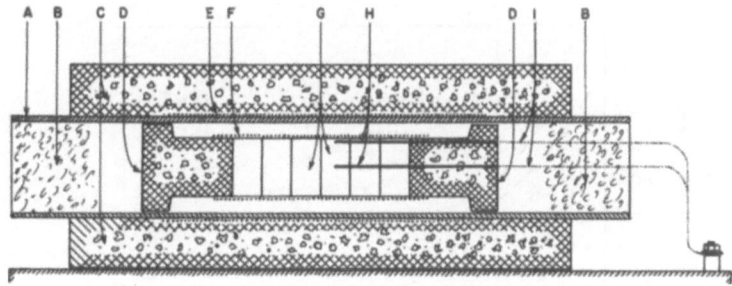

Figure 2. Schematic diagram of apparatus for measuring thermal
diffusivity.

A - mullite tube; B - fibrous ceramic insulation; C - ceramic
insulation, D - cylindrical ceramic insulator; E - furnace
heating element; F - Nichrome heater wound directly on samples;
G - sample discs; H - thermocouple holes; I - thermocouples.

heater wire and avoid electrical short-circuiting. Two thermocouple holes (H) extend through one insulator, the back-up discs
and halfway into the central sample disc. The whole assembly is
contained within a horizontal non-inductively wound electric furnace and the ends are closed with a fibrous ceramic insulation (B).

With this configuration, the radial temperature difference
was approximately 8°C during the period of measurement which lasted
some 40 sec. During this time, the sample temperature increased
12°C.

RESULTS AND DISCUSSION

The variation of thermal diffusivity with temperature as a
function of composition is given in Figure 3 and Table 1. The
smoothed curves were obtained from the original data by inspection
and represent seventy individual determinations for the 1.6 and
1.8 material and thirty five measurements for the 1.7 material.
The standard deviation of the data was determined to be 0.0001
cm^2/sec. In all cases, the precision of the data was verified by
determining the diffusivity on both the heating and cooling cycles.

As expected, it can be seen that the diffusivity decreases
with increasing temperature. However, in comparison with the data
for pure oxide ceramics reported in the literature (8), the decrease
is rather small, e.g., the 1.6 material decreases from 0.0110 to
0.0077 cm^2/sec and the 1.8 material decreases from 0.0097 to
0.0071 cm^2/sec. In addition, it can be seen that, at the lower
temperatures, the decrease is greatest for material having the
highest concentration of TiO_2, i.e., the 1.6 composition.

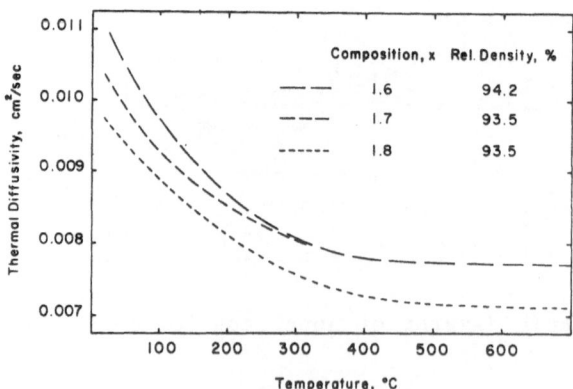

Figure 3. Thermal diffusivity of
$K_xMg_{x/2}Ti_{8-x/2}O_{16}$ as a function
of temperature and composition.

TABLE 1

Smoothed Values of Thermal Diffusivities for $K_xMg_{x/2}Ti_{8-x/2}O_{16}$

t°C	k[cm²/sec]		
	X=1.6	X=1.7	X=1.8
25	0.0110	0.0103	0.0097
50	0.0105	0.0099	0.0094
75	0.0101	0.0096	0.0092
100	0.0098	0.0093	0.0089
125	0.0095	0.0091	0.0087
150	0.0092	0.0089	0.0085
175	0.0089	0.0087	0.0083
200	0.0087	0.0085	0.0081
250	0.0084	0.0083	0.0078
300	0.0081	0.0081	0.0076
400	0.0078	0.0078	0.0073
500	0.0078	0.0078	0.0072
600	0.0077	0.0077	0.0071
700	0.0077	0.0077	0.0071

1.6   1.7   1.8

Figure 4. SEM micrographs of the fracture surface of sintered material for different values of x.

Although it is impossible to guarantee that samples will be fabricated to have exactly the same relative density, an attempt was made to eliminate the effects of porosity on diffusivity by so selecting the instrumented samples in the centre of the column that the data were obtained from specimens having essentially equal porosities. In addition, the samples were also chosen to have a similar grain size (Figure 4). With these two restrictions, it was impossible to use the specimens having the highest fired density for each composition. Consequently, a compromise was reached in which samples having a grain size of 180 ± 15 μm and a relative density of 94 ± 0.5% were used.

Although it would be expected that the highest thermal diffusivity would be obtained for material having the highest density (1.6 material, 94.2% rel. den.), the differences in diffusivity between each composition cannot be attributed to **density differences alone.** It can be seen that the data for the 1.8 and 1.7 materials are significantly different from each other yet both samples had a relative density of 93.5%. Hence, it is concluded that the differences shown in Figure 3 are not a reflection of the minor density differences but are dependent on the compositional differences.

In contrast to the data for the thermal diffusivity of $MgTi_2O_5$ recently reported by Siebeneck *et al.* (9) using the laser-flash technique, no "hysteresis" effects were observed for the present materials on cycling from room temperature to 700°C and back again. This is possibly due to a high thermal-shock resistance of these particular hollandites but it is more probably due to the different measuring technique used in the present work which does not expose the sample to such a severe thermal stress during measurement.

CONCLUSIONS

The thermal diffusivity decreases with increasing temperature and with a decreasing amount of TiO_2 in these hollandites. Over the temperature range of measurement, the diffusivity was found to be the same on both the heating and cooling cycle suggesting that the specimens did not suffer any degradation due to either the thermal cycling or the heat pulses of the individual measurements.

REFERENCES

1. R.A. Rapp and D.A. Shores, p 123 in Physicochemical Measurements in Metals Research. Part 2. Edited by R.A. Rapp. Published by Wiley-Interscience, New York. 1970.

2. W.L. Worrell, "Electrochemistry of Ceramic Electrolytes", Amer. Ceram. Soc., Bull., 425-433, 53 (1974).

3. A. Byström and A.M. Byström, "The Crystal Structure of Hollandite, the Related Manganese Oxide Minerals, and $\alpha-MnO_2$", Acta Cryst., 146-154, 3 (1950).

4. T. Takahashi and K. Kuwabara, "Preparation of Sintered Oxides with Hollandite-Type Structure ($K_xMg_{x/2}Ti_{8-x/2}O_{16}$) and Their Ionic Conduction", Nippon Kagaku Kaishi, 1883-1887, 10 (1974).

5. D.H. Quon and T.A. Wheat, "Preparation of a Ceramic Electrolyte in the System $K_2O-MgO-TiO_2$", Mineral Sciences Laboratories Report ERP/MSL 77-231(R), August 1977. Available from the Canada Centre for Mineral and Energy Technology, Department of Energy, Mines and Resources, Ottawa, Canada.

6. V.V. Mirkovich, "Thermal Diffusivity Measurement of Armco Iron by a Novel Method", Rev. Sci. Inst., 560-565, 48 (5) (1977).

7. V.V. Mirkovich, "An Apparatus for Measuring Thermal Diffusivity in Air", CANMET Report 77-21, December 1976. Available from the Canada Centre for Mineral and Energy Technology, Department of Energy, Mines and Resources, Ottawa, Canada.

8. Y.S. Touloukian, R.W. Powell, C.Y. Ho and M.C. Nicolaou, "Thermophysical Properties of Matter, Volume 10: Thermal Diffusivity", IFI/Plenum Press, New York, 1973.

9. H.J. Siebeneck, D.P.H. Hasselman, J.J. Cleveland and R.C. Bradt, "Effects of Grain Size and Microcracking on the Thermal Diffusivity of $MgTi_2O_5$", Jour. Amer. Ceram. Soc., 336-338, 60 (7-8) (1977).

THERMAL DIFFUSIVITY OF PURE IRON AND DILUTE IRON ALLOYS

FROM 500 TO 1250 K

R. Taylor and C.M. Fowler

Department of Metallurgy, University of Manchester/UMIST

Manchester, England

ABSTRACT - The thermal diffusivity has been measured from 500 -
1250K on samples of Armco iron, the N.B.S. standard reference
material S.R.M. 734 and a high purity iron designated AO.
Controlled alloying additions of Mo, Ni, Cr, V, Mn, Si, Co in the
concentration range $\frac{1}{2}$-2% have been made to the AO iron and the
diffusivities of these measured over the same temperature range.
The results are evaluated and discussed.

I. INTRODUCTION

The thermal transport properties of nominally pure iron have
been extensively measured and reviewed. Armco iron in particular
has received a great deal of attention. Powell [1] has reviewed
significant work up to 1962 and thermal conductivity data from a
round robin involving eight laboratories has been reported by
Lucks [2]. Recently Hust and Giarrantano [3] have reviewed
thermal conductivity on iron and propose N.B.S. electrolytic iron
SRM 734 as a thermal conductivity standard. Data on Armco [2],
SRM 734 and an ORNL high purity iron [4] are compared from 100 -
1000K and above about 500K the results appear to lie on the same
curve. Specimens of an Armco iron, and the SRM 734 have been
obtained and an AO iron, also of high purity, from the British
Iron and Steel Research Association (B.I.S.R.A.) was also
available.

Whilst a great deal of evaluation of the thermal conductivity
of irons of various purities has been undertaken e.g. [4] no
comparable investigation of thermal diffusivity has been made.
When the same apparatus is used for the measurements, a more valid
comparison of the properties can be made so these three different

irons were measured. Furthermore since the effect of alloying
additions is little understood samples of the AO iron were alloyed
with specific additions of a single element using concentrations
in the range typically found in low alloy steels. The diffusivity
of these alloys was measured over the same temperature range
(500 - 1250 K). It is hoped that this provides useful information
of the affect of specific impurities on the diffusivity of iron.

II. SPECIMEN CHARACTERISATION

Armco iron is a general term used to describe a fairly pure
iron made to a nominal specification. Hence small material
differences may be expected between different batches. The Armco
iron used in this investigation was in the form of a 2.5 cm
diameter bar and is presumed to have a specification similar to
that described by Lucks. The N.B.S. standard reference material
was supplied in the form of a bar 0.64 cm diameter x 10 cm long.
(SRM 734-S) Full characterisation details are listed by Hust
and Giarrantano. From each of these, specimens 0.640 cm diameter
and lengths in the range 0.140 to 0.180 cm were machined for
measurements of diffusivity. The AO high purity iron was
available in the form of a bar 5 cm square by 60 cm long.
Specimens of this material were 1 cm diameter and similar lengths.
Analysis of these three irons are listed in Table 1.

TABLE 1

Compositions of the irons measured.
Values quoted are in weight percent

Element	NBS SRM 734	Armco Iron	AO Iron
C	0.0067	0.015	0.025
Mn	0.0057	0.028	0.0056
P	0.0025	0.005	0.002
S	0.0059	0.025	0.011
Si	0.0080	0.003	0.001
Cu	0.0058	0.04	<0.01
Ni	0.041	-	<0.01
Cr	0.0072	-	<0.01
V	0.0006	-	<0.01
Mo	0.005	-	<0.01
Co	0.007	-	<0.01
Ti	0.0006	-	<0.01
As	0.0002	-	<0.01
Al	0.0007	-	<0.01
B	0.00013	-	<0.01
Pb	0.00002	-	<0.01
W	-	-	0.05
N	-	-	0.0025
O	-	-	0.105

A series of alloys were made using the AO iron as a base material in a high frequency induction furnace equipped with a melting hearth. Melting was carried out in a high purity argon atmosphere inside a pure recrystallised alumina crucible. The alloying additions were introduced by inserting the calculated weight inside a small hole in the AO rod which was then fitted with a tight fitting lid. This minimised loss of alloy. In each case the alloy was tested in the furnace cooled condition. Full analyses were not carried out on these alloys but the determinations of the alloy additions are listed in Table 2.

TABLE 2
Alloy Irons

Alloying Addition.	Co	Ni	Mn	Si	Cr	V	Mo
Concentration wt %.	1.93	0.80	1.13	2.18	3.74	1.92	0.49

III THERMAL DIFFUSIVITY APPARATUS

Thermal diffusivity was measured using the flash method [5], about which there is copious documentation which will not be repeated here. The Mark I version of the equipment has already been described [6] but the improved version will be briefly summarised here. A schematic diagram of the equipment is shown in figure 1. The sample mounted in a tungsten or graphite sample holder is held inside a graphite susceptor which is heated inside the induction coil. The heat pulse is supplied from a 100 J solid state ruby laser. Radiation from the lower face of the sample is collected using the lens and mirror system shown in the figure and focussed onto a lead sulphide detector. The amplified output from the detector is fed into a digital data acquisition system (D.D.A.S.).

The D.D.A.S. contains a minicomputer with 16,000 words of core. Signals are fed in from a high gain four channel amplifier through a 15 bit Analogue to Digital convertor. The A.D.C. record 1024 voltage readings at 1 millisecond intervals. From the input wave form which is routinely outputted into an oscilloscope, the computer is programmed to determine the datum voltage level, the maximum voltage, the $t_{\frac{1}{2}}$ value, the signal ratio at $10t_{\frac{1}{2}}$ and $t_{\frac{1}{2}}$. Because of the intense electrical interference following the laser flash the voltage level at zero time is determined before the flash by averaging the preceding 15 data points. A crystal time clock operating in conjunction with the ADC is connected to a photodiode which records the laser flash and the next sampling

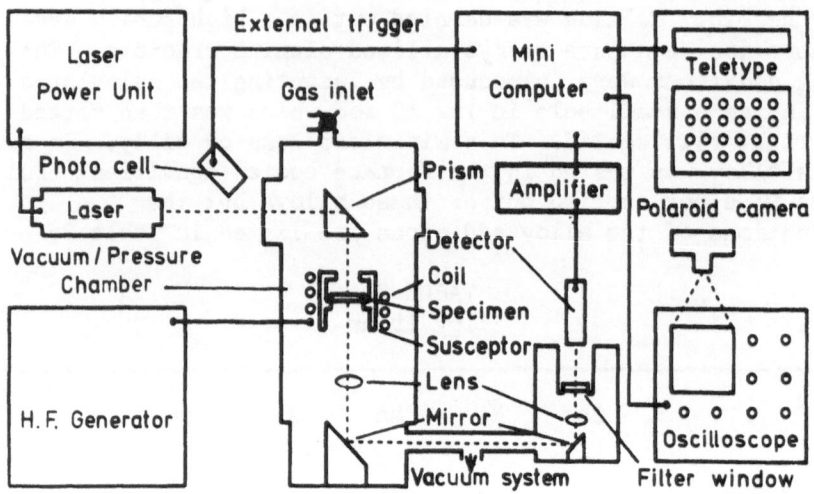

Fig. 1.   Schematic diagram of equipment.

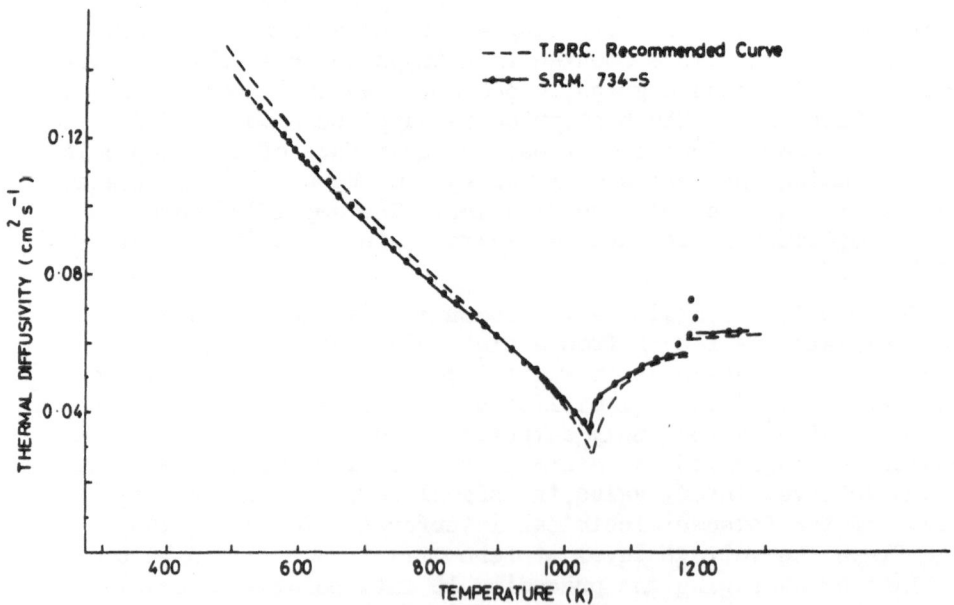

Fig. 2.   Thermal diffusivity of S.R.M. 734-S.

point, from which subsequent calculations are based. This is
added to the computed $t_\frac{1}{2}$ value. Calculated outputs from the
waveform are printed out on a teletype which can also be used to
enter instructions or programmes.

As a further addition to the data reduction the heat loss
analysis listed by Cowan [7] has been entered into the computer
which then calculates ω/π^2 from the previously determined heat
loss ratio. Finally the experimental thermal expansion for iron
was programmed and entered into the computer whereupon by typing
in the measurement temperature, via the teletype, the experimental
thermal diffusivity was calculated.

The temperature of measurement was recorded using a Chromel/
Alumal thermocouple. Diffusivity measurements were made from
500K, the lower limit of sensitivity of the PbS detector, to
1250K, the upper limit of reliable operation of the thermocouple.

A $t_\frac{1}{2}$ value is calculated to a resolution of 0.1µs and a
probable precision better than 10µs. Thus the experimental
uncertainty is 0.03% for $t_\frac{1}{2}$ > 30 ms. The length is determined to
± 0.005 mm which for a sample length of 1.5 mm corresponds to a
precision of 0.3%. Corrections to ω/π^2 because of heat losses
never amounted to more than 15% and errors in this whilst more
difficult to quantify, probably amount to less than 0.25%. Hence
the precision of measurement is better than 1%. Accuracy is more
difficult to determine since it involves the assumption of no
systematic errors but we believe it to be much less than twice the
estimated precision.

IV RESULTS AND DISCUSSION

Ten sets of data have been obtained over the temperature
range 500 - 1250K. Three of these are on 'irons' of varying
levels of purity. The remaining seven are on alloyed irons where
the effect of a specific element added to one of the aforementioned irons may be considered. These three irons we
considered together first and the alloys considered separately.

1. Pure Irons

The results for the N.B.S. SRM 734-S, Armco Iron and the A.O.
iron are shown in figures 2 and 4 respectively. Also plotted on
each curve is the T.P.R.C. recommended thermal diffusivity curve
for Armco iron [8]. Although the results for Armco iron and SRM
734-S are plotted separately in figures 2 and 3 the individual
data points for the two materials could just as readily be plotted
on the same curve so the two sets of results are, for all
practical purposes, indistinguishable. At 500K the results lie
some 4% below the T.P.R.C. curve but the difference decreases as

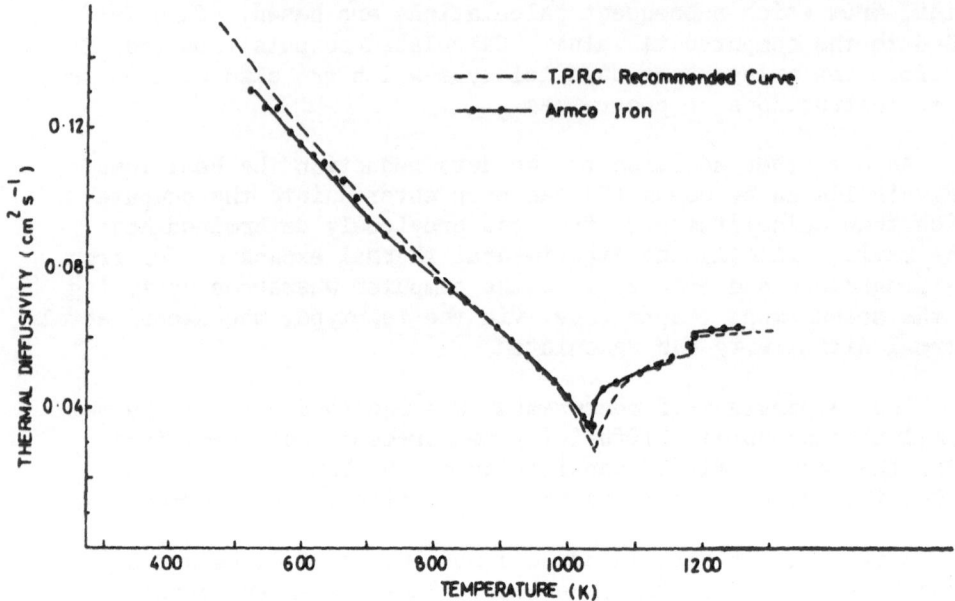

Fig. 3.   Thermal diffusivity of Armco iron.

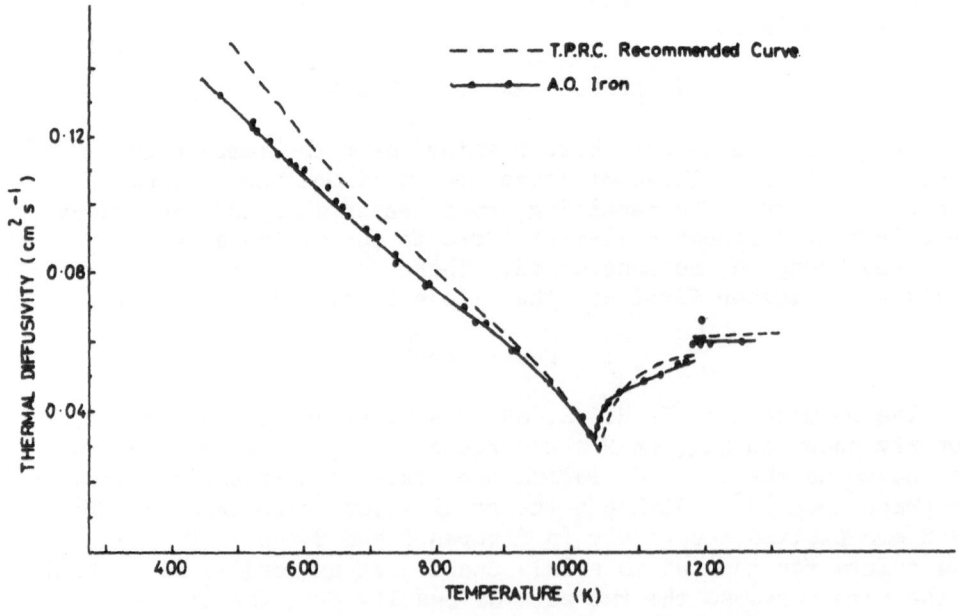

Fig. 4.   Thermal diffusivity of AO iron.

the temperature is increased until it is zero at 900K. From 900K
to the Curie point (Tc) at 1041K, the measured values lie above
the T.P.R.C. curve until at Tc the minimum in the curve occurs
at 0.0340 cm^2s^{-1} compared with a T.P.R.C. recommended value of
0.0289 cm^2s^{-1}. From 1041K to 1183, the $\alpha \rightarrow \gamma$ transition
temperature, our diffusivity results show a more pronounced
increase but the limiting value of 0.055 cm^2s^{-1} is the same.
Above 1183K our values lie some 2% above T.P.R.C. recommended
values.

The AO iron has a thermal diffusivity significantly lower
than the other two irons having a diffusivity some 7% lower at
500K and 3% lower at 900K. Above Tc this trend to a lower
diffusivity is again apparent, the limiting value at 1183K being
0.0540 cm^2s^{-1} compared with 0.0565 cm^2s^{-1} for SRM 734-S and Armco.
Even in the γ phase the diffusivity is again lower.

Armco iron is probably the most intensively investigated
material from the point of view of thermal transport properties.
The T.P.R.C. data manual records some 135 determinations of
thermal diffusivity of Armco and similar irons, with a scatter of
at least ±15% about the T.P.R.C. curve. However, in order to
delineate a curve having as many interesting features as this
some 25-30 determinations at least are required and there are
surprisingly few of these. The T.P.R.C. curve is most closely
approximated by the data of Kennedy [9] and of Cody et al [10].
Shanks et al [11][12] in a comprehensive series of investigations
obtained data some 9% below our values at 500K but their results
showed excellent agreement from 800K to 1200K. Whilst they showed
that the diffusivity of Armco is relatively insensitive to heat
treatment they achieved an increase in diffusivity by reducing
the carbon content from 0.0112 to 0.0075%. The data on Van
Craeynest et al [13] shows good agreement with our results up to
Tc but from Tc to 1183K their results are lower than ours. On
the other hand, Carter and Sidles results [14] are a good fit with
our data from 875 to 1180K but below 875K their results lie below
ours. Hence our data lie well within the scatterband of results
obtained from comparable investigations.

Some very good thermal conductivity results have been
published in the literature but any attempt to compare diffusivity
data with thermal conductivity data requires an accurate specific
heat. Such comparisons must be treated circumspectly particularly
in the temperature range 800-1250K where large changes in Cp
occur. Nevertheless using the T.P.R.C. recommended specific heat
curve [15] and converting the N.B.S. recommended thermal
conductivity curve for SRM 734-S and the measured density of
7.867 gm cm^{-3} the maximum deviation of 550K is 2.3% decreasing
to below 1% at 700K and thereafter being less than 1% up to
1000K. The N.B.S. recommended curve which is based on the low

temperature measurements of Hust and Sparks [16] for this iron
and the high temperature data of Fulkeson et al [4] on an ORNL
high purity iron is considered to be accurate to ±2% above
280K so our results show excellent agreement. Moreover our
agreement between SRM 734-S and Armco supports the comparison
drawn by Hust and Giarrantano regarding comparability at
elevated temperatures between SRM 734 and ingot iron (Armco)
as summarised by Lucks [2].

However there is a significant difference between these
two irons and the AO iron. In the vicinity of Tc the three
sets of data are similar but from Tc to 1183K the diffusivity of
the AO iron is about 6% below that of Armco and SRM 734-S and
similarly in the γ region. If we critically examine these data
in terms of the chemical analyses and metallographic structure
some reasons for the differences may be obtained. The AO iron
has a high oxygen content (0.105%) and significant levels of Mn
and S. SRM 734 has a similar Manganese content to AO and half
the sulphur level. In contrast Armco has about half the Mn and S
content of SRM 734. Unfortunately oxygen levels for Armco and
SRM 734 were not available although other more detailed analyses
of Armco [11][2] suggest oxygen levels of 0.05%. Micrographs
at x 400 of SRM 734-S, Armco iron and AO are shown in figures
5-7 respectively. All three show pools of Mn S, the largest
concentration being found in the AO iron. Armco possesses the
largest grain size and both Armco and AO iron show fine
inclusions, possibly oxide, inside the grains and at grain
boundaries. Inclusions are also present within the SRM 734
but on a much finer scale. It is difficult to quantify any
differences between the irons but is clear that the inclusion
level is higher in the AO iron than the other two. One fact
is however apparent; that even such 'pure' irons have a
substantial inclusion content which must affect the thermal
transport properties. If a diffusivity or conductivity for pure
iron is to be obtained then the impurity levels must be reduced
a great deal more.

2. Iron Alloys

Since turbulent conditions existed during melting in the
induction furnace it is assumed that the alloy was homogeneous
so the alloys were measured in the as cooled condition without
any recourse to any high temperature homogenisation anneal.
Since each alloy would also cool through the γ → α transformation
this was felt to be further reason why no heat treatment was
deemed necessary.

If we consider the alloys in terms of their respective
equilibrium diagrams [17] then they fall into three distinct
categories:

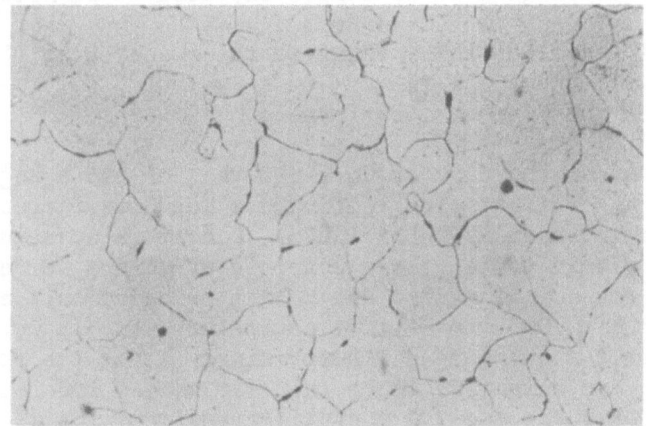

Fig. 5.   Micrograph of S.R.M. 734-S (x 500).

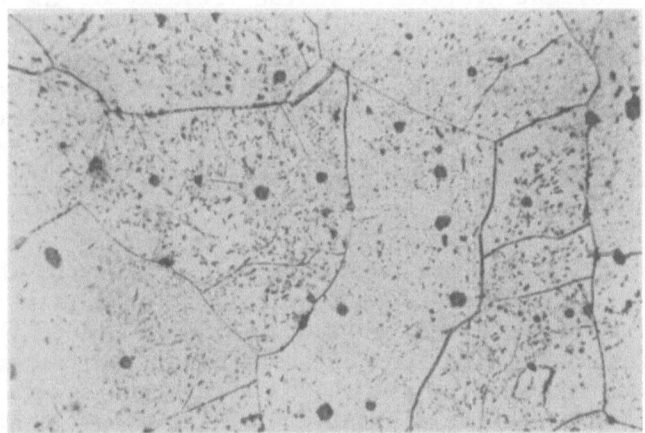

Fig. 6.   Micrograph of Armco iron (x 500).

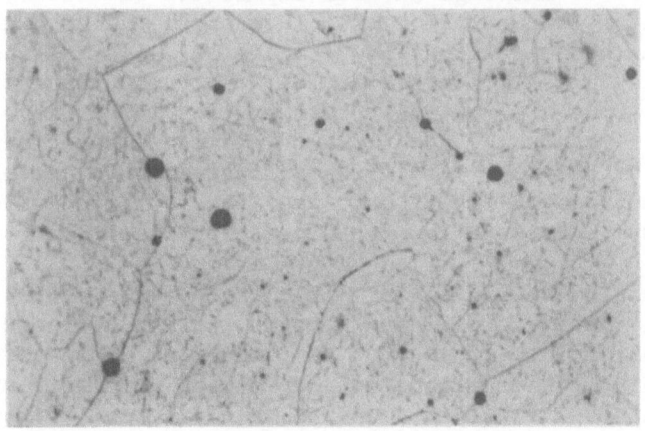

Fig. 7.   Micrograph of AO iron (x 500).

a)   Co is unique in that it is completely soluble in both α and
      γ-Fe.
b)   Mn and Ni stabilise the γ phase so the sequence of phases on
      heating will be α, α + γ, γ.
c)   Mo, Si, Cr and V are γ-loop formers.

If we consider the case of Co first then Co has a negligible
effect on the α - γ transition temperature but does increase the
Curie point Tc to about 1200K at 15%Co. Cobalt is adjacent to
iron in the periodic table, the electronic structure is $3d^7 4s^2$,
the atoms are of a similar size and Cobalt produces only a very
small increase of 0.008% in lattice parameter. [18]. Our results
shown in figure 8 show, somewhat surprisingly, that the results
during heating are slightly higher than for the AO iron by some
3% at about 600K. Because the magnetic transformation is
retarded to about 1063K, in agreement with the equilibrium
diagram, this difference increases to 7-8% at 1000K. There
appears to be a slight hysteresis in the vicinity of the α - γ
transformation which occurs at about 1183K but this again is in
accord with the equilibrium diagram. However measurements taken
during cooling from 1250K show that the diffusivity has been
reduced until at 600K it is some 4% below the AO iron curve.
Whilst it is obvious that the diffusivity of this alloy is
structure sensitive it is clear that Co has only a slight affect
on the diffusivity of iron.

Manganese and Nickel are both more soluble in γ-Fe than in
α-Fe, and of the two Nickel is more soluble in α-Fe than is
Manganese. In figure 9 we see that 0.8% Ni does reduce the
thermal diffusivity by a maximum of about 5% at low temperatures,
has no effect on Tc, and the α + γ → γ temperature occurs at
about 1155K. Yet again the diffusivity of the alloy is lower
during cooling by a maximum of about 8%. Although Ni does expand
the lattice [19][20] the magnitude of the effect is difficult to
quantify. However although Mn does expand the lattice by about
0.012% [18] at a concentration of 1.13%Mn it has a more marked
effect on the thermal diffusivity (figure 10). The diffusivity
is decreased by a maximum of about 16%, the Curie temperature is
decreased by about 22K and the transformation temperature
(α + γ → γ) is reduced by about 50°. This is as expected from
the equilibrium diagram. Again the diffusivity of this alloy is
further reduced on cooling through the transformation with some
evidence of hysteresis in the transformation itself.

The remaining four alloying elements are all more soluble in
α-Fe than in γ-Fe. Vanadium and Chromium with electronic
configurations of $3d^3 4s^2$ and $3d^4 4s^2$ are below Mn in the periodic
table, Molybdenum is a much heavier atom with an electronic
configuration $4d^s 5s$ and Silicon with an electronic configuration
$3s^2 3p^2$ has an atomic mass half the mass of the iron atom. Of

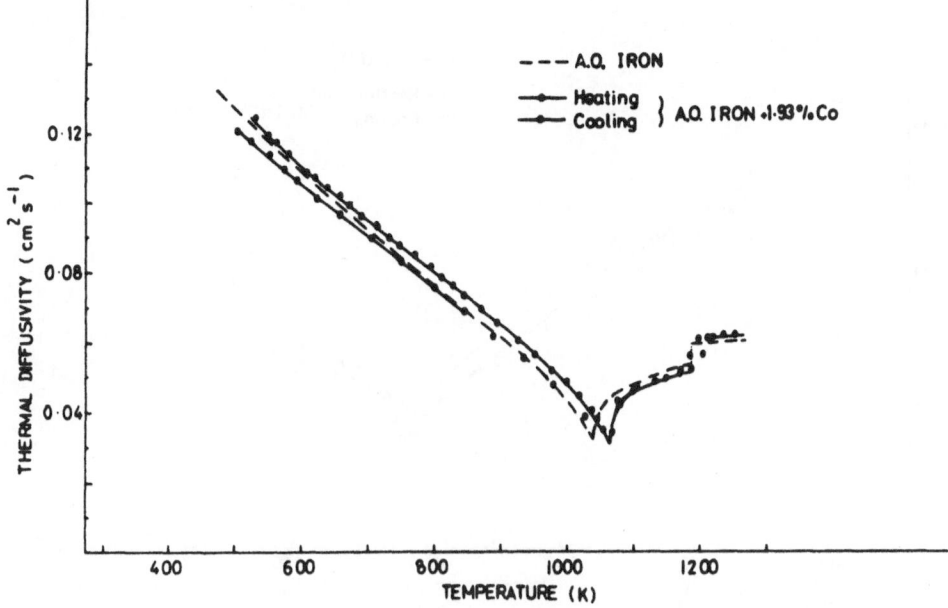

Fig. 8. Thermal diffusivity of 1.93% Co alloy.

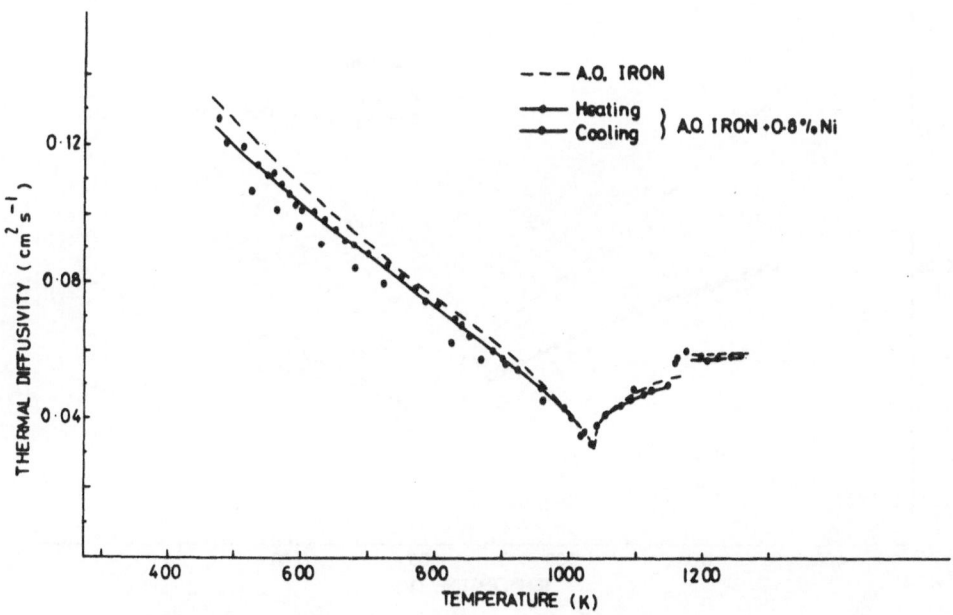

Fig. 9. Thermal diffusivity of 0.8% Ni alloy.

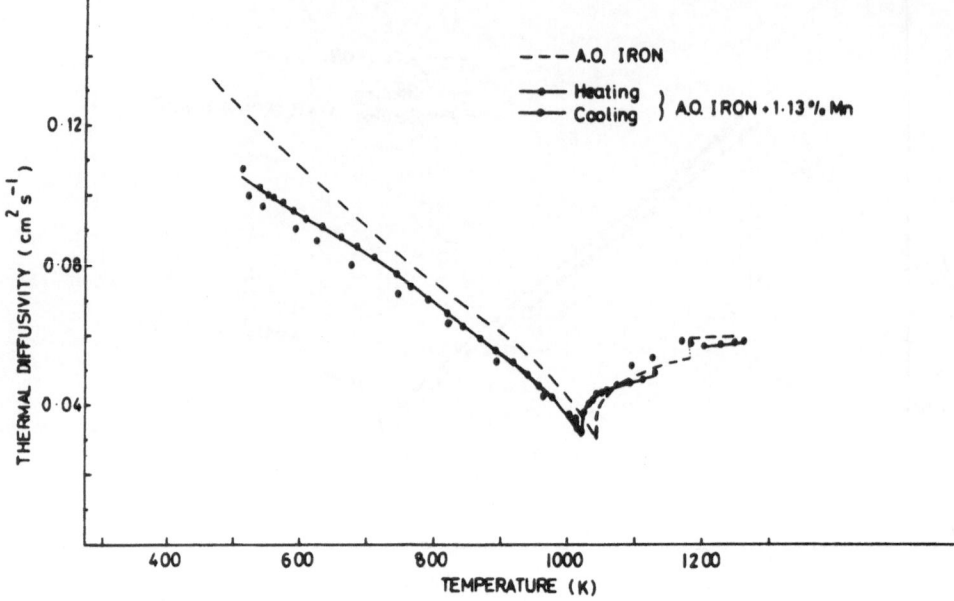

Fig. 10. Thermal diffusivity of 1.13% Mn alloy.

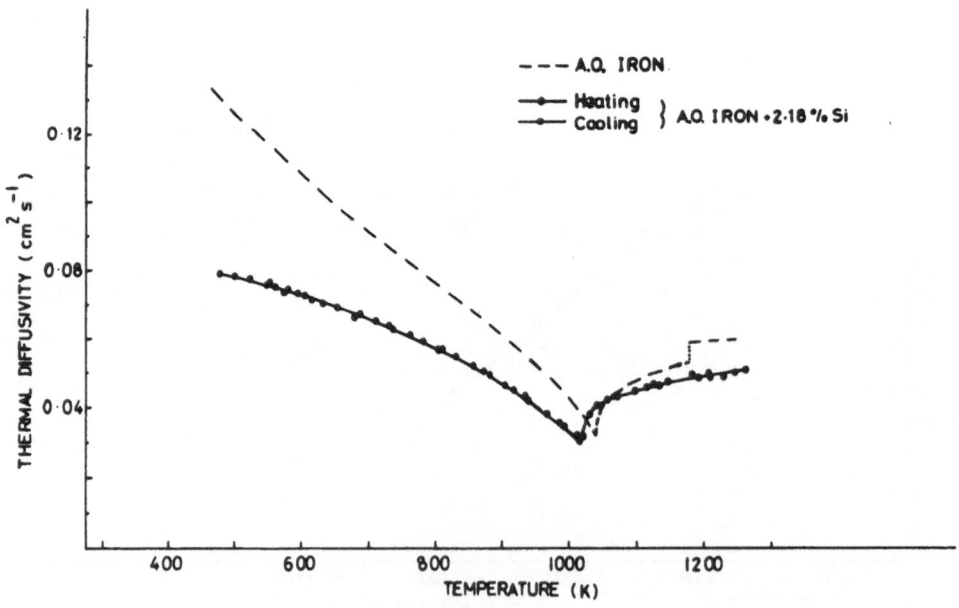

Fig. 11. Thermal diffusivity of 2.18% Si alloy.

these four elements Silicon produces the most marked change in lattice parameter of all the alloys tested being 0.11% at 2% concentration [21] and is also the only element to decrease the lattice parameter. Hence one would expect substantial changes in diffusivity and this is borne out by the results (figure 11) which show that at 500K the diffusivity is reduced 38% by the addition of 2.18% Si. The data are substantiated by equilibrium diagram considerations which show that Tc is reduced (to about 1120K) and α → γ transformation does not occur even at 1250K. The results on cooling also follow the same curve indicating no change in alloy structure. The next most marked change is shown by the addition of 3.74% Cr which reduces the diffusivity by about 28% at 500K. (figure 12) A similar reduction of 33.5% in thermal conductivity at 400K for zone refined iron with an addition of 2.96%Cr has been noted by Holder [22]. The Curie point is little changed, again, but the α → γ transition temperature is reduced to about 1125K. Both of these are in accordance with equilibrium diagram considerations but the change in the α → γ transformation temperature is perhaps 25K lower than expected. There is no change in diffusivity on cooling from 1250K. A concentration of 1.9% V is sufficient to close the γ-loop entirely so from the liquid state to room temperature a wholly body centred cubic structure will be obtained. The addition of Vanadium increases Tc to a maximum of about 1093K at 12% V. Our results (figure 13) show the low temperature diffusivity to be decreased by a maximum of 11.5% at 500K. However like the Cobalt alloy the retardation of the magnetic transformation raises the temperature of the point of inflection in the diffusivity/temperature curve so that above 900K the diffusivity is higher than that of A0 iron. Above Tc the diffusivity is lower than that of A0 iron. Results obtained during cooling follow the same curve. The addition of 0.49% Mo reduces the diffusivity by a maximum of about 7% at 500K but has a negligible effect on Tc (figure 14). Molybdenum is both a heavier and a larger atom increasing the lattice parameter by about 0.03% for this concentration [23] which may account for some of the effect. The α → α + γ → γ is retarded to 1115-1140K with some evidence of hysteresis and the diffusivity obtained during cooling is lower than that obtained during heating.

If the results for the alloys are examined in totality then a few tentative conclusions may be drawn.

1) The largest effect is shown by the addition of Silicon which shows the largest difference in electronic configuration is the only addition to decrease the lattice parameter, produces the largest change in lattice parameter and has the largest atomic mass difference ratio ($\frac{1}{2}$).

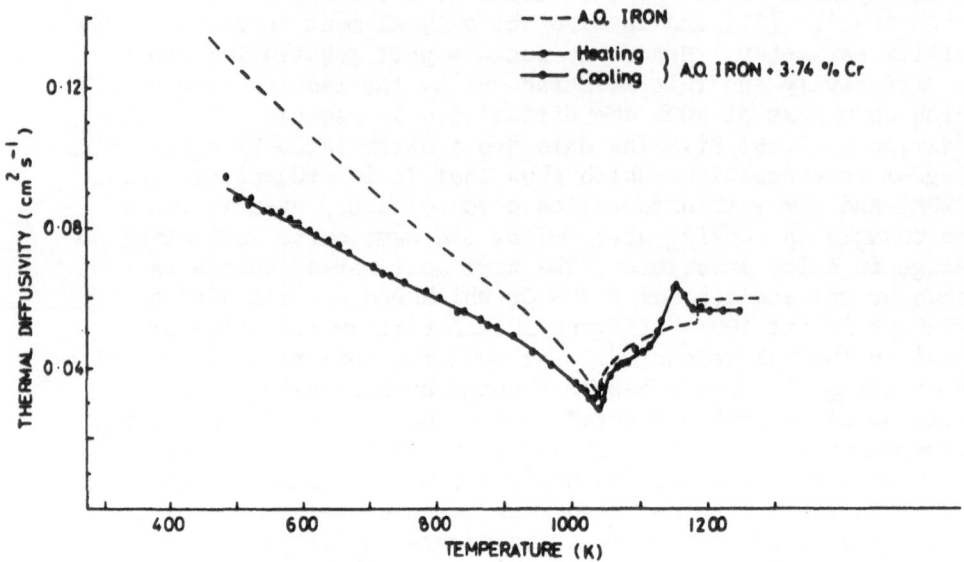

Fig. 12.   Thermal diffusivity of 3.74% Cr alloy.

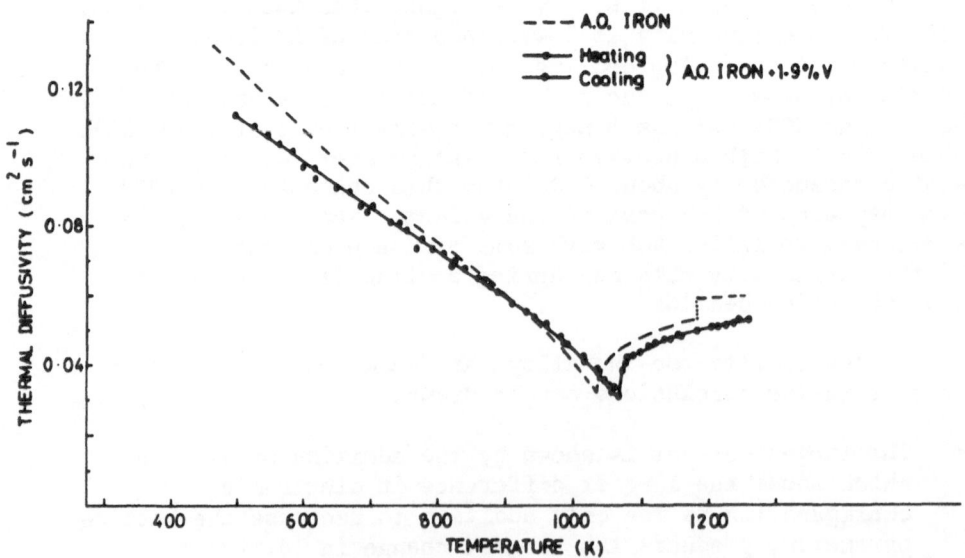

Fig. 13.   Thermal diffusivity of 1.92% V alloy.

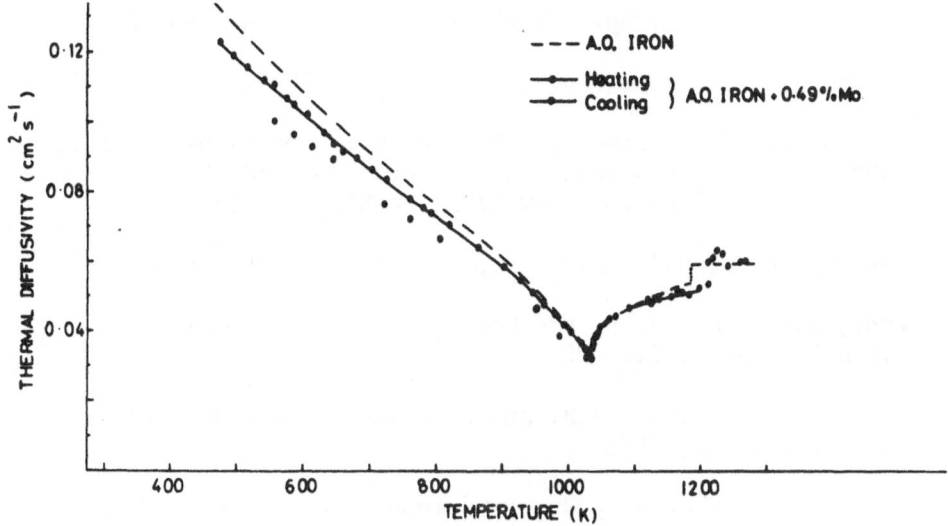

Fig. 14. Thermal diffusivity of 0.49% Mo alloy.

2)   For three of the four elements which form γ loops when
     alloyed with iron (V, Cr, Si) the heating and cooling
     measurements lie on the same curve. Conversely for the
     remaining four elements for which the measurement run did go
     through the transformation results obtained during cooling
     were always lower than those obtained during heating.

3)   Five of the elements added are adjacent to iron in the
     periodic table; Co and Ni above and Mn, Cr and V below.
     Those lying above iron appear to have a small effect than
     those lying below i.e. Co vis-a-vis Mn and Ni vis-a-vis Cr.
     However it is difficult to be precise about this since the
     alloying concentrations are different.

V   REFERENCES

1.   Powell, R.W. in 'Progress in International Research on
     Thermodynamic and Transport Properties' eds. J.F. Masi and
     D.H. Tsai (American Society of Mechanical Engineers and
     Academic Press Inc. New York, 1962) p. 454.

2.   Lucks, C.F., J. Testing and Evaluation, $\underline{1}$, 5, 422-31 (1973).

3.   Hust, J.G. and Giarrantano, P.J., N.B.S. Special Publication
     260-50, June 1975.

4.   Fulkerson, W., Moore, J.P. and McElroy, D.L., J. Appl. Phys.
     (1966) $\underline{37}$, $\underline{7}$, 2639-53.

5.   Parker, W.J., Jenkins, R.J., Butler, C.P., Abbott, G.L.,
     J. Appl. Phys. (1961) $\underline{32}$, 1679.

6.   Taylor, R., High Temp. - High Press. (1972), 4, 649-58.

7.   Cowan, R.D., J. Appl. Phys. (1963), 34, 926-7.

8.   Touloukian, Y.S., Powell, R.W., Ho, C.Y. and Nicolaou, M.C.
     'Thermophysical Properties of Matter,' Vol. 10, Thermal
     Diffusivity.1FI/Plenum, New York, Washington, (1973).

9.   Kennedy, W.L., U.S.A.E.C., report 1S-137, 1-59 (1960).

10. Cody, G.D., Abeles, B. and Beers, D.S., Trans. Met. Soc.,
     A.I.M.E. (1961), 221, 25.

11. Shanks, H.R., Klein, A.H. and Danielson, G.C., J. Appl. Phys.
     (1967), 38, (7), 2885.

12. Klein, A.H., Shanks, H.R., Danielson, G.C., Proc. 3rd Conf.
     Thermal Conductivity, October 16-18 (1963), p.747.

13. Van Craeynest, J.C., Weilbacher, J.C. and Lallemente, R.,
     U.S.A.E.C. report CEA-R-3734, 1-45 (1969).

14. Carter, R.L. and Sidles, P.H., Proc. 4th conf. on Thermal
     Conductivity, October 13-16th (1964), V-F-1/V-F-6.

15. Touloukian, Y.S. and Buyco, E.H., 'Thermophysical Properties
     of matter'. Vol. 4., Specific Heat of Metals and Alloys.
     1FI/Plenum. New York, Washington (1970).

16. Hust, J.G. and Sparks, L.L., N.B.S. Spec. Publication,
     260-31, (1971).

17. Smithells, C.J. 'Metals Reference Book, Volume 2', 4th Ed.
     Butterworths, London, (1967).

18. Sutton, A.L. and Hume-Rothery, W. Phil. Mag. (1955), 46,
     1295.

19. Jette, E.R. and Foote, F., Trans. A.I.M.M.E. (1936) 120, 259.

20. Guljeev, A.P. and Trusova, E.F., Z. Tekh. Fiz. U.S.S.R. (1950)
     20, 66.

21. Selisskii, I.P., Z. Fiz. Chim. U.S.S.R. (1946) 20, 597.

22. Holder, T.K., Oak Ridge National Laboratory report, ORNL/TM-
     5539, June 1977.

23. Bowman, F.E., Parke, R.M. and Herzig, A.J., Trans Am. Soc.
     Met. (1943) 31, 487.

THERMAL DIFFUSIVITY MEASUREMENT OF TEMPERATURE SENSITIVE MATERIALS BY AN EXTENDED PULSE TECHNIQUE[1]

A. B. Donaldson

Sandia Laboratories, Albuquerque, New Mexico 87115

B. D. Faubion

Mason & Hangar - Silas Mason Co., Amarillo, Texas 79177

Difficulty arises in applying the standard flash techniques for measuring thermal diffusivity to temperature sensitive materials because decomposition or ignition frequently results from the high front surface temperature. This difficulty can be overcome by simultaneously reducing the intensity of the energy source and increasing the exposure time. The increased exposure time permits a sufficient amount of energy to propagate to the opposite surface of the specimen so that a detectable temperature excursion results.

This paper discusses a theoretical analysis of the problem and includes extended pulse effects, radial heat conduction effects, and surface heat loss effects. The results are used to consider (1) the regime of pulse times for which data reduction can be based on the instantaneous pulse model, (2) the regime of specimen diameter/thickness ratios for which radial conduction effects can be neglected (this consideration allows minimizing explosive mass and hence, the hazard) and (3) the estimation of the front surface temperature during heating (so that ignition or decomposition temperatures are not reached).

Also, the results of experiments with the explosives HNS and RDX are discussed. Data reduction is made by finding the values of thermal diffusivity and surface heat loss coefficients which minimize the sum of the squares deviation between the experimental response curve and the theoretical response curve.

[1] This work was supported by the United States Department of Energy

Theoretical Analysis

The geometry which will be considered is that of a disk (sample) which is assumed to be well clamped on the circumference to a sample holder. The energy fluence to the sample is assumed to be constant for the time interval $o \leq t \leq \hat{t}$ and uniform with respect to radial and angular position. The energy equation and boundary and initial conditions for this problem are

Energy Equation

$$\nabla^2 T(z, r, t) = \frac{1}{\alpha} \frac{\partial T}{\partial t} (z, r, t)$$

Boundary Conditions

1. $-\frac{\partial T}{\partial z} (o, r, t) = f(t) - \frac{h_1}{\lambda} \left[T(o, r, t) - T_s \right]$

2. $-\frac{\partial T}{\partial z} (l, r, t) = \frac{h_2}{\lambda} \left[T(l, r, t) - T_s \right]$

3. $\frac{\partial T}{\partial r} (z, o, t) = 0$

4. $T(z, R, t) = T_s$

Initial Condition

$$T(z, r, o) = T_s$$

where $\quad f(t) = \begin{cases} o & t < o \\ Q/\lambda A & o \leq t \leq \hat{t} \\ o & t > \hat{t} \end{cases}$

The solution of this problem for $\hat{t} \to o$ has been given by Watt [1] and can be written in terms of nondimensional parameters as

$$\theta(\xi, \eta, \tau) = \left\{ \frac{2}{\sigma} \sum_{m=1}^{\infty} \frac{J_0(\beta_m \sigma \eta) e^{-\frac{\beta_m^2}{\sigma^2} \tau}}{\beta_m J_1(\beta_m \sigma)} \right\}.$$

$$\left\{ C + 2 \sum_{n=1}^{\infty} \frac{\gamma_n(\gamma_n^2 + H_2^2)[\gamma_n \cos(\gamma_n \xi) + H_1 \sin(\lambda_n \xi)]e^{-\gamma_n^2 \tau}}{(\gamma_n^2 + H_1^2)(\gamma_n^2 + H_2^2 + H_1) + H_1(\gamma_n^2 + H_2^2)} \right\}$$

where
$$C = \begin{cases} 0 & \text{if } H_1 \text{ or } H_2 \neq 0 \\ 1 & \text{if } H_1 = H_2 = 0 \end{cases}$$

and the eigenvalues β_m and γ_n are determined from the eigenequations

$$J_1(\beta_m \sigma) = 0$$

and

$$\tan(\gamma_n) = \frac{\gamma_n(H_1 + H_2)}{\gamma_n^2 - H_1 H_2} ,$$

respectively. This solution can now be generalized to solve the case for $\hat{t} > o$ by Duhamel's theorem:

$$\phi(\xi, \eta, \tau) = \int_{s=o}^{\tau} f(s) \cdot \frac{\partial \theta}{\partial \tau} (\xi, \eta, \tau - s) \, ds$$

Although this solution is all that is needed for reduction of experimental data, three special cases will first be considered which have a bearing on the experiment.

First, since the sample material is temperature sensitive, an estimate of the maximum temperature is useful so that degradation or ignition of the sample can be avoided. Rather than using the complete solution, an upper estimate can be obtained from the solution for the problem of a semi-infinite solid which is exposed to a constant heat flux. From Carslaw & Jaeger [2], the surface temperature rise will be

$$\Delta T(o, \hat{t}) = 2(Q/A\lambda) \cdot \sqrt{\frac{\alpha \hat{t}}{\pi}}$$

where surface heat loss has been neglected. Use of this equation requires an estimate of the magnitude of α and λ.

Second, the effect of finite pulse time on the half-time,
τ 1/2 [3], can be computed from the general solution. The case
which is treated is that of a semi-infinite slab ($\sigma = \infty$) so that
radial conduction is neglected and the nondimensional surface loss
coefficients for both front and rear surfaces are equal. Results
are shown in Figure 1 and indicate that if a half-time deviation of
less than 1% is desired, then the pulse width, τ 1/2, should be
less than ~ 0.004.

Third, the effect of variation in the sample radius-to-thickness
ratio on half-time can be computed from the general solution.
Figure 2 shows this result for zero pulse width and indicates that
if data reduction is based solely on half-time, then σ should be
greater than ~ 2.2 in order to eliminate radial conduction effects.

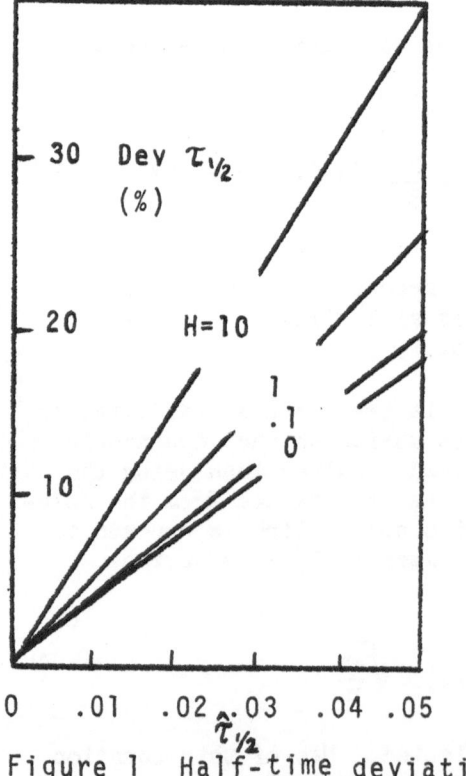

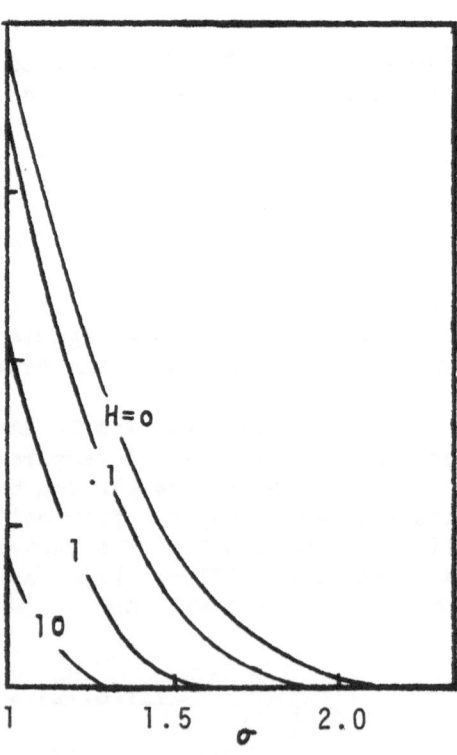

Figure 1   Half-time deviation
from value for $\hat{\tau}$=0 as a
function of $\hat{\tau}$, all for σ=∞

Figure 2   Half-time deviation
from value for σ=∞ as a
function of σ, all for $\hat{\tau}$=0

Experiment

Figure 3 illustrates the experimental configuration. A 750 mW laser is used as the power source, and a camera shutter is used to adjust the pulse width. A differential thermocouple at the rear center of the sample and the sample holder is used to measure temperature excursion. A temperature controller is used to maintain the sample at a preset temperature. The sample and sample holder environment can be evacuated, and the sample is exposed on a downward facing surface--both considerations to minimize convective heat loss.

The experimental temperature response on the rear surface is recorded on an analog recorder. Data reduction is accomplished by a computer program that finds the values of α, H_1 and H_2 which minimize the sum-of-the squares deviation between the general solution and the data at specified data points.

Figures 4 and 5 show the experimental data, the fitted solution and the input/output parameters for the explosives HNS (hexanitrostilbene) and RDX (trinitrotriazacyclohexane), respectively. Also shown is an estimate of the maximum front surface temperature for the HNS experiment and a nondimensional pulse width for the RDX experiment.

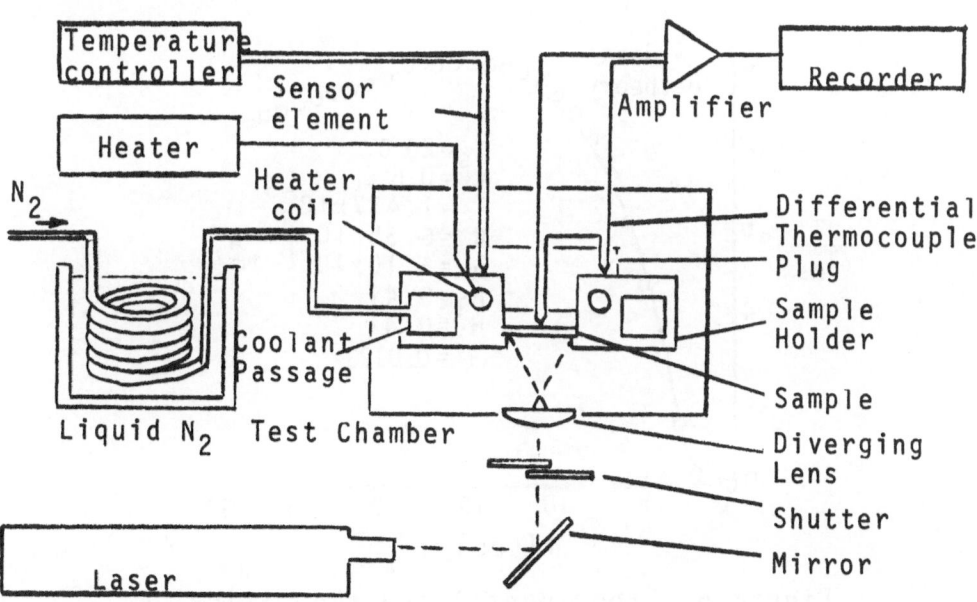

Figure 3   Schematic of apparatus

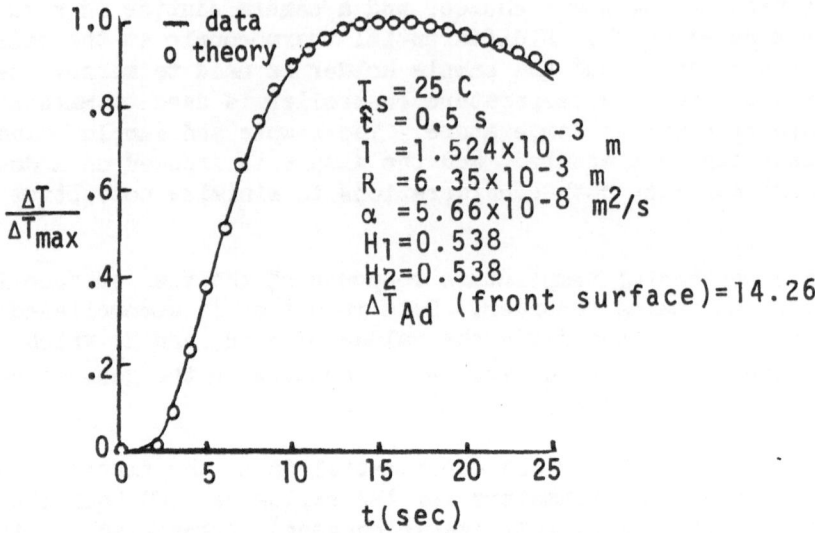

Figure 4   Experimental and theoretical
response for HNS

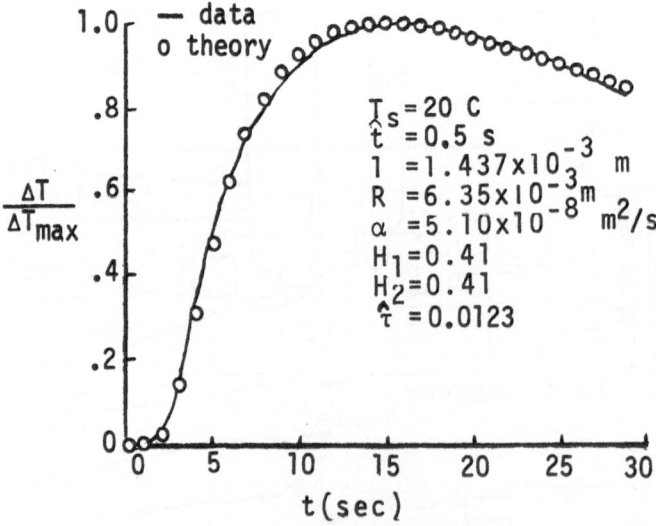

Figure 5   Experimental and theoretical
response for RDX

Conclusions

The extended pulse technique appears to be a useful way of applying the flash method to measure the thermal diffusivity of temperature sensitive materials. By properly designing an experiment, temperature excursions can be made sufficiently small, finite pulse width effects on data reduction can be minimized, and the sample diameter can be reduced to the smallest value for which radial conduction is inconsequential.

Nomenclature

H_1, H_2 front and rear surface Nusselt numbers

h_1, h_2 front and rear surface linear heat loss coefficients

J_o, J_1 ordinary Bessel functions of zeroth and first order

ℓ sample thickness

Q/A heat flux

R sample radius

r radial position variable

s dummy integration variable

T temperature

T_s surrounding temperature

t time variable

$\hat{t}$ pulse width

z axial position variable

α thermal diffusivity

η nondimensional radial position variable ($\equiv r/R$)

θ nondimensional temperature excess for zero pulse width $[\equiv (T - T_s)/(Q\ell/\lambda A)]$

λ thermal conductivity

ζ nondimensional axial position variable ($\equiv z/\ell$)

σ ratio of sample radius to thickness ($\equiv R/\ell$)

τ Fourier number ($\equiv \alpha t/\ell^2$)

$\hat{\tau}$ nondimensional pulse width ($\equiv \alpha \hat{t}/\ell^2$)

ϕ nondimensional temperature excess for finite pulse width
           [$\equiv (T - T_s)/(Q\ell/\lambda A)$]

References

1. Watt, D. A., Br. J. Appl. Phys., <u>17</u>, 231 (1966).

2. Carslaw, H. S. and Jaeger, J. C., <u>Conduction of Heat in Solids</u>,
    2nd ed., Oxford Press, London, p. 75.

3. Half-time is customarily used in data reduction, see for
    example Parker, W. J., et al, J. Appl. Phys., <u>32</u> 1679 (1961).

THERMAL DIFFUSIVITY OF DISPERSED MATERIALS*

T. Y. R. Lee

Center for Information and Numerical Data Analysis and
Synthesis, Purdue University
West Lafayette, Indiana 47907

R. E. Taylor

Properties Research Laboratory
School of Mechanical Engineering
Purdue University
West Lafayette, Indiana 47907

ABSTRACT

Measurements have been made of the effective thermal diffusivity
at room temperature of composites consisting of one phase randomly
dispersed in a second phase. The method is based on the flash
technique. Data are presented for four types of composites ranging in particle-to-matrix diffusivity ratios from 0.48 to 1137; in
volume specific heat ratios 0.04 to 1.16 and in volume fraction of
dispersed particle from zero up to 34%. The results show that the
limitations of the concept of an effective thermal diffusivity are
far beyond the stiuations to which it is currently applied in the
transient state heat conduction problems. Values of effective
diffusivities derived from values of the effective thermal conductivity calculated from the Bruggeman variable-dispersion equation
are found to agree well with the measured diffusivity values.

*This work was performed as part of an NSF grant administered by
the Engineering Division.

THERMAL DIFFUSIVITY MEASUREMENTS AT HIGH TEMPERATURES BY THE RADIAL FLASH METHOD

F. I. Chu, R. E. Taylor and A. B. Donaldson

The flash diffusivity method is extended to measure thermal diffusivity at higher temperatures using a radial heat flow technique. In this technique the energy source of radius R˙ irradiating the front surface of the sample is considerably smaller than the sample's surface. The resulting temperature rises on the rear surface are measured at two locations (1) directly opposite the center of the energy pulse and (2) at a distance r (which is greater than R). This method, which is called radial flash method is different from the conventional method (in which the entire front surface of the sample is irradiated) in several respects (1) considerable radial heat flow occurs in addition to axial heat flow (2) the radial flash method measures diffusivity in the radial direction (α_r) instead of the diffusivity in the axial direction (α_x), and (3) the radial flash method is independent of radiation heat losses from the surface. The method was tested with both an isotropic material (POCO AXM-5Q Graphite) and an anisotropic material (reactor grade graphite) at high temperatures. The results showed that thermal diffusivity in the radial direction could be measured without accounting for the very large radiation heat losses. The results are also compared with those based on the conventional flash method using smaller samples.

Work supported by National Science Foundation
F. I. Chu - RCA Picture Tube Division, Circleville, Ohio
R. E. Taylor - Properties Research Laboratory, Purdue University,
                                                    Lafayette, Indiana
A. B. Donaldson - Sandia Laboratories, Albuquerque, New Mexico

DETERMINATION OF THE THERMAL DIFFUSIVITY OF TWO-LAYER
COMPOSITE SAMPLES BY THE MODULATED HEATING BEAM METHOD[+]

R. Brandt and M. Havránek

Institut für Kernenergetik und Energiesysteme

University of Stuttgart, W. Germany

INTRODUCTION

The measurement of the thermal diffusivity of solids by the modulated heating beam method has found widespread application, since Cowan /1/ in 1961 has given a theory of determining the thermal diffusivity of small disk-like samples.

However, this theory is applicable only for single-layer samples of material, but in many situations one would like to measure samples consisting of two layers of different materials. As an example one may list coating materials on substrates, working as thermal or electrical insulators or as corrosion protectors. As an other example one could mention deposits on substrates, which are too thin or too friable, so that they could not be measured separately.

The purpose of this paper is to outline the mathematical analysis by means of which the thermal diffusivity in one layer of a two-layer composite sample can be deduced from the modulated heating beam data, provided that there exists a perfect thermal contact between the layers of the sample.

[+] This work shall be published in more details in J. Non-Equilib. Thermodyn., Vol. 3 (1978).

/1/ Cowan, R.D., "Proposed Method of Measuring Thermal Diffusivity at High Temperatures", J. Appl. Phys. 32 (1961), p. 1363-1370.

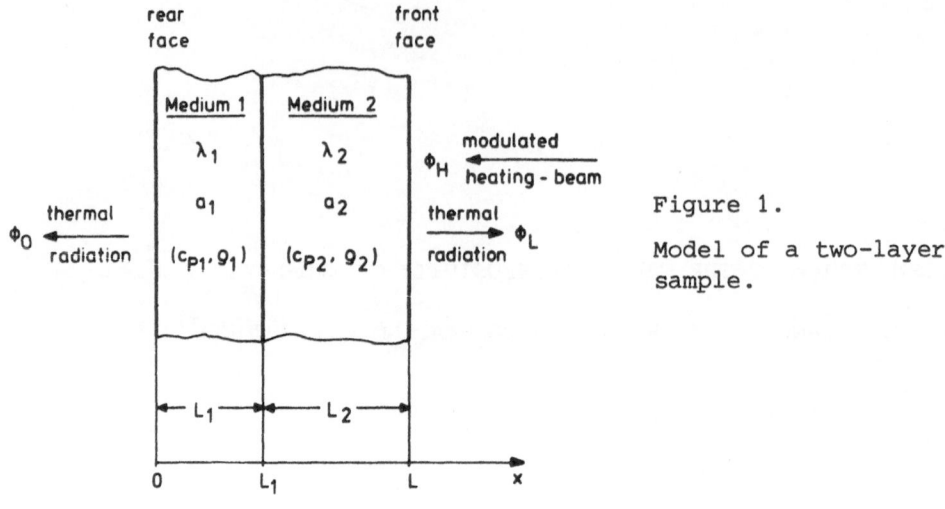

Figure 1.

Model of a two-layer sample.

THE MATHEMATICAL MODEL

The theory for the thermal diffusivity determination used in the present analysis was developed by Cowan /1/ and has been extended now to two-layer composite samples.

The sample consists of two layers with the thicknesses L_1 and L_2 (see Fig. 1), λ_1 and λ_2 are the thermal conductivities and a_1 and a_2 the thermal diffusivities of the two media. The sample is heated uniformly at the front face ($x = L$) with a sine-wave modulated heating beam Φ_H, producing temperature waves propagating through the sample. The phase shifts between the resulting temperature variations of both sample surfaces and the heating beam modulation can be measured and are related to the thermal diffusivity of the sample.

Assuming that there is no radial heat flow component through the sample and that there exists no contact resistance between the two layers, the heat flow in the sample is described by

$$a_i(\partial^2 T_i/\partial x^2) = \partial T_i/\partial x \qquad , i = 1,2 \qquad (1)$$

where T_i is the temperature of medium i, t is the time, and x is the distance from the rear face of the sample.

Eq. (1) must be satisfied by the following boundary conditions:

$$\Phi_O = \lambda_1(\partial T_1/\partial x) \qquad \text{at } x = O \qquad (2)$$

$$T_1 = T_2 \qquad \text{at } x = L_1 \qquad (3)$$

$$\lambda_1 (\partial T_1 / \partial x) = \lambda_2 (\partial T_2 / \partial x) \quad \text{at } x = L_1 \tag{4}$$

$$\Phi_H - \Phi_L = \lambda_2 (\partial T_2 / \partial x) \quad \text{at } x = L \tag{5}$$

$$\Phi_H = \Phi_H^O + A \sin(\omega t) \tag{6}$$

In the above equations Φ_O and Φ_L represent the intensities of thermal radiation at the surfaces $x = O$ and $x = L$, A is the amplitude and ω the angular velocity of the heating beam modulation.

ANALYTICAL SOLUTION

The method leading to an analytical solution of the problem represented by Eqs. (1)-(6) follows the method employed by Cowan /1/ for single-layer samples. As a solution of the problem one gets two equations, which describe the phase shifts between the temperature variations at both sample surfaces and the heating beam modulation:

$$\tan \psi_O = - \frac{KB_1 (Q_1 \tan B_2 + Q_2 \tanh B_2) + B_2 (Q_3 \tan B_2 \tanh B_2 + Q_4)}{KB_1 (Q_2 \tan B_2 - Q_1 \tanh B_2) + B_2 (Q_4 \tan B_2 \tanh B_2 - Q_3)} \tag{7}$$

$$\tan \Delta\psi = - \frac{KB_1 (Q_1 P_1 - Q_2 P_2) + B_2 (Q_3 P_3 - Q_4 P_4)}{KB_1 (Q_1 P_2 + Q_2 P_1) + B_2 (Q_3 P_4 - Q_4 P_3)} \tag{8}$$

where

$$\left. \begin{aligned}
Q_1 &= C_1 (\tan B_1 \tanh B_1 + 1) + 2B_1 \tanh B_1 \\
Q_2 &= C_1 (\tan B_1 \tanh B_1 - 1) + 2B_1 \tan B_1 \\
Q_3 &= C_1 (\tan B_1 + \tanh B_1) + 2B_1 \\
Q_4 &= C_1 (\tan B_1 - \tanh B_1) + 2B_1 \tan B_1 \tanh B_1
\end{aligned} \right\} \tag{9}$$

$$\left. \begin{aligned}
P_1 &= B_2 (\tan B_2 \tanh B_2 + 1) + C_2 \tan B_2 \\
P_2 &= B_2 (\tan B_2 \tanh B_2 - 1) - C_2 \tanh B_2 \\
P_3 &= B_2 (\tan B_2 + \tanh B_2) + C_2 \tan B_2 \tanh B_2 \\
P_4 &= B_2 (\tan B_2 - \tanh B_2) - C_2
\end{aligned} \right\} \tag{10}$$

ψ_O is the phase shift between the heating beam modulation and the rear face temperature variation ($x = O$), $\Delta\psi$ represents the phase shift between the front and rear face temperature variations. Both phase shifts are dependent on the 5 variables B_1, B_2, C_1, C_2 and K, which are defined as follows:

$$B_i = L_i \sqrt{\pi \nu / a_i} \qquad i = 1,2 \tag{11}$$

$$C_1 = ((L_1/\lambda_1)(\partial\Phi/\partial T_1))_{x = 0}$$

$$C_2 = ((L_2/\lambda_2)(\partial\Phi/\partial T_2))_{x = L}$$

$$\left.\right\} (12)$$

$$K = (\lambda_1/\lambda_2)(L_2/L_1) \hspace{2cm} (13)$$

Here, ν is the modulation frequency, C_1 and C_2 are dimensionless heat loss parameters describing the change of heat leaving the surfaces as functions of the surface temperatures.

Following Eqs. (7)-(13) one can now calculate the characteristic value B of one of the layers and from this its thermal diffusivity, when the phase shift ψ_0 or $\Delta\psi$ is measured and the value B of the other layer as well as the heat loss parameters C_1 and C_2 and the value of K are known. However, the heat loss parameters C_1 and C_2 and the thermal conductivity of the unknown layer in general would not be known when measuring the thermal diffsivity. For this reason the influence of the 5 variables B_1, B_2, C_1, C_2, and K on the phase shifts was investigated by establishing a computer program.

RESULTS

As a first result of the parameter study the computations show, that the total phase shift ψ_0 remains unaltered, when the heat flow direction in the sample is reversed. However, $\Delta\psi$ changes its value, when layer 1 is subjected to the heating beam instead of layer 2. Therefore it is to decide if it would be better to have the unknown layer at the heated or at the unheated side of the sample.

Furthermore, the computations show, that uncertainities in the knowledge of C_1, C_2, and K take the least effect, when measuring the phase shift $\Delta\psi$ instead of ψ_0 or of $\psi_L = \psi_0-\Delta\psi$, which is the phase lag between the heating beam modulation and the front face temperature variation (x = L). Therefore all further investigations are restricted to $\Delta\psi$.

As it will be shown, it is advantageous to substitute the heat loss parameters C_1 and C_2 by two other heat loss parameters called R and α in a similar way as Cowan /1/ has done for single layer samples:

$$R = C_2/(KC_1) \quad \text{and} \quad \alpha = C_1(1+R)(1+K) \hspace{1.5cm} (14)$$

In the special case, when only thermal radiation determines the heat losses at the sample surfaces ($\Phi = \varepsilon\cdot\sigma\cdot T^4$), R and α can be estimated from Eqs. (12)-(14) to:

$$R = \frac{\varepsilon_1}{\varepsilon_2} \left(\frac{T_L}{T_O}\right)^3 \quad \text{and} \quad \alpha = \left(\frac{L_1}{\lambda_1} + \frac{L_2}{\lambda_2}\right) (1+R) \, 4\varepsilon_1 \sigma T_O \tag{15}$$

Here σ is the Stefan-Boltzmann-constant, ε_1 and ε_2 are the total emissivities of both sample surfaces at temperatures T_O and T_L.

Some results of the computations concerning the influence of the heat loss parameters R and α are shown in Figures 2 and 3. The phase shift $\Delta\psi$ is plotted versus the characteristic value of the unknown layer, called B_x, for various values of α. K and R have a constant value of unity. In Fig. 2 it is assumed that the characteristic value of the known layer, called B_j, is small, i.e. the known layer is thin or is a good thermal conductor. In Fig. 2a it is assumed that the sample is mounted in the measuring device in such a way, that the unknown layer is located at the heated side of the sample, whereas in Fig. 2b the unknown layer is assumed to be at the unheated side of the sample.

For both arrangements the influence of α on $\Delta\psi$ is nearly constant in the range $2 \leq B_x/B_j \leq 20$, but generally the influence of α is about 2 or 3 times greater when the unknown layer is situated at the unheated side of the sample. For example, an uncertainity in the knowledge of α by a factor 2 can cause an error in B_x of about 12% and in a_x of about 25%, when the unknown layer is at the unheated side, whereas the location at the heated side yields an error in a_x of only about 10%. Therefore the unknown layer should be placed at the heated side of the sample.

Furthermore, the computations indicate that the influence of R on $\Delta\psi$ generally is less than a half the one of α. These different magnitudes in the influence of R and α on $\Delta\psi$ are the reason for the substitution of C_1 and C_2 by R and α. Because of the lower influence of R this parameter can be estimated with sufficient accuracy from Eq. (15). However the estimation of α from Eq. (15) requires the knowledge of the thermal conductivities. But when measuring not only the phase shift $\Delta\psi$, but simultaneously $\Delta\psi$ and ψ_O, both a_x and α can be calculated iteratively /2/.

In Fig. 3 it is assumed that the known layer is thick or is a poor conductor (great value of B_j). All other parameters are within the same values as in Fig. 2. For values $B_x/B_j > 2$ again the influence of α on $\Delta\psi$ is lower for the rear face location of the unknown layer. But for values $B_x/B_j < 2$ the curves flatten, i.e. the influence of α increases markedely and the measurements will become uncertain.

/2/ Chafik, E.M.; "Ein Verfahren zur Messung der Temperaturleitzahl fester Stoffe bei hohen Temperaturen unter Anwendung eines modulierten Lichtstrahles". Diss., University Stuttgart, 1970.

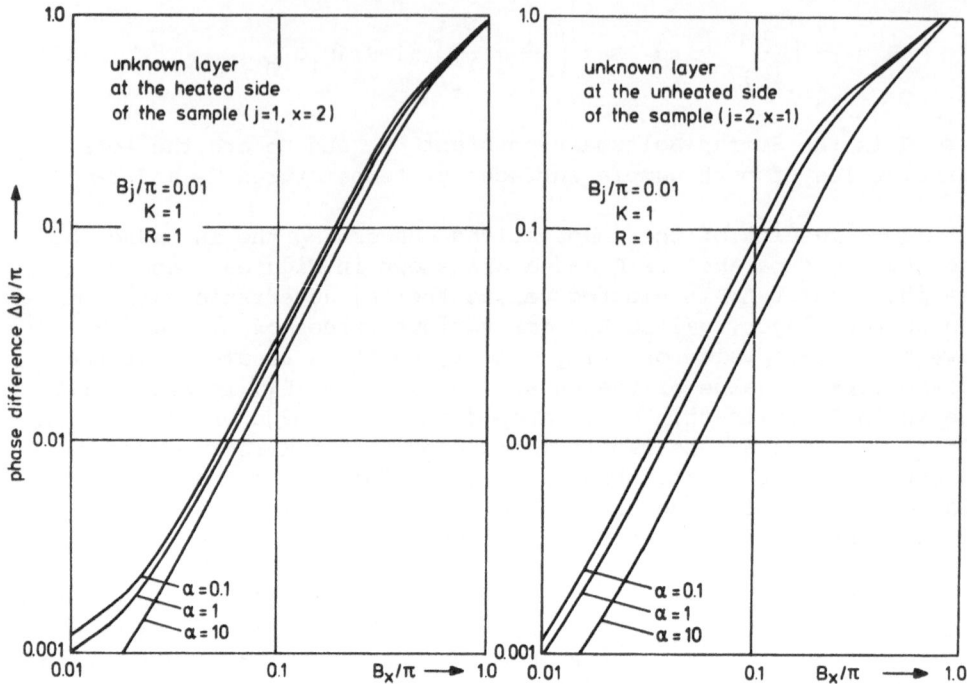

Fig. 2: Phase difference $\Delta\psi$ as a function of the unknown value B_x
for a low value of B_j and for various values of α.

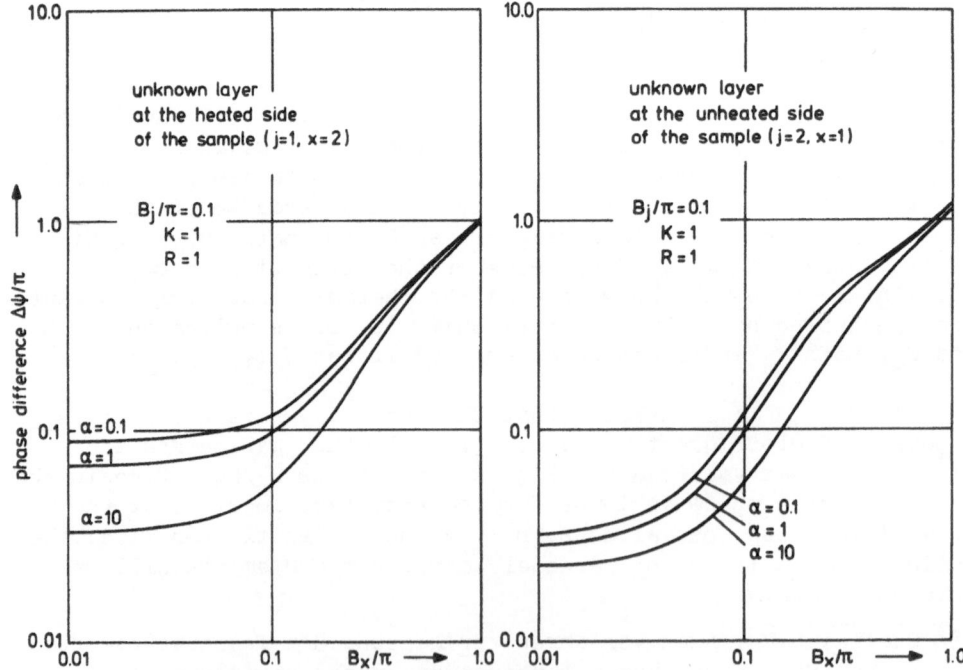

Fig. 3: Phase difference $\Delta\psi$ as a function of the unknown value B_x
for a great value of B_j and for various values of α.

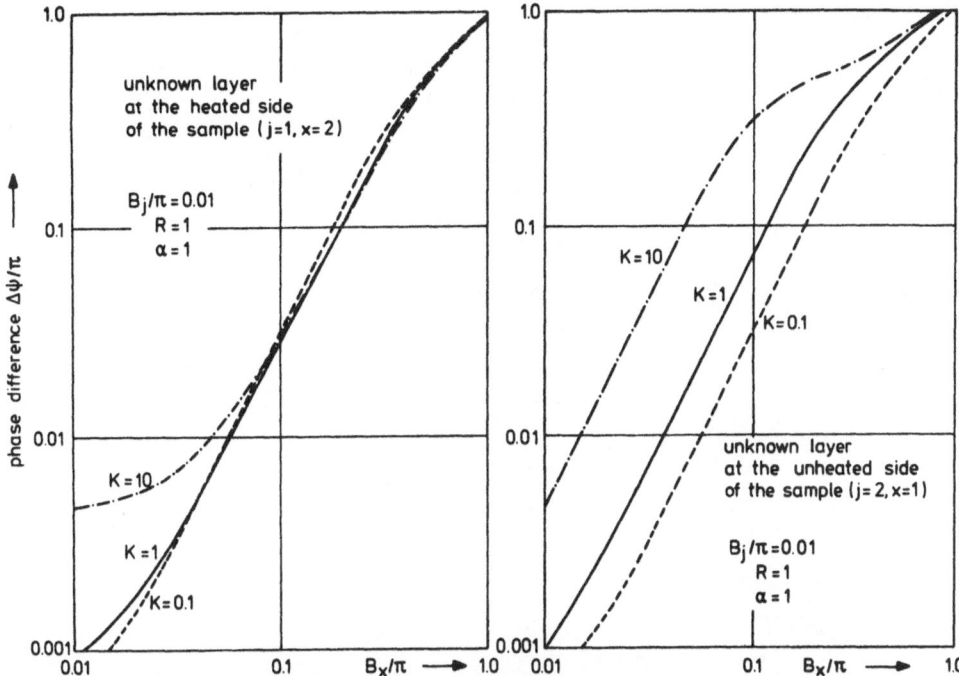

Fig. 4: Phase difference $\Delta\psi$ as a function of the unknown value B_x for a low value of B_j and for various values of K.

In Fig. 4 the influence of the parameter K on $\Delta\psi$ is shown for low values of B_j. It is evident, that there exists a considerable difference in the influence of K on $\Delta\psi$ for the two positions of the unknown layer, which even is greater than the difference in the influence of α. This fortifies the demand that the unknown layer should be placed at the unheated side. Then, from the Fig. 4a it can be seen, that for values B_x/B_j greater than about 3 the influence of K on $\Delta\psi$ is not important. An uncertainty in the knowledge of K by a factor 2 can cause an error in estimating the thermal diffusivity not greater than 3 to 5%.

CONCLUSIONS

A method of measuring the thermal diffusivity of two-layer composite samples by the modulated heating beam technique has been described. For reducing measurement uncertainties the following provisions should be made:
- The sample should be prepared in such a way that the layer of the known material is as small as possible and should be a better thermal conductor than the unknown material whenever possible.
- During the measurement the unknown layer should be subjected to the heating beam.

LIST OF AUTHORS

Ackerman, M.W.	77	Graves, R.S.	75,133,149
Acton, R.U.	107,263	Gregorio, P.	355
Afshar, R.	223		
Aleinikova, V.I.	223	Havrànek, M.	481
Aragones, M.A.	155	Heremans, J.	63
Ashworth, T.	57	Hickox, C.E.	389,423
		Ho, C.Y.	77,79
Begej, S.	115	Hsieh, C.K.	11
Bertino, J.P.	433	Hsiung, C.Y.	57
Bogaard, R.H.	77	Hust, J.G.	161
Brandt, R.	481		
Brown, W.C.	169	Issi, J.-P.	63
Chu, F.I.	479	Joshi, R.K.	223
Chu, T.K.	77,79		
Clark, R.K.	343	Klemens, P.G.	3,203
Cook, J.G.	73,187	Kopelman, I.J.	23
Cort, G.E.	433		
Cunnington, G.R. Jr.	325	Latini, G.	245
		Laubitz, M.J.	73,187
Desai, P.D.	77	Lawrence, W.E.	197
Dicus, D.L.	343	Lee, T.Y.R.	135,477
Dodson, J.G.	399	Leiser, D.B.	335
Donaldson, A.B.	135,469,479	Le Neindre, B.	217
		Levy, P.W.	289
Eatherly, W.P.	133	Lindquist, L.O.	125
Ellingson, W.A.	11	Lisagor, W.B.	343
		Llewellyn, G.H.	277
Faubion, B.D.	469		
Ferro, V.	355	McElroy, D.L.	149
Fowler, C.M.	453	Mah, R.	125
		Marchenkov, E.I.	223
Garnier, J.E.	115	Maštovský, J.	211
Garrabos, Y.	217	Mirkovich, V.V.	297,305,447
Gartling, D.K.	423	Moore, J.P.	75,133,149
Goldstein, H.E.	335	Muller, E.R.	191

489

INDEX